Ion transport and cell structure in plants

European plant biology series

Consulting Editor
Professor M. B. Wilkins
University of Glasgow

Titles in the series

Clarkson: Ion Transport and Cell Structure in Plants
Meidner: Physiology of Stomata
Robards: Electron Microscopy and Plant Ultrastructure
Robards: Dynamic Aspects of Plant Ultrastructure
Smith: Phytochrome and Photomorphogenesis
Wilkins: Physiology of Plant Growth and Development

Published by **McGRAW-HILL Book Company (UK) Limited**
MAIDENHEAD · BERKSHIRE · ENGLAND

07 084026 1

PRINTED AND BOUND IN GREAT BRITAIN

Ion transport and cell structure in plants

David T. Clarkson
Agricultural Research Council
Letcombe Laboratory

London · New York · St Louis · San Francisco · Düsseldorf · Johannesburg
Kuala Lumpur · Mexico · Montreal · New Delhi · Panama · Paris · São Paulo
Singapore · Sydney · Toronto

Preface

The range of subjects covered in this book—from the molecular architecture of membranes to the ionic relations between roots and the soil—might well have been published as a multi-author work. Instead, it has been written by a single plant physiologist with interests in fine structure and biophysical chemistry. This is meant to be neither a boast nor an apology, but to emphasize that the present work was conceived not as a compendium of information but as a text which can be read, hopefully, for general interest and instruction. I have tried to encourage the reader to think about what is actually involved, in structural terms, when ions are transported across the various barriers which have to be negotiated on entry to the cell and on subsequent movement within it. In places, this approach has involved some speculation which is, I hope, adequately identified as such.

The book is divided into two parts; the first six chapters are concerned with general properties of transport processes and the membranes across which they occur. The alga, *Hydrodictyon africanum*, has been introduced as a test case organism to provide a reference point to which the student can return and apply any knowledge or insight that he may have picked up from the various chapters. In other places I have crossed the arbitrary lines drawn between plant and animal biology and between biology and chemistry to find good evidence from phenomena which are likely to be of general occurrence. I am sorry if the plant physiologist with an interest in, say, ion uptake by plant roots, feels affronted by examples from red blood cells or slime moulds, but it is one of the basic tenets of this book that, when *his* material comes to be as well understood, it will be the details, rather than the fundamentals, of transport processes which will be its distinctive features.

The second part of the book contrasts with the first in several respects. It is almost exclusively concerned with higher plants and, in particular, with ion transport by plant roots. Hence the level of inquiry shifts from single cells to organs and intact plants; it is the multicellular nature of higher plants and the communication between cells and tissues which are judged to be the most interesting matters for consideration.

This book would not have been written without the encouragement of my wife and my colleagues at the Letcombe Laboratory or without the cooperation of the Agricultural Research Council and my Director, Dr R. Scott Russell. In particular I would like to thank Dr R. N. Crossett, Dr M. C. Drew and Dr M. G. T. Shone for their valuable discussion of the drafts to various chapters. I have also been helped in a most generous way by Dr A. W. Robards of the University of York and by Professor D. Branton, Dr B. E. S. Gunning, Dr M. C. Ledbetter and Dr W. W. Franke who have let me use their published and unpublished electronmicrographs. In this connection, I must add that speculation about the significance of structures seen in these pictures in relation to transport processes, especially where it is incorrect, must be attributed to me and not to them.

David T. Clarkson
Letcombe Regis
Nov. 1973

Contents

Part I

one

Some fundamentals, some questions and a case history

Work can be performed only if the interacting components of a system differ in energy. Heat will pass from a hotter body to a cooler one until both reach the same temperature, thus establishing an equilibrium which cannot be spontaneously disturbed. During heat transfer the energy differential between the two bodies is reduced to zero, although the total energy in the bodies plus their surroundings remains constant. At the moment we first observed the system, the hotter body possessed some useful heat energy which would be potentially available to do work. In the final situation, when both blocks reach the same temperature, neither possesses any useful or free energy. The heat energy has come to the most random distribution in the system, and thus degraded is no longer available for work.

The second law of thermodynamics, essentially a statistical law, predicts that the total energy of the universe will come to be more and more randomly distributed until all forms of energy transfer have been completed and universal equilibrium is reached. All spontaneous changes of state are in the direction of increasing chaos, and the chaos is more stable than order. Perhaps we should think a little more about this paradox and start by examining this statement, 'highly ordered systems invariably have more free energy than less ordered ones'. We can put this idea into a biological context by thinking about the relative energy content of 1 gram . mole of glucose in comparison with the equivalent number of free molecules of carbon dioxide and water at the same temperature and pressure. The biological oxidation of glucose in respiration proceeds with a great release of potential energy which can be used to perform muscular work, provide body heat, transport ions, etc. The assembled atoms of carbon, hydrogen and oxygen possess this energy by virtue of the orderly structure in which they are placed, but when the glucose is respired they are recombined into less ordered structures such as carbon dioxide and water. Throughout the universe ordered structures are broken down into less ordered ones in a similar way. This process will continue until the universe is reduced to an inert chaos.

Randomization of solutes by diffusion

In this book we shall be studying the processes whereby plants and plant cells accumulate solutes from the external environment. In all solutions there is a tendency for the solute to become randomly distributed throughout the available volume; spontaneous gradients of solute concentration cannot be maintained because the process of diffusion causes the movement of solute from areas of high to low concentration. We may think of diffusion, therefore, as a process which brings about the most random deployment of solute molecules. We can illustrate this by considering Fig. 1.1 in which two compartments, A and B, are separated by a thin barrier which allows both the solute and the solvent to pass through it. Initially, the solute concentration in A is greater than in B but, with time, there will be a net movement of solute molecules into B so that eventually the concentrations will become equal. As long as a concentration difference exists across the barrier we can predict the spontaneous direction of flow of solute. When the concentrations become equal the movement of solute molecules will continue between the two compartments by random thermal motion, but there will be no longer any net flow. Once randomization has been achieved the system cannot return spontaneously to the original condition.

Chemical potential as a driving force in solute movement

Let us now consider what driving forces act on solute molecules. When a substance is dissolved in a solvent its chemical potential is

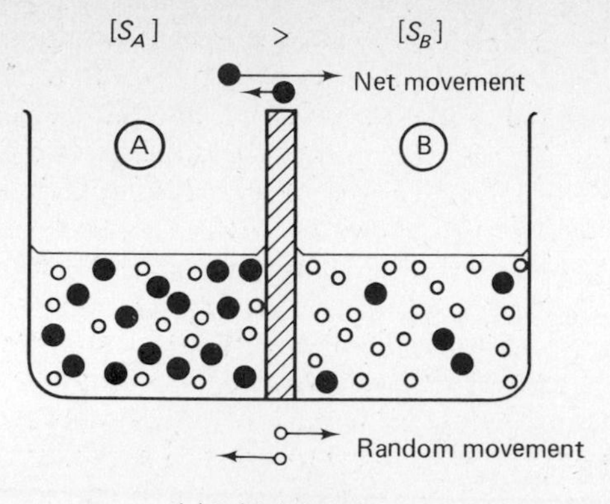

(a) Initial situation

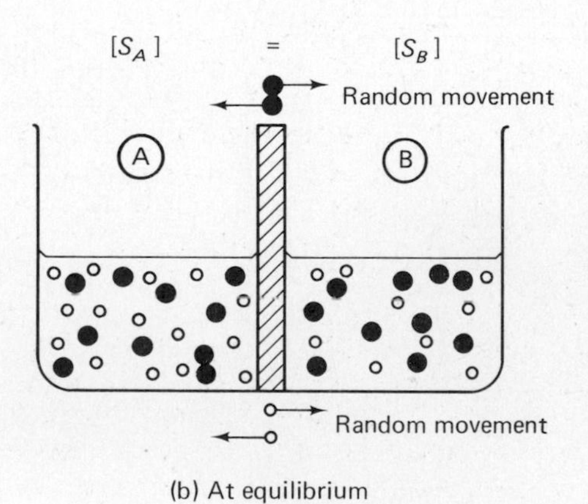

(b) At equilibrium

***FIG. 1.1** Randomization of solute molecules by diffusion across a permeable membrane. (S_A) and (S_B) = the concentration of solute in compartments A and B. Solid circles represent solute molecules, open circles water molecules. In the initial situation, there will be a net flux of solute molecules across the membrane; at equilibrium movement between the compartments is at random and there is no net flux.*

related to its concentration (or strictly, its activity) in the following way:

$$\mu = \mu^* + RT \ln a_s \tag{1.1}$$

where μ = the chemical potential in joules mole^{-1}.

μ^* = the chemical potential of the solute in its standard state (usually taken as 1 atmosphere pressure, where activity is unity, no reactions of solute with solid or gas–liquid interfaces and zero electric potential). It is therefore a constant for a given solute.

R = the gas content.

T = the absolute temperature.

$\ln a_s$ = the natural logarithm ($\log_e$) of activity of the solute in g-moles litre^{-1}.

Except in very dilute solutions, intermolecular attraction between solute molecules causes them to exert less than their potential effect. The solution does not therefore behave in an ideal manner and the effective concentration of solute is less than the actual concentration; the ratio of effective and actual concentration is the molar activity ratio of the solute and for electrolytes is usually designated, f.

For electrolytes the activity of ions in solution can be precisely inferred from the total electrolyte concentration in the solution (Debye–Hückel equation), but for non-electrolytes, particularly if they are large molecules, such calculations cannot be made. It is possible to derive a practical activity coefficient for non-electrolytes but this does not usually differ greatly from unity unless the concentration exceeds 10^{-2} M. The value of f for di- and polyvalent electrolytes may differ markedly from 1 even in quite dilute solutions. For our present discussion we will assume that two solutions, A and B, are appreciably less concentrated than 10^{-2} M so that we may assume that the practical activity coefficient is unity; we can then rewrite eq. (1.1) in a more familiar form

$$\mu = \mu^* + RT \ln C_s \tag{1.2}$$

where C_s = concentration of the solute. If the concentration of the solute in A is greater than in B and the temperature in both compartments is the same, then there will exist a difference in chemical potential between them:

$$\mu A - \mu B = \Delta\mu \tag{1.3}$$

The movement of a solute by diffusion is dependent, then, on the existence of a gradient of chemical potential, $d\mu/dx$, between the two compartments.

If we now consider the situation in Fig. 1.2 we can see that if the membrane is semi-permeable, i.e., impermeable to the solute but allowing the passage of the solvent, the chemical potential difference can be made use of to perform work. Water will move from B into A

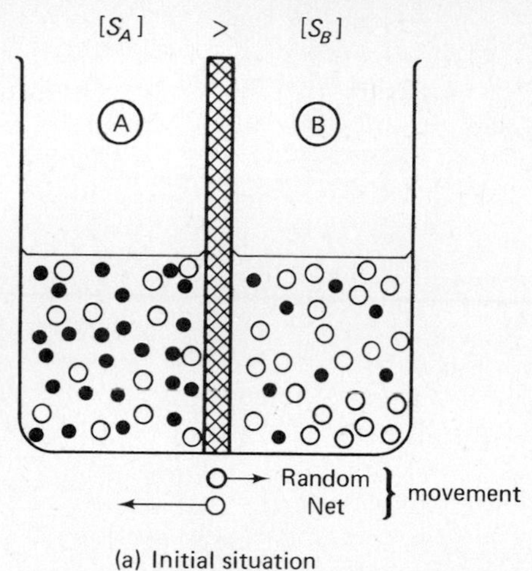

(a) Initial situation

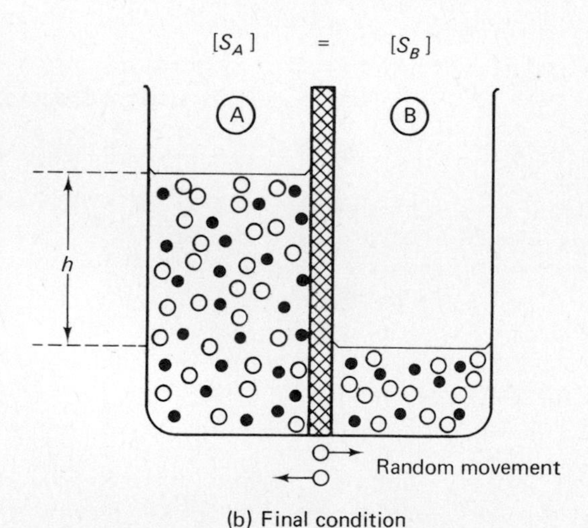

(b) Final condition

***FIG. 1.2** Movement of water molecules through a semi-permeable membrane separating two compartments initially containing a solute (solid circles) at different concentrations. There is a net movement of water molecules into A because the membrane is impermeable to the solute. When a final steady state is reached there is a head of water,* h, *proportional to the original difference in solute concentration (the osmotic potential).*

until the concentration of solute is equalized, but in so doing a head of water is built up the size of which depends on the original difference in chemical potential of the solute in the two compartments and certain frictional interactions between the solvent and the membrane. This process could be used as a basis for a machine; for instance, the rising water level in A could be used to move a piston. The force developed in A is again related to the difference in chemical or *osmotic potential*. To prevent the movement of solvent into A we must apply a force, the *osmotic pressure*, equal to the osmotic potential of the solution in A.

We have considered the driving force of the solute but what is it that causes the solvent molecules to pass from B into A? The molecules of the solvent, like everything else, move in response to a force which again results from a difference in chemical potential. When we dissolve a solute in a solvent the chemical potential of the solvent is lowered, thus the greater concentration of solute molecules in A exists in a solvent of lower chemical potential than those in the more dilute compartment. Solvent molecules, therefore, move 'downhill' thermodynamically from compartment B into compartment A. In more formal terms the solute affects the colligative properties of the solvent causing a group of related changes in its properties, e.g., the depression of freezing point, the elevation of boiling point, the depression of vapour pressure of the solvent and the creation of osmotic potential.

Diffusion gradients and the plant cell

The plant cell invariably has a higher total solute concentration in its sap than is present in the external solution; if it had not, water would pass from the cell causing its death. The cell is permeable to water and the chemical potential of water in the cell is lower than outside, so unless some specific provision is made to limit the entry of water, the protoplast will swell and eventually dilute itself out of existence or burst. The cellulose wall surrounding the plant cells prevents this happening and at the same time provides a framework of considerable rigidity because the protoplast exerts a pressure on the wall (the *wall pressure* in earlier texts) when the cell is fully turgid.

Since the concentration of most solutes within cells is usually higher than in the surrounding medium, diffusion is potentially capable of destroying the cell by dispersing its contents. It is necessary therefore for the cell to surround itself with some surface barrier which will limit this process; the barrier is known as the plasma membrane or plasmalemma and it is with this structure that much of this book is either directly or indirectly concerned. We shall see, however, that cells are able to accumulate certain solutes, especially ions, against gradients of concentration in a thermodynamically 'uphill' direction. Such movements are the thermodynamic equivalent of heat passing from a cooler to a warmer body and can only be achieved if energy is consumed by the system and chemical work is performed. At this point in the discussion we are ignoring certain factors such as the electrical potential difference which is usually found across the plasma membrane, and the very important influence which this has in determining whether ions move uphill or downhill on their journey into cell. These matters will be discussed fully in chapter 3.

Creation of order by organisms

Organisms, especially those capable of photosynthesis, are able to synthesize complex molecules from simpler ones in their environment. The synthesis of highly ordered polysaccharides (e.g., starch) by plants from carbon dioxide and water leads to a decrease in the randomness of carbon in the system. This can be achieved only by the absorption of energy from the surroundings because the atoms of carbon are moved from a lower to a higher chemical potential. The source of energy is light emitted from the fusion reactions in the sun and trapped by chlorophyll in the photosynthetic apparatus. If we consider now the whole or universal thermodynamic system we may ask whether the increase in order resulting from the synthesis of starch is greater or less than the increase in randomness in the solar system resulting from the transfer of radiant energy to the plant. Because the conversion of radiant energy into chemical energy by the plant is an inefficient process, more potential energy is dissipated by the sun than is created by the temporary reconstruction of order from chaos. Thus, the randomness of the system as a whole increases. A system containing sun, plants and their chemical environment is a rather large and unwieldly one, but the example makes the general point that the decrease in potential energy in one component of a system may, by the performance of work, serve to increase the potential or free energy of another.

Free energy, work and entropy

When it is in equilibrium every system contains an intrinsic amount of energy, U, and if some change of state occurs in the system the change of its intrinsic energy is given by

$$U_i - U_f = \Delta U \tag{1.4}$$

where U_i and U_f represent the intrinsic energy in the initial and final states. If the conditions are so arranged that this change of state occurs at constant pressure and temperature, some energy (heat) will be gained from or lost to the surroundings; if we ensure that no work is performed during this change of state the heat absorbed, the *enthalpy change*, ΔH is given by

$$\Delta H = \Delta U + P\,\Delta V \tag{1.5}$$

where P = pressure in atmospheres
and ΔV = change in volume of the system.

In most changes of state which are of interest in biochemistry the term $P\,\Delta V$ is negligibly small so that

$$\Delta H = \Delta U \tag{1.6}$$

The sign of the enthalpy change may be positive, in which case heat will be absorbed by the system and an *endothermic* reaction will have occurred, or may be negative as in the case of an *exothermic* reaction.

If the system is to be put to useful work, the sign of the enthalpy change must be negative and energy (heat) must be liberated by the change of state. Is all the heat liberated in the enthalpy change available for work? The answer to this question is no. A certain amount of energy is consumed by the system itself during the change of state. The amount of useful energy liberated by the system is called the Helmholtz free energy, F, and is not the same as the change in enthalpy. We can symbolize these statements as follows:

Since $\Delta F \neq \Delta H$
and $\Delta F < \Delta H$
then $\Delta F = \Delta H - (x)$ (1.7)

The quantity of energy (x) is not available for work because it has been consumed in the change of state; this quantity is the *entropy change* which depends on the absolute temperature of the system and is defined

$$Q = T\,\Delta S \tag{1.8}$$

where Q = heat necessary to bring about the change of state, and S = entropy. $T\,\Delta S$ has the units of calories mole^{-1} degree^{-1}.

We can now rewrite eq. (1.7) in a more conventional form

$$\Delta F = \Delta H - T\,\Delta S$$

The maximum amount of work, W_{max}, which can be performed by a change of state is simply related to the change in free energy

$$W_{max} = -\Delta F \tag{1.9}$$

Note that work can only be performed if ΔF is negative, i.e., there has been a *decline in the free energy of the system*.

Physiologists and biochemists use a slightly different definition of free energy, the Gibbs free energy, G, which is related to the Helmholtz free energy

$$G = H - TS + PV \quad (1.10)$$

In fact G and F are practically identical except in changes of state which involve significant changes in volume against a constant environmental pressure. Such situations rarely concern the physiologist so that the distinction between F and G need not concern us greatly.

All changes of state involve changes in the entropy of the system and where work is performed by the system this leads to an increase in entropy. Entropy is a measure of the randomness and disorder mentioned in the introductory paragraphs of this chapter. In eq. (1.11) entropy is defined in a somewhat different manner by relating it to a thermodynamic probability function, B

$$S = \text{constant} \ln B \quad (1.11)$$

According to this relationship, systems increase in entropy when they move from a lower to a higher state of probability and hence become more stable. We may therefore predict, with a high degree of probability, the increasing disorder in the universe.

Consumption of free energy in solute transport

There are many components in a cell so that a decrease in free energy in one component may lead to an increase in the free energy of another. Let us illustrate this by considering the relationship between respiration and solute accumulation in the cell. The oxidation of 1 g-mole of glucose in aerobic respiration to form carbon dioxide and water proceeds with a decrease in the free energy of the system of −673 000 cal.mole. Some of this energy becomes available, through processes which we need not specify at this stage, for solute accumulation in the cell, so that the free energy released is used to make the cell more concentrated in a particular solute than the surroundings. Using classical thermodynamic theory, and for the moment ignoring certain important properties of the cell membrane, we can calculate the amount of free energy which must be consumed in moving a solute from the dilute surroundings to the concentrated interior of the cell using eq. (1.12)

$$\Delta G = RT \ln \frac{C_i}{C_o} \quad (1.12)$$

where R equals the gas constant, T the absolute temperature and C_i and C_o the concentration of the solute inside and outside the cell respectively. For instance, if the concentration of an uncharged solute in the cell is 10^{-2} M and in the external medium is 10^{-4} M, then the *minimum* amount of energy which must be expended to move 1 g-mole of that solute against the concentration gradient at 20 °C may be worked out as follows:

$$G = 1{\cdot}98\ (R) \times 293\ (T) \times 2{\cdot}303 \log_{10} \frac{10^{-2}}{10^{-4}}$$

$$= 1{\cdot}98 \times 293 \times 2{\cdot}303 \times 2$$

$$= 2667\ \text{cal.mole}^{-1}$$

Thus, the oxidation of 1 mole of glucose should be capable of transporting 673 000/2667 moles of solute up the concentration gradient into the cell. In practice, however, we would find that much more energy has to be invested in the transport of the solute into the cell than is given by the equation above. This is because of the properties of the cell membrane which we chose to ignore in our first approximation. The equation above (1.12) is one which is used in equilibrium thermodynamics and its inability to describe completely a biological situation points to a very important feature of active transport systems requiring an input of energy to drive them. The membranes of cells are leaky and, even though work is going on to concentrate the solutes, there is a continual flow back into the surrounding medium. Thus, when we observe a steady state in the distribution of a solute between the cell and its surroundings, we are not looking at a static situation but a dynamic one in which rates of movement into and out of the cell are balanced. In chapter 3 we shall be considering these *fluxes* of solute in more detail. Thus the amount of energy necessary to transport 1 mole of solute against the one hundredfold concentration step is likely to be much greater than is given by eq. (1.12).

Basic cell structure

Before we can understand how metabolic processes are linked with transport systems and how the cell selectively absorbs essential nutrients from its environment we should have a clear but simple idea of the structure of a plant cell and the functional relationships between its various components.

The plant cell is bounded by two structures, the outer being a cellulose wall with a central amorphous matrix of long polymeric molecules, collectively described as pectin, which carry fixed negative charges. One important feature of this wall is that, in most situations, the interstitial spaces between the chain-like polymers of cellulose and pectin are very large in comparison with the size of dissolved solute molecules. What appears to us down the light microscope as a dense barrier is, relative to most solutes and all inorganic ions, a huge, very loosely formed net. Once through this wall ions encounter the plasma membrane, a much more compactly organized structure which does not have large holes in it. This membrane encloses a layer of cytoplasm which can be regarded as a complex arrangement of fibrillar protein molecules carrying fixed charges in which are suspended numerous organelles each surrounded by its own membrane. The cytoplasm is, therefore, a system of many compartments in which the concentration of ions may not be uniform. The large number of sites in the cytoplasm carrying fixed charges are neutralized by electrostatic binding of small ions of opposite sign which are transported from the outside solution. Thus, a proportion of the ions in the cytoplasm is fixed and not free to diffuse. The inner boundary of the cytoplasm is marked by a membrane, known as the *tonoplast*, which surrounds the vacuole. The vacuole is a concentrated solution largely inorganic in nature in which ions are in true solution. The question is at once raised whether the 'barrier' properties of the tonoplast and the plasmalemma differ from one another and whether the relative concentration of solutes in the vacuole differs from that of the cytoplasm. We can forecast, however, that the total concentration of its solutes (its osmolality) is similar to that of the cytoplasm. If this were not so the cytoplasm would either become crushed against the cell wall or the cell would lose its turgor. It is not, however, in direct communication with the external medium. To what extent does the presence of the cytoplasm and the plasma membrane influence the properties of the tonoplast? Is the ionic concentration of the vacuole very different from that of the cytoplasm? We can partly answer the latter question by examining examples of ionic composition of giant algal cells, one of which will be referred to frequently in later chapters. Algal cells may seem a remote starting point for those who want to understand transport processes in the higher plants, but they have been chosen because more is known about the factors controlling ion transport in them than in any other type of plant cell.

Giant algal cells

Certain families of the algae contain species with very large cells. These species have attracted the attention of research workers because they are sufficiently large to allow accurate, direct determinations of the ionic composition in the various compartments of the cell. Such measurements are much more difficult, if not impossible, with the small cells of higher plants. The volume of the vacuole in a cortical cell from a barley root (*Hordeum vulgare*) is approximately 7×10^{-6} ml, while that in the alga *Hydrodictyon africanum* is approximately 4×10^{-2} ml. The algal vacuole is, therefore, about

5000 times larger and contains enough sap for measurements of solute concentration to be made on single cells. Single internode cells of the alga *Nitella translucens* contain enough cytoplasm for direct measurement of its average ionic composition (MacRobbie, 1964). Gutknecht (1966) developed a technique whereby the vacuole of a marine alga, *Valonia ventricosa*, can be perfused with solutions of predetermined ionic composition and was thus able to make direct observations on the influence of sap composition on ion movements across the plasma membrane and the tonoplast.

The comparatively large volume of the cells in these algae facilitates measurement of the electrical potential differences which exist across the plasma membrane and the tonoplast. These membrane potentials are measured by inserting glass micro-electrodes into the cells and comparing the electrical potential with that of the external solution by connecting them through a millivoltmeter with sufficiently high impedance to prevent a short circuit between the electrodes. The measurements, although very simple in character, are complicated by a number of mainly technical difficulties. These will be discussed in detail in chapter 3.

Although several algae will be used in later chapters to illustrate specific matters one of them, *Hydrodictyon africanum*, will be more extensively discussed. The ionic regulation in this organism, which has been intensively studied by Raven (1967a, b; 1968; 1969a, b; 1970; 1971a, b), will be used as a case history to which the theory developed in later chapters will be applied.

Hydrodictyon africanum: a case history

Hydrodictyon africanum, a member of the Chlorophyceae, family Chlorococcales, is a brackish water species which is found in low lying flooded coastal flats in South Africa where it may experience and adapt to considerable fluctuations in the ionic composition of its environment (Pocock, 1960). It may be cultured in synthetic media and under laboratory conditions will go through the rather unusual features of its morphogenesis. After fertilization, the zygote undergoes successive divisions to form 256 or 512 daughter cells. These cells grow rapidly in size and form themselves into a loose net-like structure which is called the *coenobium*. Although they are physically joined in this way there appears to be no protoplasmic connection between the individual cells (coencytes), and when they reach a mature stage in their development the coenobium breaks up yielding 256 or 512 spherical coenocytes of similar age and identical genetic composition. The coenocytes are two to four months old at this stage and may be 3–6 mm in diameter, but they continue to swell up into spheres 1 cm or more in diameter (Fritsch, 1935). The coenocytes have a large central vacuole and a thin layer of cytoplasm much of which is filled by a highly complex irregular chloroplast which appears to be a continuous but convoluted sheet with numerous holes (fenestrations) in it.

If coencytes of *Hydrodictyon africanum* are maintained in a dilute solution of sodium and potassium chloride for a few hours a steady state is established in which there is a marked asymmetry in the distribution of the ions across the plasma membrane (Table 1.1), all three ions being at a higher concentration in the vacuole and the cytoplasm than in the outside solution. We should not conclude that work must be done in transporting all three ions.

***Table 1.1** Ionic composition of* **Hydrodictyon africanum** *in relation to the outside solution*

Ion	Concentration (millimoles/litre)		
	Outside solution	Inside	
		Cytoplasm	Vacuole
K	0·1	93	40
Na	1·0	51	17
Cl	1·3	58	38

Measurements made when cells in equilibrium at 14 °C in the light. *(Data taken from Raven (1967a) with kind permission.)*

Although there is little evidence to suggest that an anion and a cation are transported in company with one another (as a molecule of salt) the movements of ions of opposite charge interact so as to maintain approximately the electrical neutrality of the system. The membrane is generally more permeable to cations than anions but both types of ion move across the membrane at comparable rates. We can illustrate this by thinking about a membrane which separates two solutions of KCl at different strengths; the membrane is more permeable to K and as salt begins to diffuse across it the K^+ begins to move ahead of the Cl^-. This tendency begins to separate the charges so that an electric field is set up which is positive at the advancing K^+ front; this will impede the progress of the K^+ by charge repulsion. Conversely, Cl^- will be attracted to the positively charged zone and its migration will be speeded up under the influence of this additional driving force. Eventually, both ions will move at an approximately comparable rate so that little net charge separation occurs.

This simplified example does not illustrate exactly what happens but it makes the valid point that the movement of one ion may modify the movement of another by producing an electric driving force.

It may not be necessary to expend free energy on the transport of both anions and cations since the transport of one may passively induce the transport of the other. We shall be returning to this matter in chapter 3.

Table 1.1 also shows that the ratio of sodium to potassium changes from 10:1 in the outside solution to approximately 0·5:1 in the cytoplasm and the vacuole; thus the membrane discriminates in favour of potassium. When the cell is in a steady state, the rates of inward and outward movement of ions are balanced and the coenocyte is said to be in a state of *flux equilibrium*. Measurement of membrane potentials and flux rates at equilibrium and after physical or biochemical perturbation of the system will provide us with a basis for deciding which of the fluxes require direct inputs of energy to drive them (*active transport*) and those which do not (*passive diffusion*). Some understanding of the relationships which may exist between active transport and cellular metabolism can be reached by selectively blocking certain metabolic processes with inhibitors.

If as a result of what we learn in the subsequent chapters we can begin to understand how *Hydrodictyon* regulates its internal ionic composition then we shall have a secure base from which we can explore the processes governing ion transport in all other types of plant cell.

Priorities

We have raised a number of questions in these introductory remarks but let us be sure that we have a clear idea of the most important ones before we become involved with too much detail. The membranes of the cell are clearly key structures determining the asymmetric distribution of ions between the cell interior and the surroundings. It is essential therefore that, before we do anything else, we learn something of their detailed structure. The second important matter is to consider how ions actually cross the membrane and to understand the driving forces involved in their movement. When work has to be performed to transport an ion into the cell we would like to know how this work is coupled with the ion-transporting apparatus both biochemically and structurally; and, lastly, we should ask how the apparatus in the membrane involved in ion transport is able to distinguish between various species of ion.

Bibliography

a Further reading

Dainty, J. (1969) The water relations of plants, ch. 12 in *The Physiology of Plant Growth and Development*, Ed. M. B. Wilkins. McGraw-Hill, London.

A readable and simplified account of water movement based on thermodynamic principles.

Slatyer, R. O. (1967) *Plant–Water Relationships*. Academic Press, N.Y. and London.

A more rigorous treatment of water relations which would, perhaps, be a rather ambitious starting point for the elementary reader. It is, however, one of the most valuable books available to students equipped with some basic knowledge of the subject.

Lehninger, A. L. (1966) *Bioenergetics*. Benjamin, N.Y.

This book provides a stimulating and lucid introductory account of thermodynamic principles in biochemistry and physiology.

Spanner, D. C. (1964) *Introduction to Thermodynamics*. Academic Press, N.Y. and London.

Written for biologists this book contains a particularly good account of the relationship between free energy, work and entropy and also provides an introduction to the thermodynamics of irreversible processes.

Morris, J. G. (1968) *A Biologist's Physical Chemistry*. Edward Arnold Ltd, London.

A useful book for the student to have beside him to consult any point of physical chemistry not adequately explained in this text.

Clowes, F. A. L. and Juniper, B. E. (1969) *Plant Cells*. Blackwell, Oxford.

A large number of cell types are considered in great detail and the book contains profuse illustration.

b References in the text

Fritsch, F. E. (1935) *The Structure and Reproduction of the Algae*, Vol. I, p. 171. Cambridge.

Gutknecht, J. (1966) Sodium, potassium and chloride transport and membrane potentials in *Valonia ventricosa*. *Biol. Bull.* **130**, 131–137.

MacRobbie, E. A. C. (1964) Factors affecting fluxes of potassium and chloride ions in *Nitella translucens*. *J. Gen. Physiol.* **47**, 859–877.

Pocock, M. A. (1960) *Hydrodictyon*: a comparative biological study. *J. South African Bot.* **26**, 167–204.

Raven, J. A. (1967a) Ion transport in *Hydrodictyon africanum*. *J. Gen. Physiol.* **50**, 1607–1625.

(1967b) Light stimulation of active ion transport in *Hydrodictyon africanum*. *J. Gen. Physiol.* **50**, 1627–1640.

(1968) The linkage of light-stimulated Cl influx to K and Na influxes in *Hydrodictyon africanum*. *J. Exp. Bot.* **19**, 233–253.

(1969a) Action spectra for photosynthesis and light stimulated ion transport process in *Hydrodictyon africanum*. *New Phytol.* **68**, 45–62.

(1969b) Effects of inhibitors on photosynthesis and the active influxes of K and Cl in *Hydrodictyon africanum*. *New Phytol.* **68**, 1089–1113.

(1970a) The role of cyclic and pseudocyclic photophosphorylation in photosynthetic $^{14}CO_2$ fixation in *Hydrodictyon africanum*. *J. Exp. Bot.* **21**, 1–16.

(1971a) Cyclic and non-cyclic photophosphylation as energy sources for active K influx in *Hydrodictyon africanum*. *J. Exp. Bot.* **71**, 420–433.
(1971b) Inhibitor effects on photosynthesis, respiration and active ion transport in *Hydrodictyon africanum*. *J. Membrane Biol.* **6**, 89–107.

two

The structure of cell membranes

At the present time, much of the molecular organization of biological membranes is imperfectly understood and the precise relationship between structure and physiological function is not known. The advances in knowledge in recent years have, however, been most impressive and it may not be long before the functional architecture of the membrane can be described at the molecular level.

The evidence of this chapter is not set out as a chronology of research discoveries since we shall begin by considering the gross morphology of the cell membranes revealed in the electron microscope. It should be realized, however, that there were inspired hypotheses about membrane structure well in advance of the time when it became possible to see membranes in the electron microscope. On the basis of chemical and physical studies in the late 'thirties Danielli and Davson suggested a general picture of membrane structure which has subsequently been confirmed by direct observation; this work stands as one of the highest achievements in modern biology. But having paid this tribute, we will take advantage of recent technical advances which allow us to look at the membrane directly. Before we do this it is perhaps important to realize that the membrane as described by different kinds of investigator may not always be exactly the same thing; the biochemist, the anatomist and the biophysicist have combined their talents most effectively in an

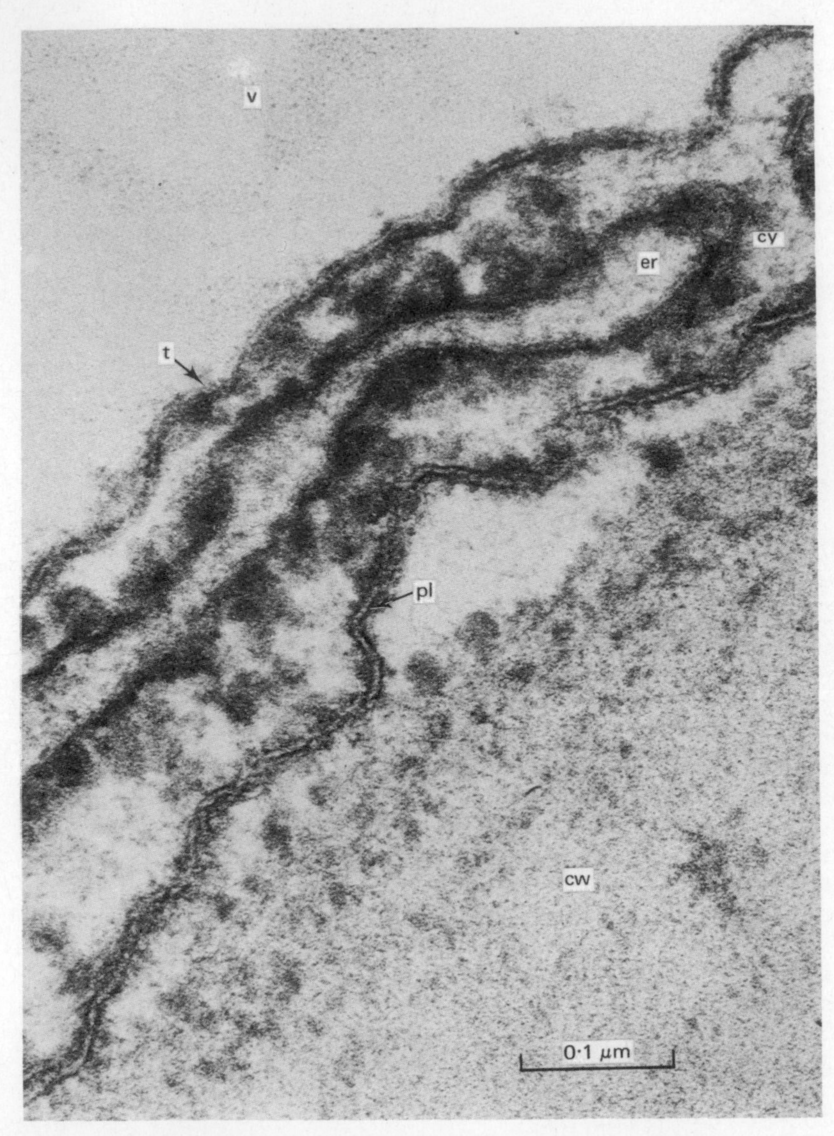

***FIG. 2.1** Electron micrograph showing unit membranes in a cell from the root tip of* **Juniperus chinensis.** *Notice that the plasmalemma (pl) is more darkly stained than the tonoplast (t). Between these membranes is the cytoplasm (cy) and a long profile of endoplasmic reticulum (er). The amorphous regions are, top left, the vacuole (v) and, bottom right, the cell wall (cw). Magnification about ×100 000. (I am indebted to Dr Myron Ledbetter for this print.)*

effort to understand membrane structure in relation to those functions which interest the physiologist, but it is not yet certain whether what we can see and what we can infer from these studies is the whole story.

Membrane structure seen in electron micrographs

The electron micrograph in Fig. 2.1 is taken from a cell in the root tip of *Juniperus* and two separate membranes can be clearly distinguished. The plasmalemma, which limits the entire protoplast, and the tonoplast, which encloses the vacuole, consist of two darkly staining bands which have adsorbed the electron dense ions of osmium, separated by a layer of low electron density. The plasmalemma appears to be somewhat thicker and is stained more deeply—this is often the case in other tissues. Careful investigation of these bands from *Juniperus* and a range of cells as diverse as those from yeast and mammalian nerves shows that the outer and inner bands are approximately 2·5 nm in width and that the central unstained band is 2·5–3·5 nm. With rare exceptions all electron micrographs, prepared from tissues which have been embedded and viewed in thin section, show this structure irrespective of the cell type or function. The three layers are integral parts of the membrane and this arrangement has become known as the *unit membrane*.

In Fig. 2.1 a portion of the endoplasmic reticulum, ER, occurs in the cytoplasm; it is seen here to consist of a sack-like structure surrounded by a rather thin intensely staining unit membrane. The view which we get from thin sections largely obscures the fact that the endoplasmic reticulum is a continuous network of cavities bounded by unit membrane which ramifies throughout the cytoplasm. The true picture of the ER was built up by painstaking reconstruction of serial sections and the variations of the design of the ER in various types of animal cells has been described by Palade (1956). One of the features of the ER is that it provides a relatively large internal surface for biochemical reactions, e.g., protein assembly on ribosomes, and for transport of solutes to its interior, i.e., the cavity. There is usually very little structural detail to be seen inside the cavity suggesting that it contains a relatively homogeneous simple solution.

In much the same way as the ER can be regarded as an envelope bounded by unit membrane, the nucleus is surrounded by a double membrane which is usually called the *nuclear envelope*. One of the features of the nuclear envelope which distinguishes it from the plasmalemma and the tonoplast is that, unlike them, it is liberally peppered with large holes or pores. These can be seen in Fig. 2.2, which shows the nuclear envelope in a cell from the root tip of

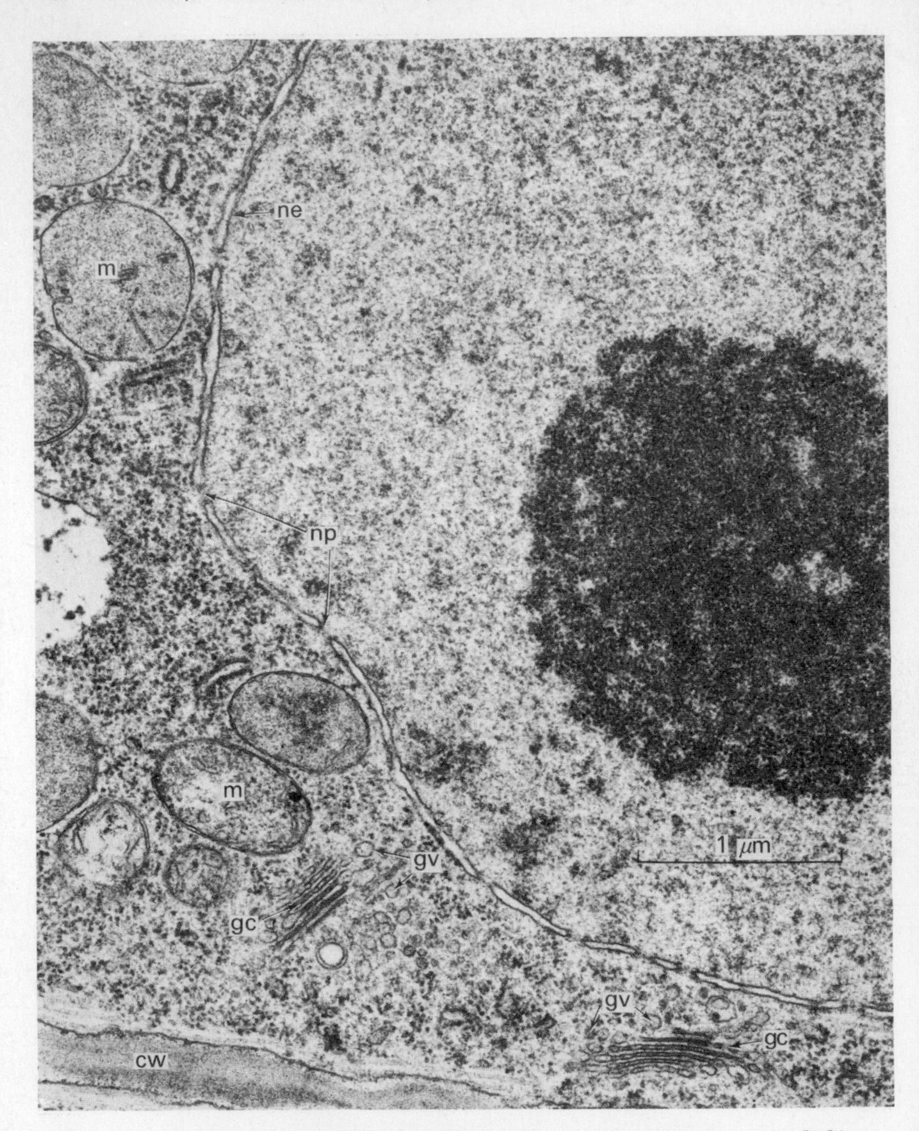

FIG. 2.2 Electron micrograph of a cell from the root tip of Raphanus *(radish) sp. Most of the area shown is occupied by the nucleus; the darker spot at its centre is the nucleolus. The nuclear envelope (ne) is frequently interrupted by pores (np). The granular appearance of the cytoplasm is due to ribosomes. Golgi cisternae (gc) and vesicles (gv) originating from them are seen towards the bottom of the picture; some of these vesicles may discharge their contents across the plasmalemma and thus add material to the developing cell wall (cw). Several mitochondria (m) are seen with their cristae only partly developed. Magnification about ×25 000. (I am indebted to Dr Myron Ledbetter for this print.)*

Raphanus sativus (radish), and also in surface view in the freeze-etched surface replica in Fig. 2.19 (page 45). The pores are very large 40–100 nm in width and on close examination have a complex structure associated with them which Franke (1970) has found to be broadly similar in a number of plant and animal cells. Figure 2.3 shows a pore in the nucleus of a meristematic cell in *Allium cepa* (onion); on both inner and outer faces of the envelope there are granular bodies apparently linked by fibrils to a fibrillar ring within the pore itself. The structure is illustrated in the diagram (Fig. 2.4) in which there is a central granular structure which Dr Franke believes to be ribonuclear-protein material *in transit* between the nucleus and the cytoplasm. The association of this material with the peripheral structures suggests that the pores do more than simply provide a large number of perforations in the nucleus; it seems possible that there may be some control of the rate of passage of such molecules as messenger RNA through the pores. The pores are, however, too large and too numerous to maintain an effective diffusion barrier to ions and other small solutes and it is unlikely, therefore, that the solute concentration or composition of the nucleus differs very markedly from that of the cytoplasm.

In other prominent cell organelles, the mitochondria and chloroplasts, macro-pores of the kind described above do not occur. Like

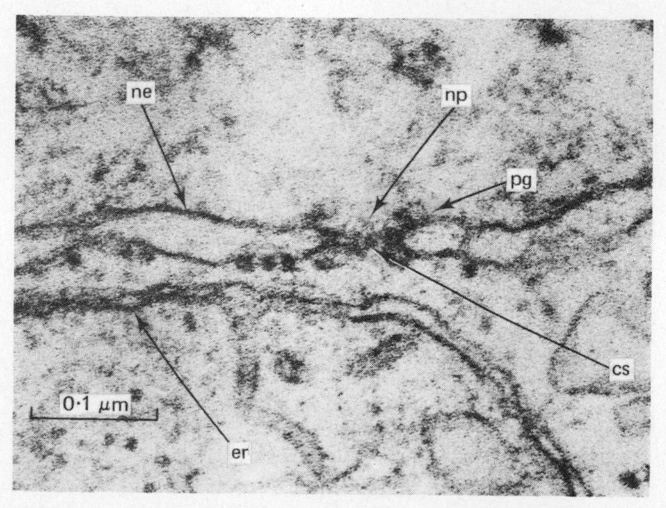

FIG. 2.3 Electronmicrograph showing structural detail of a pore in the nuclear envelope of Allium cepa *(onion). For discussion see text. Symbols: ne = nuclear envelope; np = nuclear pore; pg = peripheral granule; cs = central structure; er = endoplasmic reticulum. Magnification about ×160 000. (From Franke, 1970, by kind permission of the author and Springer-Verlag, Berlin.)*

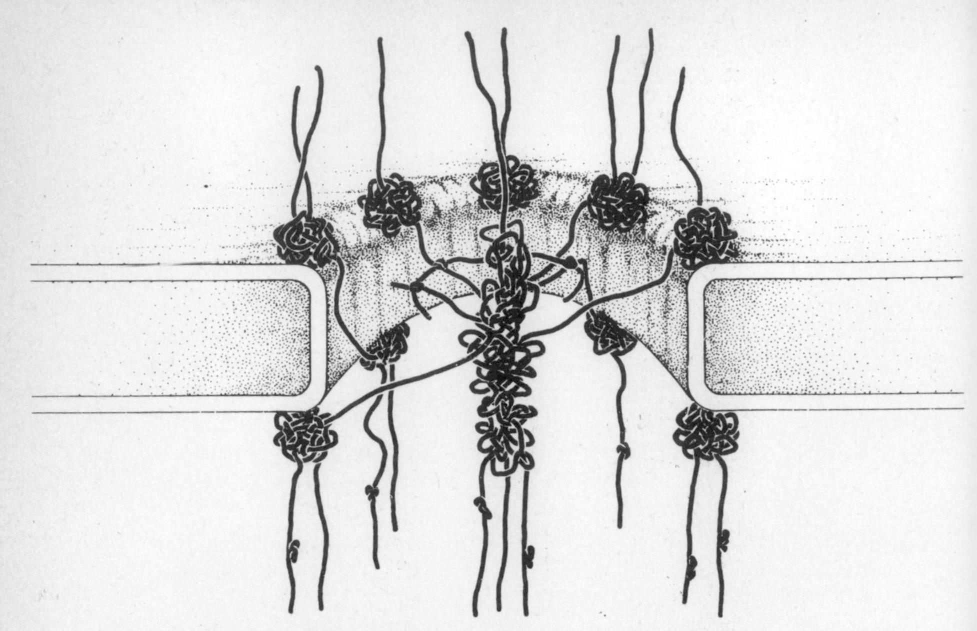

FIG. 2.4 Diagram of hypothetical ultrastructure of the nuclear pore. This represents a three-dimensional cross-section. The peripheral granules appear to have links with the central structure. (From Franke, 1970, by kind permission of the author and Springer-Verlag, Berlin.)

the nucleus, however, both are surrounded by a double membrane system or envelope, the inner layer of which is greatly elaborated (Fig. 2.5). There is much evidence to show that the internal solute composition of both mitochondria and chloroplasts does differ markedly from the cytoplasm and that the permeability and transport capabilities of their membranes are very different from those of the plasmalemma. They do, therefore, represent separate ionic compartments which can exist in a steady state with respect to the cytoplasm. We shall see that adjustments of this steady state promoted by metabolic activity may have major consequences at the cell's principal permeability barrier, the plasmalemma. The membranes of the chloroplast and the outer mitochondrial membrane appear to have the familiar trilayered structure about 7·5 nm wide when seen in transverse section, but the inner mitochondrial membrane suggests some major departure from the ubiquitous 'sandwich-cake' arrangement. Sjöstrand (1963) showed that this membrane has a ladder-like appearance in which the rungs are deeply staining bands which traverse the electron-transparent interior. It has been further established that there are projections from the membrane surface which contain specific assemblies of enzymes which have been the subject

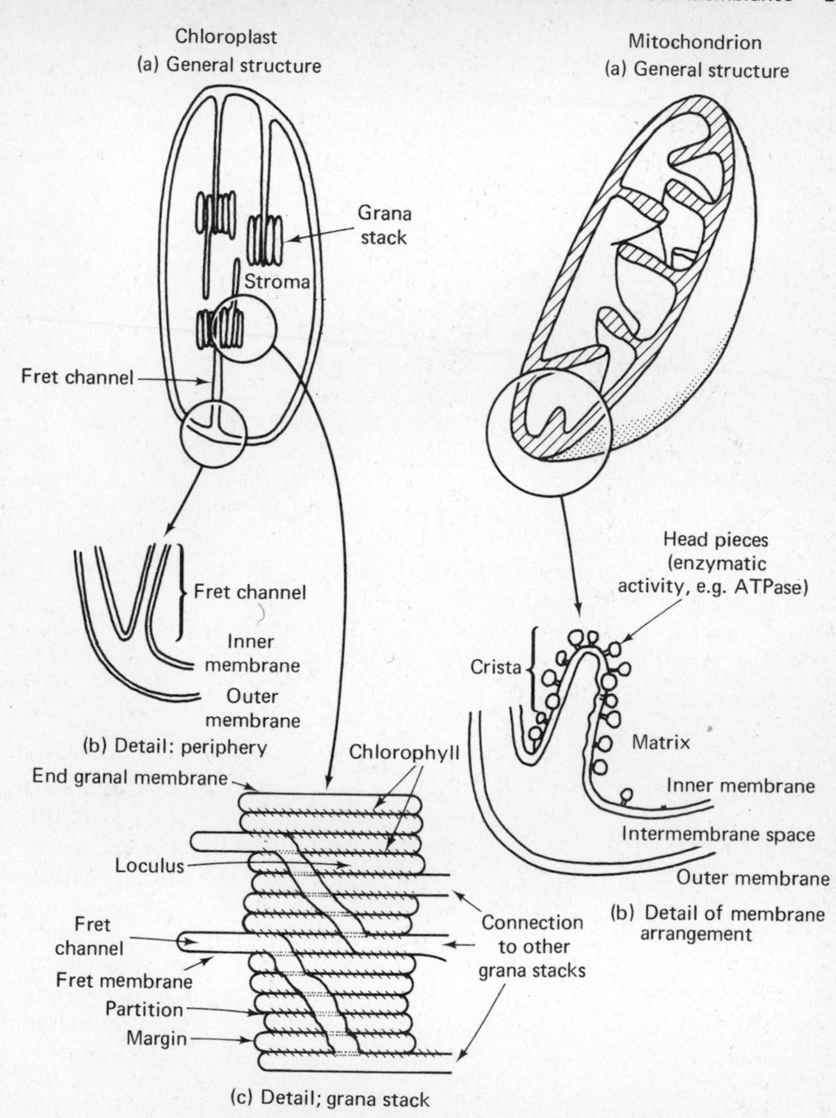

FIG. 2.5 Diagram of the double membrane system of mitochondria and chloroplasts. Both types of organelle are bounded by an outer membrane of relatively simple design and an inner one greatly elaborated for its functional role in electron transport and coupled phosphorylation.

of much research and controversy, some of which will be discussed later in the chapter.

Apart from this exception most membranes, even those which differ radically in their function, appear to be put together in much the same way. This should make us suspect that what we have seen so far in the electron microscope is merely a starting point for an enquiry into membrane organization. The tonoplast of the alga *Hydrodictyon africanum*, which is highly permeable to ions, the plasmalemma of the Schwann cell of nerve myelin, which functions primarily as an electrical insulator having a low ionic permeability (conductivity) and the membranes of chloroplasts which are concerned with energy trapping and electron transport can hardly be put together in exactly the same way, even if at first they appear to be. More recent electron microscopic evidence derived from freeze-fracture and freeze-etching studies, indicates quite clearly that there is a membrane matrix common to most, if not all, membranes which corresponds with the structure we have considered so far, but that there is also a substructure which may determine the characteristic function of a given membrane. We shall consider these two aspects of membrane structure separately.

The membrane matrix

Evidence from permeability studies

One of the earliest predictions of the composition of the plasma membrane was made by Overton (1899) who studied the rate of permeation of root hair cells of *Hydrocharis* by a range of non-electrolytes. The rate appeared to be a function of the lipid solubility of the solute tested and from this it was concluded that the membrane was predominantly lipoidal in character. This view was further advanced in the classic studies of Collander and Barlund (1933) and Collander (1949) who found that the rate of permeation of the giant internodal cells of the alga *Chara ceratophylla* by a number of organic solutes could be predicted from their molecular dimensions and their partition coefficient between olive oil and water. Figure 2.6 is a simplified version of a graph presented by Collander (1949) and only a few of the compounds tested by him are shown. The units of P (the permeability) which appear on the ordinate are in $cm\,.\,h^{-1}$; $M^{1/2}$ is the square root of the molecular weight. The graph shows, as we would expect from Overton's work, that molecules of high lipid solubility have high values for the parameter $PM^{1/2}$. The data also lead to the prediction that molecules of similar size differing only in their lipid solubility cross the plasmalemma and enter the cell at different rates, e.g., thiourea and propylene glycol both have a molecular weight of 76 and both have rather similar molecular

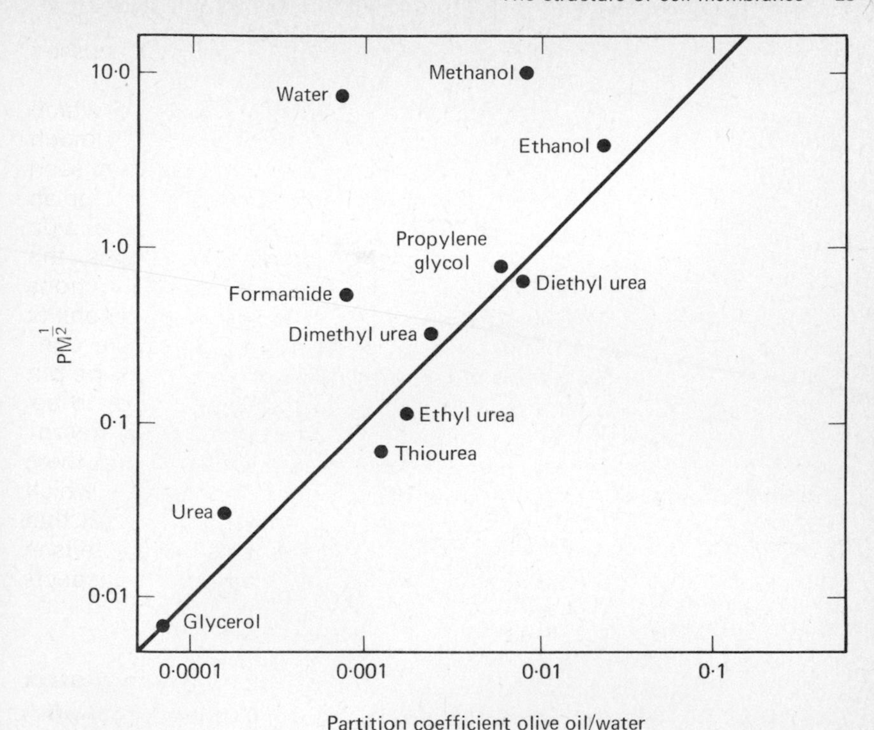

FIG. 2.6 The permeability of internodal cells of **Chara ceratophylla** *to organic non-electrolytes differing in their solubility in olive oil. The ordinate* $PM^{1/2}$*, indicates the permeability of the cell,* P = *the permeability in cm* h^{-1} *and* M = *the molecular weight of the solute under consideration. Note that three substances depart markedly from the linear relationship. (Adapted from Collander, 1949.)*

dimensions in solution; the latter, however, diffuses across the membrane ten times faster than thiourea since it is appreciably more soluble in lipid. There are, however, some non-electrolytes which do not fit this simple relationship. Figure 2.6 shows that the position of water is anomalous. The lipid solubilities of water and thiourea are similar and their molecular weights differ only by a factor of 4 and yet the value of P for thiourea is 0·09 cm . h^{-1} while that of water is 200 times greater, 1·88 cm . h^{-1}. The cell membrane appears, therefore, to be much more permeable to water than the lipid solubility of water would suggest. There are a few other anomalous molecules, e.g., methanol, formamide, but in each case they are small ones; this led Collander to propose that the plasmalemma had sieve-like pores in it, large enough to permit the passage of very small molecules but sufficiently small to ensure that larger ones have to diffuse through the lipid in order to enter the cell.

When a solute is present in an aqueous solution a number of hydrogen bonds will be formed with water molecules in its vicinity. The number of these hydrogen bonds depends on the chemical groups present in the molecule. In Table 2.1 the numbers of H-bonds

Table 2.1 Hydrogen-bonding functions and N values for various chemical groups

Function	Group in which present	N value
—OH	Alcohols, sugars, glycols, carboxyllic acids	2
H—O—H	(Water)	4
$—NH_2$	Primary amines and amides	2
—CO—	Carboxyllic acids, amides, aldehydes	1
	Esters	0·5

(Data abridged from Stein, 1967, with kind permission.)

for various types of group are shown. When the solute leaves the aqueous phase (the external solution) and enters the lipid phase of the membrane, the hydrogen bonds which it makes with water must be broken. Clearly this must involve some force, and the force will be greater in the case of molecules forming larger numbers of H-bonds. Stein has reconsidered Collander's data and shown that $PM^{1/2}$ can be replotted against the number of hydrogen bonds (the N value) formed with water and a somewhat better fit with the observed permeabilities is achieved (Fig. 2.7). But in this treatment also, the permeability of water remains anomalously high, adding weight to the proposal that there must be some special provision for getting water through the membrane. If there are pores in the membrane which would allow the rapid passage of water, how large would they have to be? The diameter of a single water molecule is about 0·24 nm which sets a lower limit for the diameter of these hypothetical pores. Much experimental work, principally with red blood cells, has been directed to determining the size of the pores and most estimates give a radius in the order of 0·4 nm (see Goldstein and Solomon, 1960, and Stein, 1967, for full discussion). Could we hope to see pores as small as this in the electron microscope if, indeed, they exist? Unfortunately the present limitations of resolution of the electron microscope, especially in sections which are 20–50 nm thick, do not permit visualization of such small structures. Evidence for their existence is, therefore, circumstantial, but there is general agreement that a pore at any place in the membrane may not be a permanent structure and may be best thought of as a transient gap or imperfection in the membrane resulting from a number of different molecular movements and rearrangements of the membrane matrix.

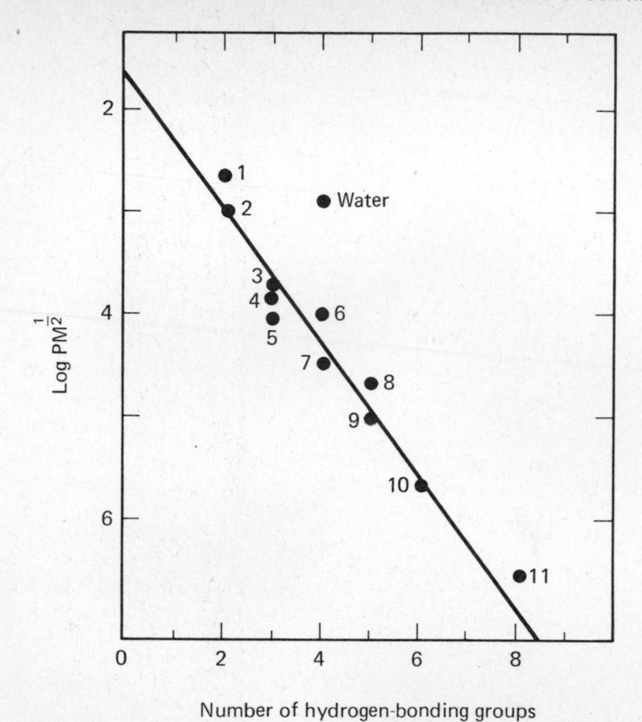

FIG. 2.7 Permeability data for Chara ceratophylla *replotted to show the variation of log* $PM^{1/2}$ *with the number of hydrogen bond forming groups in a number of non-electrolytes. The solutes have been numbered as follows: 1. methanol; 2. ethanol; 3. diethyl urea; 4. formamide; 5. dimethyl urea; 6. ethylene glycol; 7. ethyl urea; 8. thiourea; 9. urea; 10. glycerol; 11. erythritol. Note that in this relationship methanol and formamide fall on the same line as the other solutes (cf. Fig. 2.6) but that the position of water is still anomalous. (Data taken from Fig. 3.6 in Stein, 1967, with the kind permission of the author.)*

Chemical composition of the matrix

Isolation of membranes for chemical analysis was not achieved for some years after the early predictions of lipid character of the membrane but analyses have shown that these predictions were substantially correct. A proportion of the dry weight of all membranes is made up of lipid and phospholipid; there is in addition a variable amount of protein. In membranes, such as nerve myelin, which has a very low ionic conductance 76 per cent of the dry weight is lipid and only 20 per cent is protein. At the other extreme, an analysis of isolated plasmalemma from rat liver had 85 per cent protein and only 10 per cent lipid (Emmelot *et al.*, 1964). A range of intermediate situations from bacterial cells, animal and plant tissues has been described. A third group of substances associated with some membranes are mucopolysaccharides which are present with the protein in the peripheral layers. The antigenic characteristics of cells are probably determined by these molecules.

Orientation of components of the membrane

All membranes contain phospholipid and a consideration of the structure of these molecules gives an important insight into the way a membrane is constructed. Phospholipids possess a charged 'head' region with which phosphorus is associated and a long uncharged 'tail' of hydrocarbon. Figure 2.8 illustrates the structure of one of the

Lecithin: phosphatidyl choline

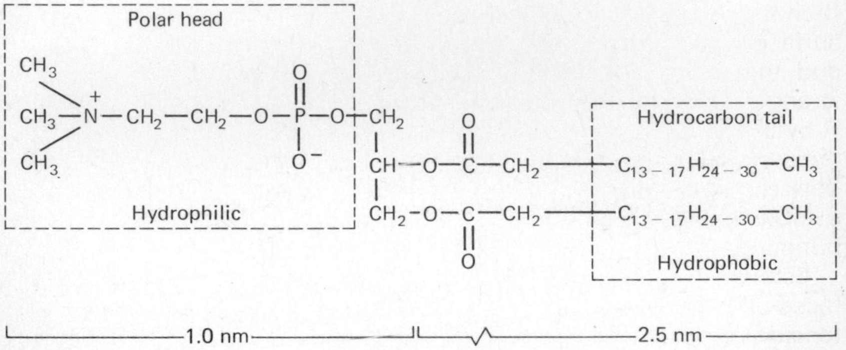

Polar head

$$(CH_3)_3\overset{+}{N}-CH_2-CH_2-O-\overset{O}{\overset{\|}{P}}(O^-)-O-CH_2$$

$$CH-O-\overset{O}{\overset{\|}{C}}-CH_2-C_{13-17}H_{24-30}-CH_3$$

$$CH_2-O-C(=O)-CH_2-C_{13-17}H_{24-30}-CH_3$$

Hydrophilic

Hydrocarbon tail

Hydrophobic

1.0 nm — 2.5 nm

FIG. 2.8 Chemical structure of phosphatidyl choline (lecithin), a phospholipid.

most commonly occurring phospholipids, lecithin or phosphatidyl choline. The polar head region forms hydrogen bonds with water and is therefore hydrophilic; it is approximately 1 nm in length. The long tail is composed of hydrocarbon chains which are uncharged and are hydrophobic. The two phases which are separated by the cell membrane are both aqueous; we might expect, therefore, that the hydrophilic parts of the phospholipid are directed towards the aqueous phases on either side of the membrane and the hydrophobic tails away from them. The phospholipid molecules thus form a bilayer, or bimolecular layer as it was first described by Gorter and Grendel (1925).

Neutral lipid molecules, such as cholesterol do not have a polar region and are associated with the hydrophobic core of the membrane where they are probably inserted between adjacent phospholipid molecules where they tend to increase the structural order and reduce the fluidity of the hydrocarbon phase of the membrane. The significance of this effect will emerge in later sections.

The analysis of membranes shows that they contain often large quantities of protein; how are these molecules to be incorporated into the hypothetical structure we are building up? Protein can become associated with the polar head region of phospholipid by electrostatic bonding but can also undergo hydrophobic bonding with the hydrocarbon. Evidence from several widely different types of investigation helps us to establish how a major portion of the protein is bound to the membrane.

The orientation of the components of the membrane was suggested by Danielli after a consideration of the surface tension of cells. At any liquid interface, the surface behaves as if it were a skin of molecules. A tension is developed in this skin because molecules in the surface layer are more strongly attracted to molecules in the interior than they are to those in the other phase. Any attempt to expand the surface of the interface is resisted by this inwardly directed tension and the magnitude of the interfacial tension will depend on the strength of the mutual attraction of the molecules of the two phases. A hydrophobic surface of an oil droplet has, therefore, a high surface tension when dispersed in water since there is little mutual attraction between the molecules of oil and water. The inclusion of a material, soluble in both phases, breaks down this surface tension; this is the principle involved in the emulsifying properties of certain molecules.

Estimates of the surface tension of oil or neutral lipid/water interfaces usually lie in the range 10–50 dyn cm^{-1} at 20 °C. The surface tension of cells is generally much lower than this, estimates ranging from 0·1 to 2·0 dyn cm^{-1} (Frey Wyssling, 1953). This reduction is explained partly by the fact that the polar heads of the phospholipids themselves lower the surface tension (see Table 4.4, page 104) and partly by the presence of protein on the outer surface of the cell membrane. Protein is capable of bonding both with the lipid components of the membrane and the surrounding water molecules; it is therefore a surfactant.

Support for this view comes from studies on the morphology of synthetic lipid bilayers. These synthetic membranes can be formed when mixtures of lipid and phospholipid are dispersed on the surface of water (see also chapter 4). At first, lipid tends to form a monomolecular layer with the polar regions of the molecules facing the aqueous phase. Lateral compression or crowding of the phospholipid molecules on the surface causes folding and a characteristic series of bilayers is produced. The morphology of these bilayers can be seen in the electron microscope after preparations have been fixed and treated with osmium. They resemble membranes seen in the electron micrographs from cells (Figs. 2.1 and 2.2) in that they consist of a repeating series of three bands. They differ from natural membranes principally in the fact that the outer bands are thinner, 1–1·5 nm as opposed to 2·5 nm in natural membranes, and that the bands are less electron dense, having adsorbed less osmium. The central 'unstained' region can be about the same width as, but is frequently seen to be somewhat thicker than, biological membranes. Capacitance and electron microscope studies of bilayers and biological membranes indicate that the thicker central band in the former is often a product of the hydrocarbon solvent used to disperse the lipid (Fettiplace, Andrews and Haydon, 1971). (It should be noticed that the thickness of the faintly staining outer bands is about that given for the polar region of lecithin in Fig. 2.8.) The inclusion of a suitable protein in the lipid dispersion forms bilayers in which the peripheral bands are thicker (2–2·5 nm), and stained more intensively with osmium—the central region remains unstained and the same width. This bilayer now resembles closely the characteristic appearance of the unit membrane seen in living cells. The inescapable conclusion is that protein has become attached to the polar heads of the phospholipid and is thus on the outside of the membrane.

The third line of evidence for the superficial bonding of protein in the membrane comes from studies on red blood cells from which as much as 70 per cent of the protein of the membrane can be solubilized in distilled water provided that the membrane surface is continuously flushed to minimize ionic concentrations (Branton, 1969). It is difficult to imagine how this could occur if the protein were bonded to the hydrocarbon region of the lipid and phospholipid (see Branton, 1969, and Branton and Deamer, 1972, for a discussion of opposing views on the membrane structure). Perhaps some of the confusion results from the way in which investigators have set about stripping the protein from the lipid matrix. It is possible that drastic treatments of red cell and other types of membranes with detergents and solutions of high ionic strength actually reduced the amount of protein extractable, suggesting that some rearrangement of protein binding might result from such treatments thus giving a specious indication of hydrophobic binding.

Having stressed the evidence for the superficial binding of membrane protein it must be admitted nevertheless that there is good evidence for hydrophobic interactions between protein and lipid (Richardson *et al.*, 1963; Lenaz, 1970). The fact that a number of proteins cannot be readily solubilized from membrane preparations in aqueous media (particularly from mitochondria) unless detergents are included but are extractable with organic solvents supports the idea that, in certain membranes, hydrophobic interactions may be extensive. Does this mean that such proteins are embedded in lipid and are not at the membrane surface? Stoeckenius and Engleman (1969) suggest that this need not be the case and propose that non-polar parts of the protein may project through the phospholipid heads into the hydrocarbon interior of the membrane—the remaining part may, thus, remain at the membrane surface. I have employed this compromise in the model of the membrane matrix in Fig. 2.9.

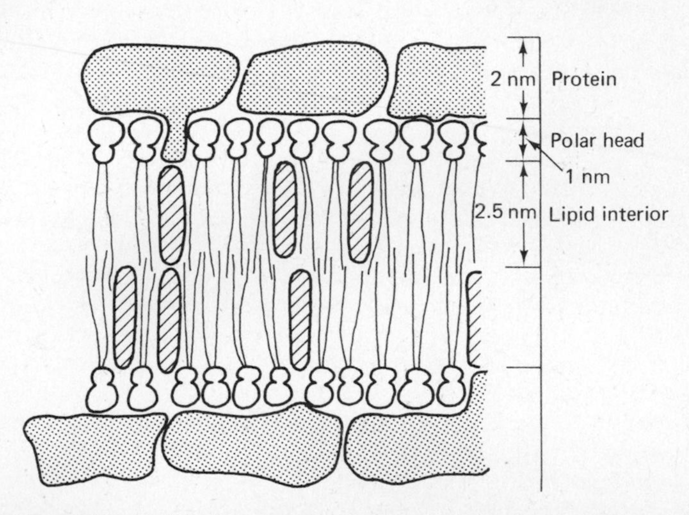

FIG. 2.9 Tentative model of the membrane matrix. This model embodies most of the features which we have discussed so far. The protein on the membrane surfaces is largely in association with the polar heads of phospholipids, but in one instance (top left) it is shown intruding into the hydrophobic lipid interior. The hatched, rod-shaped areas represent lipid molecules such as cholesterol.

A model for the membrane matrix

The conclusions we have drawn so far about the organization of the membrane matrix are embodied in Fig. 2.9 which is substantially the same as the model proposed by Davson and Danielli in 1935. Where membranes differ in gross morphology it is usually due to dark staining peripheral bands which may vary in thickness according perhaps to the amount of protein or polysaccharide which is associated with the polar region of the phospholipid.

How can we test whether this model really represents the natural membrane? The most obvious approach has been to compare the properties of synthetic lipid bilayers, whose structure is more predictable, with accumulated information on the properties of natural membranes. This matter will be dealt with in detail in chapter 4, but for our present purpose we can say that a number of properties of synthetic membranes, even when they lack any protein component, are similar to those of natural membranes. The most striking differences lie in the electrical resistance and surface tension which are at least an order of magnitude greater in synthetic bilayers (see Table 4.4, page 104). The water permeability of bilayers falls in the lower part of the range spanned by biological membranes. This and the larger activation energy for water transport may be at least partly dependent on the higher surface tension of bilayers which in turn is probably due to the absence of peripheral bands of protein.

Water and membrane structure

We have seen that, in an aqueous system, mixtures of lipid and phospholipid assume a bilayered structure—to this extent, therefore, water determines the underlying structure of the cell membrane. There are, however, more specific associations of water molecules and membranes whose importance in physiological matters can only be guessed at.

There is a measurable water content in all membranes which is usually found to be more than 30 per cent of the dry weight of the other components. This water does not seem to be affected by freezing which suggests that it is already fully immobilized or 'bound' in hydration shells. It has been suggested that the interactions between proteins and lipids are dependent on this water of hydration which, therefore, contributes to the stability of the membrane; water of hydration is also likely to be necessary for the normal activity of enzymes which may be present as part of the total membrane protein.

At the surface of the membrane, where the charged groups of protein are exposed, there will be a layer of immobilized water molecules in structured layers. These layers have physical properties more akin to ice than water as we commonly think of it. For instance, the average viscosity of liquid–crystalline water layers is 39 times that of pure water at 23 °C (Schultz and Assunmaa, 1970). It is clear that water organized in this way may add significantly to the diffusion-barrier properties of a membrane, as well as contributing to its mechanical stability.

Some authors claim that some of the permeability properties of membranes may be determined by rather large pores in the lipid components, the core of which contains structured water with a small central channel of about 0·40 nm in diameter. The dimensions of this central channel could be modified by changes in the degree of structuring of the water in the bulk of the pore volume due to the addition of drugs or electrical and ionic gradients (see Schultz and Assunmaa, 1970, for discussion). Acceptance of such views seems to depend on the membrane being composed of hexagonal lipid-protein sub-units; this structure is by no means agreed to be the general case.

Variations in the form of the membrane matrix

The bilayer structure of membranes is one of high stability and probably represents the condition of most of the membrane matrix. There may be situations, however, in which the lipid components of the membrane are arranged as globular structures called *micelles*. Figure 2.10 illustrates a micelle showing interdigitated molecules of cholesterol and lecithin. The polar groups are directed towards the outside and the hydrophobic groups are buried in the centre of the micelle. Certain phospholipids form structures of this kind spontaneously and one in particular, lysolecithin, has proved very useful in studies on the fusion of cell membranes (Lucy, 1970). The molecules of lysolecithin are more wedge-shaped than those of lecithin so that when they are closely packed together there is a strong tendency for them to form into a globular structure. This molecule has an interesting property of inducing micelle formation in both synthetic and natural membranes which are predominantly laminar by forming an additive complex with the existing phospholipid. The transition of the membrane matrix from a predominantly laminar bilayer into a predominantly micellar structure causes great instability and the membrane may break up into a series of vesicles or, where two membranes are in juxtaposition, promote their fusion into a single mem-

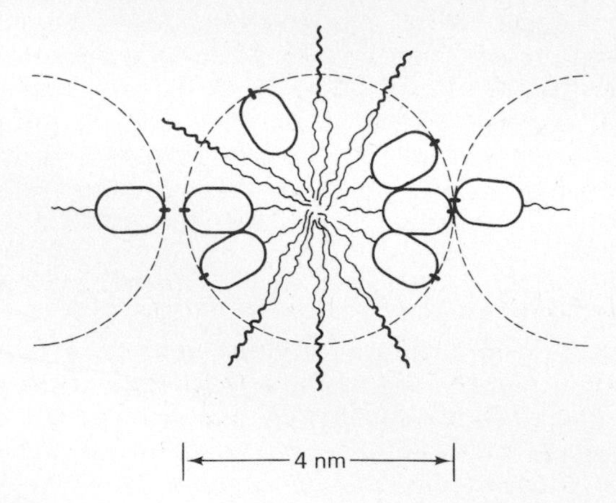

FIG. 2.10 Possible arrangement of molecules in a globular lipid micelle composed of cholesterol (ovoid structures) and lecithin the polar groups of which are at the surface of the globule. (Redrawn from Lucy and Glauert, 1964, by kind permission of the authors.)

brane (Fig. 2.11). Poole *et al.* (1970) have suggested that localized synthesis or addition of molecules like lysolecithin to the cell membrane and their subsequent degradation by the enzyme phospholipase could introduce a measure of transient instability into the membrane which would go a long way to explaining the pinching off of the small vesicles which are observed in pinocytosis (see page 49) and the coalescing of intracellular vesicles with the plasmalemma in secretory processes.

Lucy and Glauert (1964) pointed out that if globular micelles are closely packed together in synthetic preparations, spaces of about 0·5 nm in diameter are found between the globules, thus giving a pore in the lipid membrane of approximately the same diameter as that proposed for natural membranes to account for their high permeability to water. The possibility exists therefore that transient changes in parts of the membrane between bilayer and lipid micelle could give rise to pores of the appropriate size to permit the passage of water and certain other small molecules in solution. Furthermore, if we recall that in adjacent micelles the polar groups are directed towards one another then the pore is hydrophilic.

There are many structures within the cell in which the unit membrane may have to undergo fairly sharp curvature, e.g., the inner membrane system of the cristae in mitochondria. Robertson (1964) has shown that the lipid bilayer cannot be curved through 360 degrees with an external diameter of less than 30 nm without loss of stability. Figure 2.12 shows us why. The minimum distance between the centre of the polar heads of adjacent phospholipid molecules at the inner surface of curvature is 1 nm (see page 28); if the polar regions of phospholipid on the other surface of curvature are separated by more than 2 nm the hydrophobic core of the bilayer becomes exposed to the aqueous phase causing instability in the membrane. Stein (1967) has suggested that the sharp bends and convolutions which are seen in structures such as the endoplasmic reticulum may be made up from a combination of bilayer and micelles.

Conformational isomers in the matrix hydrocarbon layer

Structural transitions and defects in the hydrocarbon interior of the membrane matrix may provide a pathway whereby water can diffuse rapidly from the outside solution to the cell interior. In a most interesting theory, Träuble (1971) has pointed out that the hydrocarbon tails of phospholipids may be expected at all times to have a fairly large number of conformational isomers, the so-called 'kink' isomers. These are formed on a straight hydrocarbon chain by the rotation of a C—C bond through 120 degrees followed by the rotation of the adjacent C—C bond through −120 degrees (i.e., in the

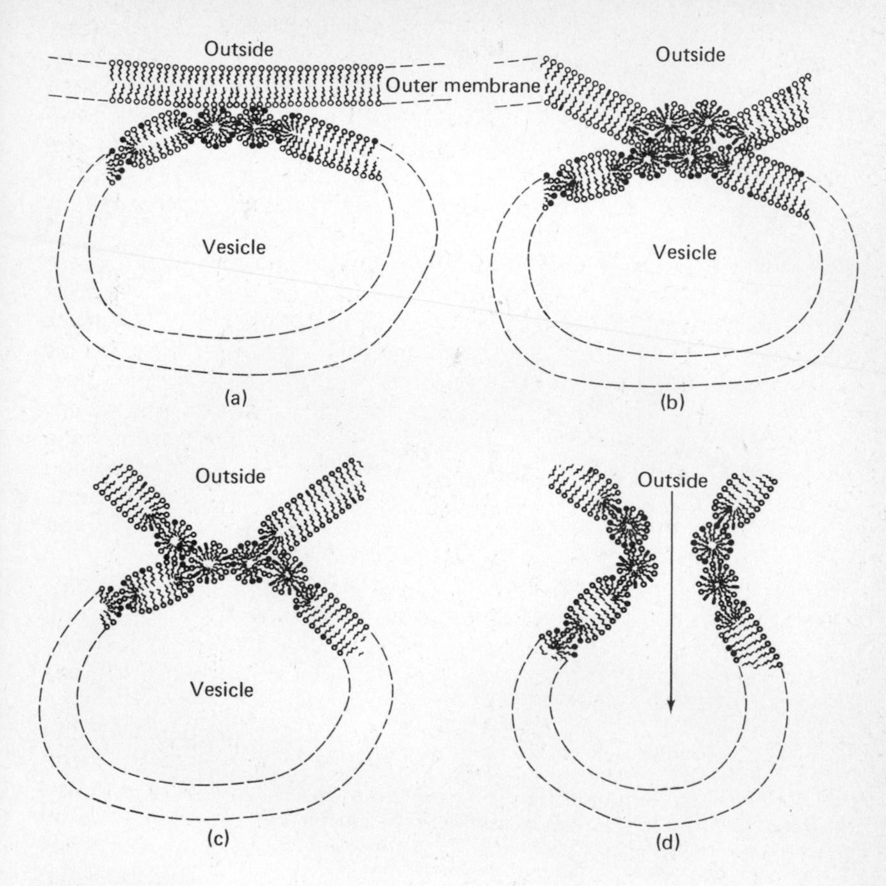

FIG. 2.11 Diagram to illustrate the possible involvement of micellar conformations of lipid molecules in the fusion of membranes. (a) Cross-sectional view of the adjacent parts of two membranes. The lipid molecules of the upper membrane are wholly in a bilayer arrangement while those in the approaching vesicle are partly in globular micelles: specific, endogenous molecules, which may be either lipid (as illustrated ●〰) or protein, are responsible for the stability of the globular micelles in the vesicle. It is postulated that fusion between the two membranes will not occur if either (or both) of the two adjacent membranes remains in a predominantly bilayer configuration. (b) As a result of some exogenous or endogenous perturbing molecule (e.g., lysolecithin), represented by ■〰, a transition has occurred in the orientation of lipid molecules in the outer membrane so that localized regions of globular micelles are formed. It is postulated that in these conditions membrane fusion can occur. (c) Interdigitation of the globular micelles of the two adjacent membranes leads to fusion locally into a single entity. (d) Breakdown of the junction now incorporates the vesicle membrane into the structure of the outer membrane, the contents of the vesicle continuous with the outside solution. (Diagram and Legend taken from Lucy, 1969, and slightly modified by kind permission of the author and North Holland Publishing Company.)

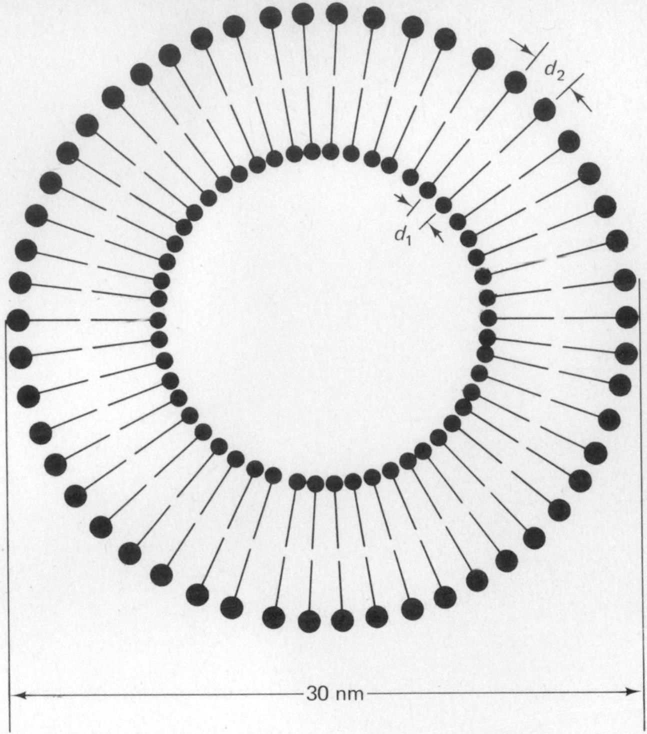

***FIG. 2.12** Molecular diagram of a unit membrane bounding a spherical vesicle with outer diameter of 30 nm. Lipid heads are denoted by black dots and are shown to scale with a diameter of 1 nm. When the heads are packed as close as possible the distance between the centres will be d_1 = 1·0 nm. The phospholipids on the outside will be more loosely packed and the distance between the heads d_2 = 2·0 nm. Any further separation of the outer heads would expose the hydrophobic interior of the bilayer to the aqueous environment which would induce stability. It is believed that this is the reason why membrane bound vesicles of less than 30 nm diameter have not been seen. (Taken from Robertson, 1964, by kind permission of Academic Press, N.Y.)*

opposite direction); they arise because of the thermal motion of the hydrocarbon chains and resemble the situation illustrated in the molecular model (Fig. 2.13). The important feature of the 'kink' is that it is mobile, moving from one end of the hydrocarbon chain to the other. As the model shows, the presence of a 'kink' creates a small free volume which migrates along the hydrocarbon with the 'kink'. The models in Fig. 2.14 show that the expected dimensions of the free volume are large enough to accommodate water molecules and possibly small solutes as well. In Träuble's model the rate of diffusion of water molecules would be intimately related to the

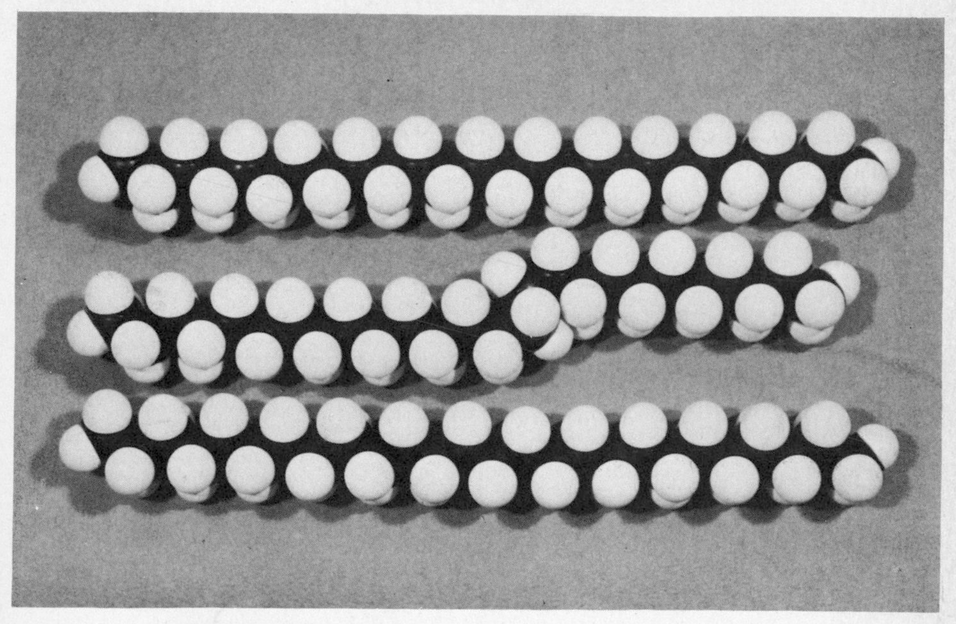

FIG. 2.13 A 'kink' in a hydrocarbon chain. This molecular model shows a kink formed by two 120 degree rotations about C—C bonds which are separated by one unit on the hydrocarbon chain. By the formation of this kink the chain is shorted by a length equivalent to one CH_2 unit. (From Träuble, 1971, by kind permission of the author and Springer-Verlag N.Y., Inc.)

number of 'kinks', their volume and their rate of movement along the hydrocarbon chain. Calculations based on reasonable assumptions of the number of 'kinks' and their rate of movement indicate that diffusion of the free volume associated with the 'kinks' would give rise to a flow of water through a bilayer membrane which is compatible with the measured values for diffusive permeability.

The 'kinks' differ from the transient pores we described earlier in several respects. The most important is that water molecules and solutes will travel across the membrane in discrete packets, or quanta; thus their rate of diffusion is determined by the properties of the 'kinks'. Once molecules are enclosed in the space formed by the 'kink' they are no longer in direct contact with molecules in the bulk aqueous phases separated by the membrane, and will not be directly influenced by events occurring in those phases. If, for instance, a hydrostatic pressure were applied on one side of the membrane we would not expect this to have any direct effect on the movement of water in the 'kink' spaces. Now compare the situation in which two aqueous phases are linked directly by minute water filled pores. The

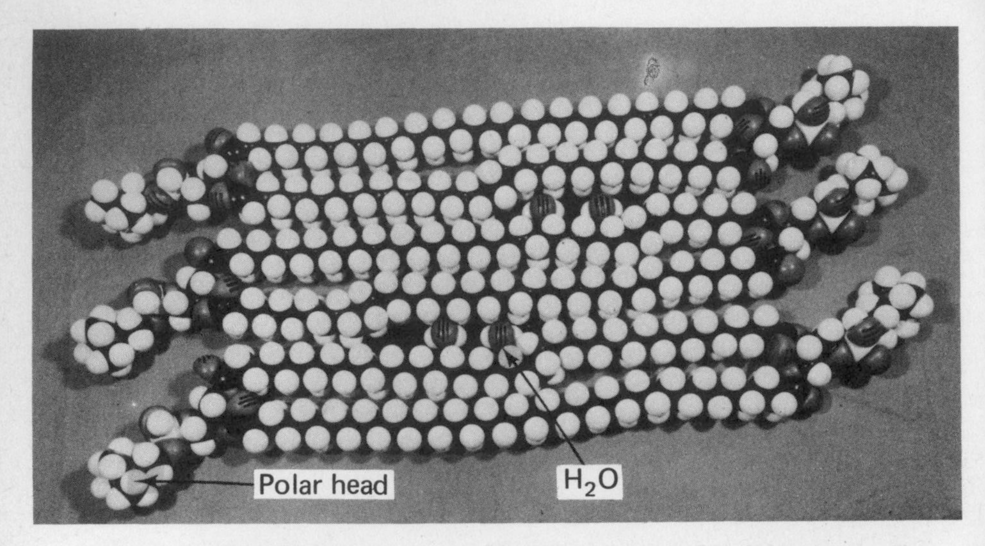

FIG. 2.14 Phospholipid bilayer model with voids caused by kink isomers. Two water molecules are shown to fit neatly in the void but somewhat larger molecules might also be accommodated in this way. (From Träuble, 1971, by kind permission of the author and Springer-Verlag, N.Y., Inc.)

application of a hydrostatic pressure would tend to push water molecules through the pore, i.e., the pressure would be communicated from one water molecule to another because they form a cohesive column. In this situation, water would pass through the membrane faster than it would by diffusion alone under no pressure gradient. In pure phospholipid bilayers, the application of hydraulic pressure or osmotic pressure does not generally increase the rate of water movement through them (Holtz and Finkelstein, 1970), but in natural membranes there is an appreciable deviation of the diffusive permeability coefficient, P_d, and the osmotic or hydraulic permeability, L_p, with increasing pressure (Dainty, 1963). It seems, therefore, that even if some water crosses biological membranes as Träuble suggests, there must be some water filled pores in addition.

Alternative models of membranes built from repeating sub-units

In 1963, Sjöstrand observed that the membranes of cristae of mitochondria appeared to be composed of globular sub-units separated by septa of about 1 nm width. Subsequently, such structures were seen by other workers and evidence of this kind was used to advance the view that membranes, in common with most other macromolecular structures in nature, are made up from repeating sub-units

(Green and Perdue, 1966). In this type of membrane the characteristic sub-unit shape is determined not by the molecules of phospholipid but by a skeleton of protein (Fig. 2.15). This is clearly a radical alternative to the model we have proposed and it is worthwhile reviewing the evidence on which it is based and to see if some of the features can be accommodated with the bilayer concept which we have developed. Green and his co-workers were primarily concerned with the inner mitochondrial membrane on which the enzymes and cytochromes of the electron transport pathway are arranged. They had already found that a treatment of mitochondria with the detergent sodium lauryl sulphate (SLS) released a range of particles with molecular weights 0·5–1 $\times$ 10^6 containing all the membrane bound apparatus involved in electron transport. Biochemically, therefore, there seems to be good evidence for the existence of some kind of sub-unit in the mitochondria. When SLS was removed from the solution by dialysis, the sub-units reaggregated to form membrane-like structures.

Powerful support for the idea that, in mitochondria, the basic skeleton of the membrane was composed of protein came from studies in which lipid was extracted from mitochondria with aqueous acetone (Fleischer *et al.*, 1962). Chemical investigation showed that most of the lipid was removed by this procedure. Subsequent examination of these preparations in the electron microscope showed that the inner membrane of the mitochondrion closely resembled that in untreated mitochondria which contained the full complement of

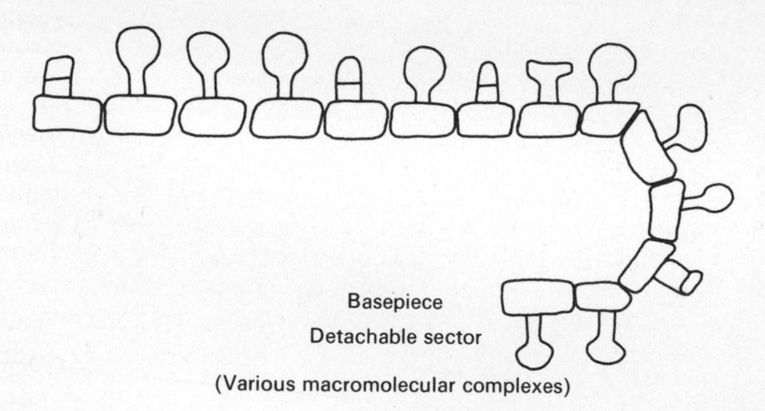

FIG. 2.15 A model of a membrane from a mitochondrion composed of sub-units. The headpieces can be removed by sonication and are composed of several macromolecular complexes differing in size and enzymatic activity. The basepiece consists of identical repeating sub-units of lipid associated with a protein framework, the reverse of the situation proposed in Fig. 2.9. (After Green and Perdue, 1966.)

lipid. This is very surprising because in the unit membrane protein bands are separated by lipid and one might expect the characteristic spacing to collapse or appear as a single dark band of about 4 nm or so in width if the lipid were removed. The fact that this did not occur makes it unlikely that the simple picture of a membrane matrix which we have constructed in the preceding pages can represent the condition in all membranes.

The plasmalemma clearly differs from the cristae of mitochondria in a functional sense; it is not, for instance, specially designed for electron transport neither has there been much microscopic evidence for sub-units of the type described by Sjöstrand (1963). It is, however, possible to reduce the plasma membrane of the bacterium *Micrococcus lysodeikticus* to a series of lipoprotein sub-units by treatment with SLS (Butler *et al.*, 1967; Grula *et al.*, 1967). But is the presence of sub-units in the extract really evidence of their existence in the intact membrane? Synthetic lipid bilayers totally lacking in protein can be broken down into discrete globular micelles resembling sub-units by extensive sonication, by certain ionic detergents and by lysolecithin, but this does not mean that the membrane was comprised originally of micelles. Similarly, the reaggregation of trilaminar membrane structures from the sub-units does not provide conclusive evidence for the sub-unit nature of the original membrane because, as Branton (1969) points out, the addition of protein, clearly unrelated to membrane protein, to dispersions of lipid causes the formation of vesicles bounded by membrane-like structures on reaggregation. The possibility remains therefore that the lipoprotein particles formed by disruption of membranes consist of a rather non-specific association of protein and lipid.

An extremely detailed and complex model of the possible arrangement of the biochemically distinct sub-units of the inner mitochondrial membrane has been constructed by Sjöstrand and Barajas (1970). The sub-units differ considerably in their composition and molecular weight and it is not possible to get a tight, repeating fit into the model which has, as a consequence, a rather untidy appearance. This model to some extent represents a compromise between the two rather orderly views of membrane structure embodied in the unit membrane and sub-unit membrane proposals, because the spacing between the macromolecular protein units is provided by a bilayer to which the proteins may be bonded either electrostatically or hydrophobically (Fig. 2.16).

Readers may be puzzled by the conflicting interpretations which have appeared in the scientific literature on this matter. But such conflicts illustrate the fact that new information in rapidly expanding areas of knowledge is eagerly seized by investigators and frequently interpreted according to their special interests. This is in no sense a

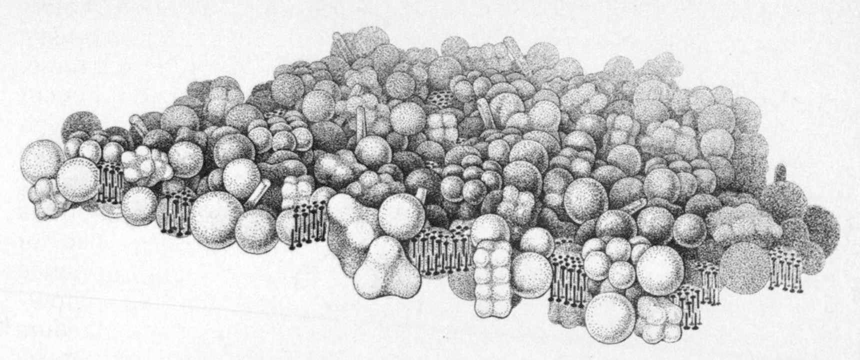

FIG. 2.16 A model of the inner membrane of a mitochondrion. This complex drawing shows a transverse section through a slice of the model. The macromolecular complexes, which can be separated from the membrane, are represented in proportion to their molecular weight. They are embedded in, but project through, a conventional phospholipid bilayer (represented by the 'nails'). The model differs from that in Fig. 2.15 in that no regular repeating pattern is developed. The complex with rod-shaped projections represents respiratory chain cytochromes and succinic dehydrogenase. (Taken from Sjöstrand and Barajas, 1970, with the kind permission of the authors and Academic Press, N.Y., Inc.)

criticism of the protagonists of the various points of view, but students should be aware that such conflicts arise frequently and reserve their judgement until they have consulted several sources of information.

As frequently happens, the exploitation of a new technique allows deeper insight into the nature of things. At least some of the controversy surrounding lipid bilayer versus sub-unit membranes has been resolved by the discovery, in replicas of freeze-fractured membranes, of a particulate sub-structure in the membrane matrix. This we will now consider.

Membrane sub-structure

We have concluded already that there are difficulties in explaining the specific functions of different membranes if all our ideas of membrane structure are based on the simple unit membrane model. The discovery of sub-units in membranes has indicated where their specificity may reside.

Freeze-fracture and freeze-etching

It has been shown that cells of yeast and other organisms may be rapidly frozen to a very low temperature (−196 °C) so that the water in the cell sets to something like a glass in which ice crystals are very small. Large sharp ice crystals normally disrupt cellular organelles and their membranes. Cells frozen in this way can be thawed out and a high proportion of them remain alive and resume their normal functions. When in the frozen condition cells and tissues may be fractured by cutting them with a glass knife (Fig. 2.17). The surface left after the passage of the knife may be coated with carbon and gold and a replica formed which can be examined in the electron microscope. (A full description of these techniques appears in a recent book by Branton and Deamer, 1972.) Among the many attractive features of this technique is one of prime importance. Since it can be shown that frozen cells are in a state of suspended animation it is reasonable to conclude that the structure we observe is a structure which must exist in a functional organism. This is certainly not always the case in standard electron microscope techniques where chemical fixatives are known to cause a number of artefacts in the tissue.

If the fractured specimen is left in a vacuum while still frozen, water will sublime from its aqueous parts and etch the surface so that non-aqueous structures are thrown into more prominent relief when a replica is eventually prepared (Fig. 2.17).

At first sight, the variety of structures revealed in the fracture planes is rather bewildering since the cells may be fractured at any point—fractures close to the surface may show quite large portions of cell wall and plasmalemma in surface view; fractures which cut obliquely across the cell may reveal an interesting array of surface and oblique transections of membrane-bound structures. Examples of these two types of fracture, kindly provided by Professor Branton, are shown in Figs. 2.18 and 2.19.

Figure 2.18 shows a surface view of the plasmalemma of a cell from the root tip of onion. The preparation is freeze-fractured and what we are looking at is the inner face of the membrane exposing the hydrocarbon interior of the lipid as a smooth background on which numerous small projecting particles are arranged. The particles are arranged either at random or in linear files which are sometimes cross-connected. The larger structures sticking out from the matrix are plasmodesmata which formerly connected the cell we are looking at, through the wall (bottom left), and the adjacent cell (see page 217 for a description of the fine structure of plasmodesmata). The grouped plasmodesmata will determine the location of pits in the secondary wall of the cell.

Figure 2.19 shows a field from a fracture of a cell interior in which numerous membranes are exposed. In these young cells there is no central vacuole but a number of smaller vacuoles and, in each one of these, particles of more or less uniform size can be seen against a smooth background of membrane matrix. The frequency of the particles is greater than in the plasma membrane. A small portion of the

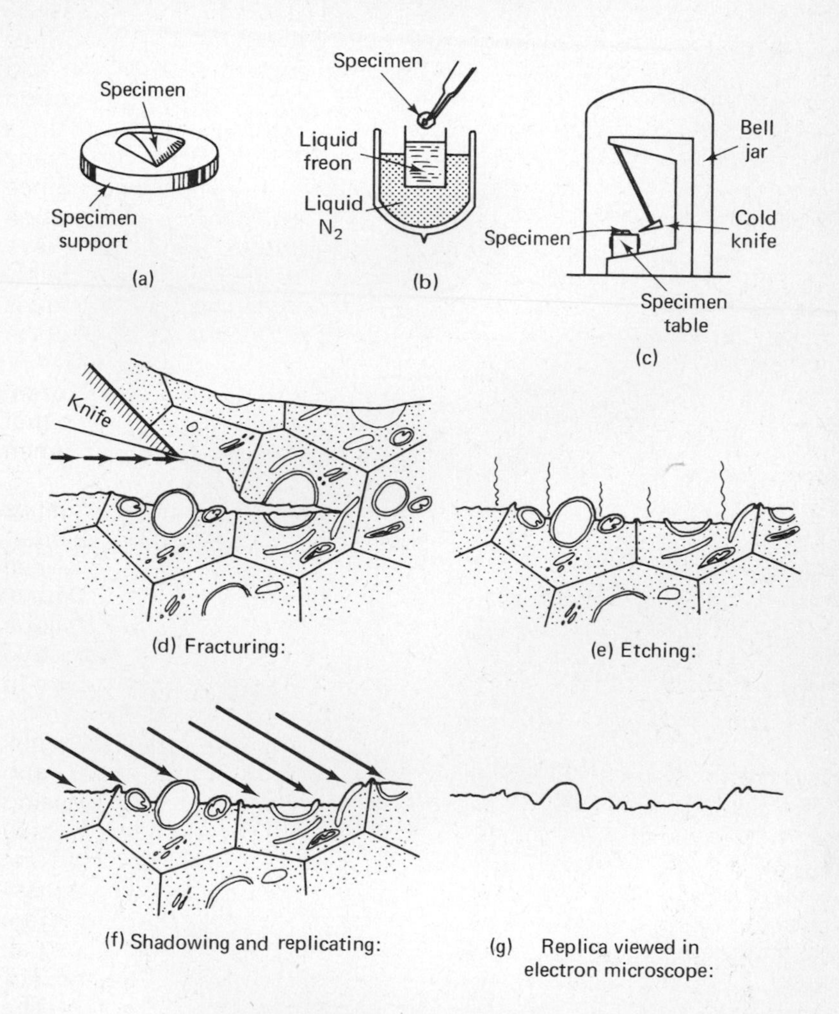

FIG. 2.17 Steps involved in freeze-etching. The fresh specimen is (a) placed on a copper disk, (b) rapidly frozen in liquid Freon 22 cooled by liquid nitrogen (−196°C), (c) placed in a vacuum chamber with a cooled specimen table (d) specimen fractured during the passage of a cooled microtome knife and (e) the freshly fractured face is left under vacuum while water sublimes, thus etching the surface; (f) the surface is shadowed and replicated with platinum and carbon. After dissolving away the specimen the remaining replica, (g) is observed in the electron microscope. (Taken from Branton and Deamer, 1972, by kind permission of the authors and Springer-Verlag, N.Y.)

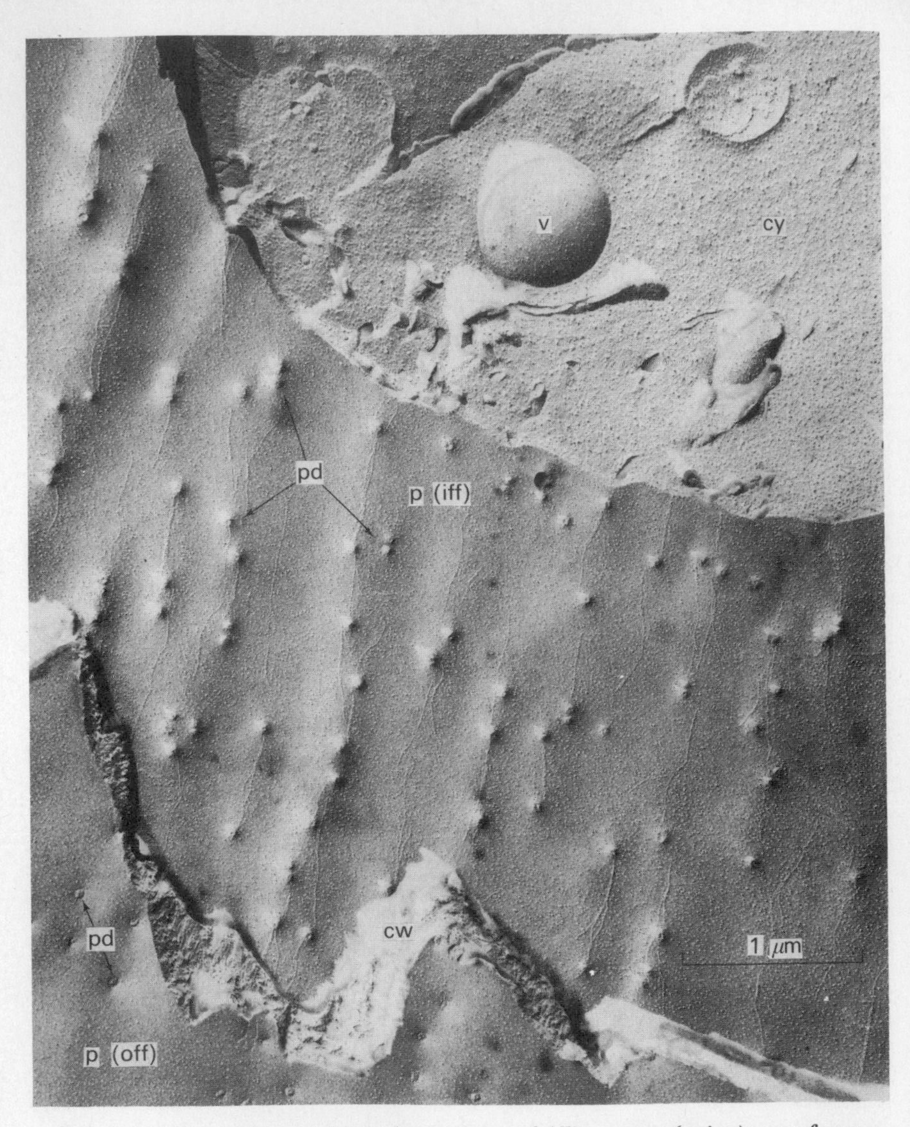

FIG. 2.18 Plasmalemma from the root tip of Allium cepa, *(onion) seen from a freeze etched preparation. For explanation see text. Symbols: P = plasmalemma, OFF and IFF represent the outer and inner fracture faces respectively of the plasmalemma, (notice that the density of particles is much greater on the IFF than on the OFF, and also that regular files of particles are seen running between the plasmodesmata on the IFF). CW = cell wall, pd = plasmodesmata seen in disrupted transverse section (note that they are clumped into areas which will become primary pit-fields), V = small vacuole, cy = cytoplasm. Magnification about ×36 000. (I am indebted to Professor Daniel Branton for this print.)*

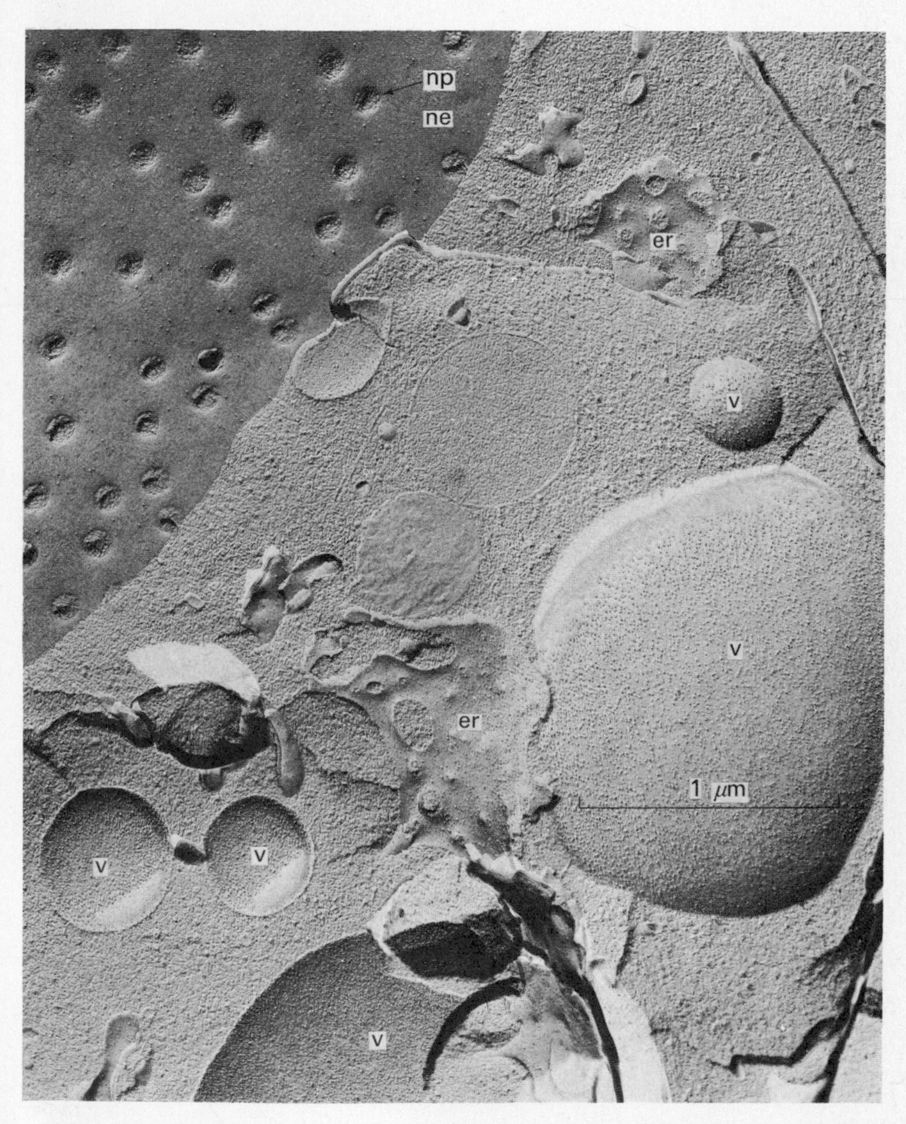

FIG. 2.19 Cytoplasmic detail from cell in the root tip of Allium cepa *(onion) seen in a freeze etched preparation. For description see text. Symbols: v, small cytoplasmic vacuole, er, endoplasmic reticulum, ne, nuclear envelope, np, nuclear pore. Magnification about ×40 000. (I am indebted to Professor Daniel Branton for this print.)*

membrane enclosing the endoplasmic reticulum can also be seen in which the particle density is about the same as that in plasmalemma but the pattern of distribution of particles and their size is not the same. The nuclear envelope is seen as a very smooth sheet of membrane matrix with very few particles but is perforated by a rather regular array of nuclear pores (see page 21).

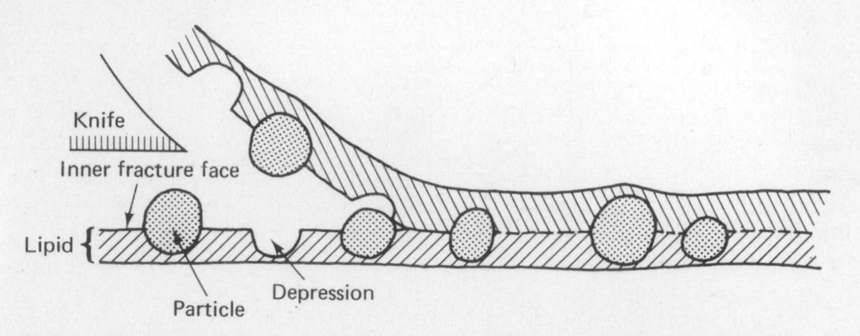

FIG. 2.20 Diagram to interpret the plane of fracture in membranes. The hydrophobic central region of the bilayer represents a line of weakness in the membrane along which cleavage occurs. The particles exposed, therefore, must be embedded in the matrix as shown.

When freeze-fracture preparations of membranes were first observed there was some controversy about what it was that actually appeared in replica; some workers believed that the true face or surface of the membrane was exposed by the fracture, but Branton (1966) argued that it was the inner face of the membrane which was exposed by cleavage along the plane of the hydrocarbon tails of the phospholipid. In other words, there is a line of weakness running along the centre of the lipid bilayer where, of course, there is little water in the form of ice. Thus, in Fig. 2.18 we see that the inner face of the membrane is exposed and what we are actually looking down on is the hydrophobic tail region of the lipids. The diagram in Fig. 2.20 shows quite clearly that the sub-units are embedded in the membrane matrix rather than superimposed on it, as was suggested by some other workers. Pinto da Silva and Branton (1970) devised an elegant experiment to confirm this. When the protein ferritin is introduced into a suspension of red blood cells, particles of ferritin (which are easily recognized) become attached to the surface of the plasma membrane, thus acting as a marker for the outer face. Oblique freeze-fractures followed by deep-freeze etching clearly distinguished between the superficial particles and those embedded in the membrane matrix. When covered by the outer layer of membrane the profile of the particulate sub-units appeared smooth as if they were covered by a blanket. Sub-units exposed along the fracture plane, i.e., the inner face, were quite sharply defined and no ferritin was seen associated with them.

At this time, the chemical composition of sub-units can only be guessed at but there seems little doubt that protein accounts for much of the particle. Support for this suggestion comes from experiments by Engstrom (1970, and cited in Branton and Deamer, 1972), in which red blood cell membranes were treated with the enzyme pronase. It could be shown that pronase extracted protein from the red blood cells and that, concomitant with this extraction, particulate sub-units in the membrane gradually declined in frequency and finally disappeared. It is noteworthy that after this pronase treatment the membrane remained intact indicating that the matrix is continuous and that the removal of the sub-units does not destroy the structure of the membrane even though its function may be altered greatly.

Distribution of sub-units in different membranes

Branton (1969) has shown that functionally different membranes appear structurally different when examined by freeze-etching. For example, nerve myelin membranes appeared as extensive smooth sheets, but other membrane systems appear as smooth sheets in which particles are embedded (average diameter 8·5 nm). Moreover, the density and distribution pattern of these particles seems to be a fixed characteristic of a given membrane type. The density of particles is greatest in membranes which are most active physiologically and have a large capacity for rapid ion transport, e.g., chloroplast membranes, and least in inactive membranes like nerve myelin which function primarily as inert electrical insulators around the nerve axon. In Table 2.2 observations of this kind are summarized. When the membrane fractures, it seems to be consistently the case that more particles are associated with one of the faces than with the other.

An interesting observation by Scott, Carter and Kidwell (1971) indicates that the number of 7 nm particles associated with the inner and outer fracture faces of the cell membrane of synchronously dividing cells in tissue culture is at a minimum after mitosis and builds up during G_1 and S stages of the mitotic cycle (Fig. 2.21).

Membranes as dynamic structures

In organisms such as *Amoeba*, which change continuously their shape and surface area, the plasma membrane must not only be flexible but must be capable of rapid assembly and disassembly. The model of the membrane in Fig. 2.9 shows why this capability is

Table 2.2 Particle density on fractured membrane faces

Type of membrane	Number of particles μm^{-2}		Membrane area covered with particles %
	Densely populated face	Thinly populated face	
Lecithin myelin forms	0	0	0
Nerve myelin sheath	0	0	0
Mithochondrial inner membrane	2700	—	—
Endoplasmic reticulum (root tip)	1700	380	12
Nuclear envelope (root tip)	1790	420	12
Tonoplast (root tip)	3300	2480	32
Plasmalemma (root tip)	2030	550	15
Plasmalemma (red blood cell)	2800	1400	23
Chloroplast lamellae	3860	1800	80

Based on data collected by Branton and Deamer (1972).

required. Phospholipid molecules are closely packed together so that there is little scope for shrinkage of a membrane unless phospholipid molecules are removed. Similarly, expansion is limited since gaps between adjacent phospholipid molecules greater than 1 nm begin to expose the hydrophobic core of the membrane to the aqueous phase and induce instability. Expansion of the membrane depends, therefore, on the assembly of new areas of membrane matrix. This process can occur in some cases with surprising rapidity. Protoplasts of higher plant cells may be shrunk or plasmolysed by placing them in a solution of high concentration of an impermeable solute which causes water to move out of the cells; if the solution is replaced by one of lower osmotic strength the cells absorb water rapidly and regain their normal size in a matter of seconds. During these processes of shrinkage and expansion it has been estimated that 23 μm^2 of membrane surface may be disassembled or reassembled per second (Stadelmann, 1969). This rate is impressive if we realize that there are approximately two million phospholipid molecules packed into each square micron.

These prodigious rates of synthesis and degradation should remind us that membranes are not static but are dynamic structures. We have already considered (page 34) how transient changes in the orien-

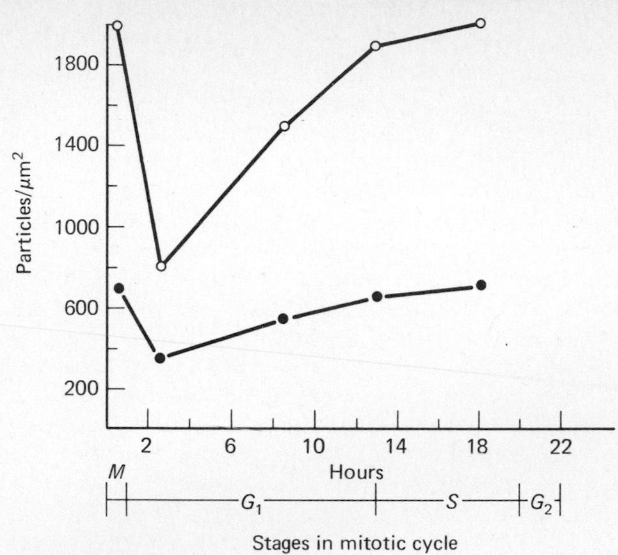

FIG. 2.21 Density analysis of 7 nm particles associated with the fracture faces of the plasmalemma of cultured 'L' cells which have been synchronized with respect to their division cycle. Open circles = inner fracture face; solid circles = outer fracture face. Stages in mitotic cycle; M = mitosis, G_1 = precursor synthesis, S = DNA synthesis, G_2 delay before mitosis. (From Scott, Carter and Kidwell, 1971, by kind permission of Macmillan (Journals) Ltd.)

tation of phospholipid micelles may form temporary pores in the membrane which could facilitate the entry of water, small solutes and ions into the cell but there is another process whereby this may be achieved—*pinocytosis*.

Pinocytosis

Pinocytosis was first described in macrophages from rat plasma by Lewis (1931) who observed that infoldings of the plasma membrane became pinched off into vesicles. Subsequently, the process has been observed in a variety of organisms, including higher plant protoplasts. The vesicles are likely to contain a solution of similar composition to the external aqueous medium. But in addition to solutes, Mayo and Cocking (1969) have shown that particles of latex (0·088 μm mean diameter) can enter isolated tomato fruit protoplasts from which the cellulose walls have been moved by enzymic digestion. Some of the particles inside the cell were surrounded by membrane while others were not. The latter finding suggests that the contents of the pinocytotic vesicles can be released into the cytoplasm after degradation of the membrane. Other observations suggest that the vesicles formed at the plasmalemma can migrate through the cell and fuse with other membrane-bound compartments in the cell. Where such fusion occurs the membrane structure may change transiently from the laminar bilayer to the micelle orientation of lipids (Lucy, 1970). Lipid bilayers and natural membranes are inherently stable structures and may be formed into large series of concentric lamellae without showing any tendency for one membrane to fuse with another. It has been shown, however, that the induction of micelle formation at their point of contact causes membranes to fuse and then to break down, mixing the contents of the vesicle and the cellular compartment (Fig. 2.11, page 35).

The implications of pinocytosis in transport processes are considerable, since compartments within the cell may have an almost direct communication with the external medium; solutes and suspended particles entering them do not have to pass through the plasmalemma in the normal sense. This is thought to be the mode of entry of some virus particles into plant and animal cells and recently MacRobbie (1969) has described the transfer of chloride to the vacuole of the giant alga *Nitella translucens* in similar terms (see page 162).

Membrane flow

The processes leading to membrane fusion described by Lucy (1970) are clearly supported by mounting evidence that the formation of certain membranes is accomplished by the physical transfer of membrane material from one cell component to another. Although vigorous proof is lacking, circumstantial evidence points to the endoplasmic reticulum, ER, as the source of membrane building blocks which ultimately become the plasmalemma. This may be achieved by budding of vesicles from the Golgi cisternae. The transition between the ER-type and the plasmalemma-type membrane probably takes place in the Golgi apparatus (Franke *et al.*, 1971). Franke *et al.* (1971)

***Table 2.3** Estimated time constants for the initial phase of the incorporation of L-(guanido-^{14}C)-arginine into membrane fractions from rat liver*

Fraction	Labelling time (min.)	
	Half max.	Max.
Endoplasmic reticulum	5·5–6·5	10
Golgi apparatus	8·5	30
Plasmalemma	>60	>180

have obtained evidence for the pattern of labelling of membrane-protein in rat liver which strongly supports this idea. They measured the incorporation of a labelled amino acid, L-(guanido-^{14}C)-arginine, into the protein of various membrane fractions following a single injection of the tracer into a living rat. Each membrane showed a peak of labelling after the injection but the maximal activity of the ER and golgi apparatus appeared well in advance of that of the plasmalemma (Table 2.3).

Tentative model for the structure of biological membranes

We can now summarize our ideas about membrane structure in the diagram in Fig. 2.22. The matrix of the membrane is most likely to be a bilayer with polar groups of phospholipid directed towards the outside. Embedded in the matrix are particulate substructures which are represented as protein and as stabilized lipid micelles. Covering the whole surface we have protein and mucopolysaccharide which is hydrogen bonded to the polar groups of the phospholipid or to the protein of the sub-units.

It has been pointed out by Stadelmann (1969) that the structure of the membrane matrix may be variable between one organism and another. This cautionary attitude is based on observations of the permeability of a number of non-electrolytes into the cells of a wide variety of plant species. In these studies, it was found that the permeability of the cells to malonamide, urea and methyl urea relative to their permeability to glycerol varied over an extremely wide range.

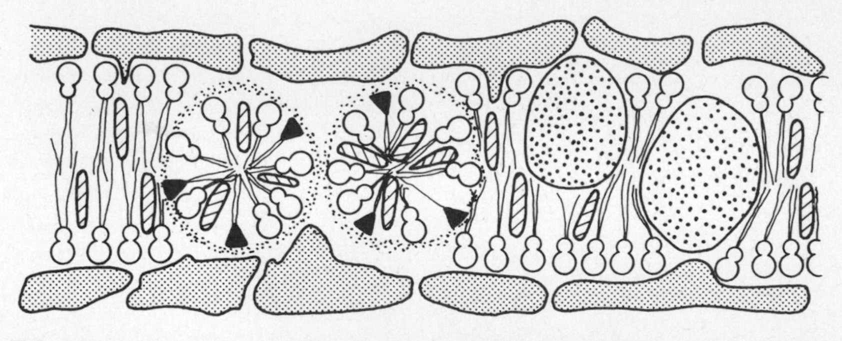

FIG. 2.22 A generalized model of plasma membrane structure. This model is developed from the one in Fig. 2.9 to include lipid micelles and membrane particles containing protein. The micelles are stabilized by a wedge-shaped phospholipid such as lysolecithin. The superficial protein is probably fibrillar rather than globular as envisaged in early models of membrane structure. It should be stressed that several other kinds of model are possible, especially in mitochondria and chloroplasts where specialized biochemical processes are associated with membrane-bound macromolecules.

For instance, in the giant algal cells *Chara ceratophylla* the ratio was 0·15, whereas in another algal cell of *Oedogonium* the ratio was about 5. The exact significance of these differences is not clear but they may indicate that the lipid composition and substructure of the membrane matrix varies between organisms. Such wide variations suggest that it may prove necessary, in the long run, to modify our ideas about the permeability of non-electrolytes through the lipid phase of the membrane.

We have said sufficient to appreciate that although the model is thermodynamically a stable arrangement of molecules, any part of it may have only a transient existence. Large sheets of synthetic bilayer assemble themselves spontaneously and at great speed where appropriate mixtures of lipid and phospholipid are dispersed in water. The fact that biological membranes can be formed rapidly is not, therefore, surprising but suggests that components of the membrane stored in compartments in the cytoplasm near the membrane surface may be readily available for assembly.

What are the implications of the proposed model for the transport of solutes and, in particular, of ions? It can be seen in the model that, in the region of the membrane particles, there is essentially a ladder of protein crossing the hydrophobic core of the lipid bilayer. Molecules which are small enough, e.g., water, formamide, etc., may find this a hydrophilic pathway across the membrane which allows them to bypass the hydrophobic core.

The observant student will have noticed that up to this point I have avoided discussing the problems involved in the transport of ions across our simple model membranes; we must now consider these problems. In aqueous solutions, all ions are surrounded by tightly bound shell water molecules. As a general rule the smaller the non-hydrated radius of an ion, the larger its hydration shell will be (see chapter 4, page 99). Each water molecule in the hydration shell has a tendency to form before hydrogen bonds with the water molecules in its vicinity. It follows that an ion with a large number of water molecules in its hydration shell will make a large number of hydrogen bonds which will have to be broken if the ion is to diffuse from the aqueous phase into the hydrophobic lipid phase. In other words, hydrated ions have large N values and their permeability through bilayers might be expected to be quite low. In fact, we know that the passage of many ions through biological membranes is quite rapid.

To explain this it is usually assumed that the passive diffusion of ions is facilitated by some kind of carrier molecule which has a higher solubility in the hydrocarbon phase of the membrane than the hydrated ion. The nature of these carriers will be considered in chapter 4.

The sub-units which we can see in membranes are probably not concerned with this passive permeation but may be of particular importance in the active transport of ions. Active transport of ions against thermodynamic gradients is often thought to involve enzyme reactions. It is, therefore, extremely interesting to find that, embedded in the membrane matrix, there are protein substructures which are arranged in characteristic patterns and with frequencies which correspond broadly with the transport capabilities of the membrane.

Before we can go on to consider how enzyme systems may transport ions actively across the membrane, we must determine from first principles which of the ions entering the cell require directly the expenditure of energy and which do not. This is the subject of the next chapter.

Bibliography

a Further reading

Branton, D. and Deamer, D. W. (1972) *Membrane Structure*. Springer-Verlag, Vienna.

This book is an extended and clearly written review in which the evidence for the different proposals for the ultrastructure of the membrane matrix is discussed. It summarizes concisely Branton's own work on membrane substructure revealed by freeze-etching techniques.

Hendler, R. W. (1970) Biological membrane ultrastructure. *Physiological Reviews* **51**, 66–97.

Reviews the evidence for and against the unit membrane hypothesis in a systematic and precise way. Very highly recommended to the student reading this subject in depth.

Singer, S. J. and Nicholson, G. L. (1972) The fluid mosaic model of the structure of cell membranes. *Science* **175**, 720–731.

Reviews and opinions about membrane structure are very frequent occurrences in the literature at the present time. This is a very interesting one advancing the view that alternating globular proteins in a phospholipid bilayer is the only membrane model consistent with thermodynamic restrictions.

Stein, W. D. (1967) *The Movement of Molecules across Cell Membranes*. Academic Press, N.Y.

Although this book adopts a rigorous theoretical approach to the subject it is very clearly written and widely regarded as one of the finest monographs available. Chapter 3 is particularly helpful in providing a molecular basis for the diffusion of solutes across membranes. Nearly all of the examples discussed are of transport in animal cells.

b References in the text

Branton, D. (1966) Fracture faces of frozen membranes. *Proc. Natl. Acad. Sci.* (US) **55**, 1048.

Branton, D. (1969) Membrane Structure. *Ann. Rev. Pl. Physiol.* **20**, 209–238.

Branton, D. and Deamer, D. W. (1972) *Membrane Structure*. Springer-Verlag, Vienna.

Butler, T. F., Smith, G. L. and Grula, E. A. (1967) Bacterial cell membranes. I. Reaggregation of membrane sub-units from *Micrococcus lysodeikticus*. *Canad. J. Microbiol.* **13**, 1471–1479.

Collander, R. (1949) The permeability of plant protoplasts to small molecules. *Physiol. Plant.* **2**, 300–311.

Collander, R. and Barlund, H. (1933) Permeabilitatsstudien an *Chara ceratophylla*. *Acta Botan. Fennica* **11**, 1–14.

Dainty, J. (1963) Water relations of plant cells. *Adv. Bot. Res.* **1**, 279–326.

Danielli, J. F. and Davson, H. (1935) A contribution to the theory of permeability of thin films. *J. Cellular Comp. Physiol.* **5**, 495–508.

Emmelot, P., Bos, C. J., Benedetti, E. L. and Rumke, P. L. (1964) Studies on plasma membranes. I. Chemical composition and enzyme content of plasma membranes isolated from rat liver. *Biochim. Biophys. Acta* **90**, 126–145.

Engstrom, L. H. (1970) Structure in the erythrocyte membrane. Ph.D. Dissertation, University of California, Berkeley.

Fettiplace, R., Andrews, D. M. and Haydon, D. A. (1971) The thickness, composition and structure of some lipid bilayers and natural membranes. *J. Membrane Biol.* **5**, 277–296.

Fleischer, S., Brierley, G., Klouwen, H. and Slautterbach, D. B. (1962) Studies of the electron-transfer system. XLVII. The role of phospholipids in electron transfer. *J. Biol. Chem.* **237**, 3264–3272.

Franke, W. W. (1970) On the universality of nuclear pore complex structure. *Z. Zellforsch.* **105**, 405–429.

Franke, W. W., Morré, D. J., Deumling, B., Cheetham, R. D., Kartenbeck, J., Jarasch, E-D. and Zentgraf, H-W. (1971) Synthesis and turnover of membrane proteins in rat liver: an examination of the membrane flow hypothesis. *Z. Naturforsch.* **26b**, 1031–1039.

Frey-Wyssling, A. (1953) *Submicroscopic Morphology of Protoplasm*, 2nd Edition. Elsevier, Amsterdam.

Goldstein, D. A. and Solomon, A. K. (1960) Determination of equivalent pore radius for human red cells by osmotic pressure measurements. *J. Gen. Physiol.* **44**, 11–17.

Gorter, E. and Grendel, F. (1925) Bimolecular layers of lipoids on chromocytes of blood. *J. Exptl. Med.* **41**, 439–443.

Green, D. E. and Perdue, J. F. (1966) Membranes as expressions of repeating sub-units. *Proc. Natl. Acad. Sci.* (U.S.) **55**, 1295–1302.

Grula, E. A., Butler, T. F., King, R. D. and Smith, G. L. (1967) Bacterial cell membranes. II. Possible structure of the basal membrane continuum of *Micrococcus lysodeikticus*. *Canad. J. Microbiol.* **13**, 1499–1507.

Holtz, R. and Finkelstein, A. (1970) Water and non-electrolyte permeability induced in thin lipid membranes by the polyene antibiotics Nystatin and Amphotericin B. *J. Gen. Physiol.* **56**, 125–145.

Lenaz, G. (1970) Studies on the structure of the mitochondrial membranes. *Ital. J. Biochem.* **19**, 54–82.

Lewis, W. H. (1931) Pinocytosis. *Bull. Johns Hopkins Hosp.* **49**, 17–27.

Lucy, J. A. (1970) The fusion of biological membranes. *Nature* **227**, 814–817.

Lucy, J. A. and Glauert, A. M. (1964) The structure and assembly of macromolecular lipid complexes composed of globular micelles. *J. Mol. Biol.* **8**, 727–748.

MacRobbie, E. A. C. (1969) Ion fluxes to the vacuole of *Nitella translucens*. *J. Exp. Bot.* **20**, 236–256.

Mayo, M. A. and Cocking, E. C. (1969) Pinocytic uptake of polystyrene latex particles by isolated tomato fruit protoplasts. *Protoplasma* **68**, 223–230.

Overton, E. (1899) Über die allgemeinen osmotischen Eigenschaften der Zelle, ihre vermutlichen Ursachen und ihre Bedeutung für die Physiologie. *Vierteljahrsschr. Naturforsch. Ges. Zürich* **44**, 88–107.

Palade, G. E. (1956) The endoplasmic reticulum. *J. Biophys. Biochem. Cytol.* **2**, Suppl. 85–125.

Pinto da Silva, P. and Branton, D. (1970) Membrane splitting in freeze-etching. Covalently bound ferritin as a membrane marker. *J. Cell. Biol.* **45**, 598–605.

Poole, A. R., Howell, J. I. and Lucy, J. A. (1970) Lysolecithin and cell fusion. *Nature* **227**, 810–813.

Richardson, S. H., Hultin, H. O. and Green, D. E. (1963) Structural proteins of membrane systems. *Proc. Natl. Acad. Sci.* (U.S.) **50**, 821–826.

Robertson, J. D. (1964) Unit membranes: A review with recent new studies of experimental alternations and new sub-unit structure in synoptic membranes. In *Cellular Membranes and Development* (M. Locke, ed.), pages 1–81, Academic Press, N.Y.

Schultz, R. D. and Assunmaa, S. K. (1970) Ordered water and the ultrastructure of the cellular plasma membrane. *Rec. Prog. Surface Sci.* **3**, 291–332.

Scott, R. E., Carter, R. L. and Kidwell, W. R. (1971) Structural changes in membranes of synchronized cells demonstrated by freeze-cleavage. *Nature, New Biol.* **233**, 219–220.

Sjöstrand, F. S. (1963) A new ultrastructural element of the membranes in mitochondria and of some cytoplasmic membranes. *J. Ultrastruct. Res.* **9**, 340–361.

Sjöstrand, F. S. and Barajas, L. (1970) A new model for mitochondrial membranes based on structural and biochemical information. *J. Ultrastruct. Res.* **32**, 293–306.

Stadelmann, E. J. (1969) Permeability of the plant cell. *Ann. Rev. Pl. Physiol.* **20**, 585–606.

Stein, W. D. (1967) *The Movement of Molecules across Cell Membranes*. Academic Press, N.Y. and London.

Stoeckenius, W. and Engleman, D. M. (1969) Current models for the structure of biological membranes. *J. Cell. Biol.* **42**, 613–616.

Träuble, H. (1971) The movement of molecules across lipid membranes: a molecular theory. *J. Membrane Biol.* **4**, 193–208.

three

Driving forces and the transport of ions across membranes

This chapter is divided into a theoretical section and a section in which practical methods are described; the student may find it helpful to read them in parallel especially when evidence supporting a theory depends on methods which may be unfamiliar.

Electrochemical potential gradients as driving forces on ions

We showed in chapter 1 that an uncharged solute (a non-electrolyte) will diffuse from a region of high concentration to one of lower concentration and that the physical driving force in this movement was the gradient of chemical potential of the solute in the two regions. The chemical potential of the solute is a function of its concentration (strictly, activity)

$$\mu = \mu^* + RT \ln C_s \qquad (3.1)$$

hence, if the temperature in two compartments is the same, solute molecules will move down a concentration gradient. When the solute we are considering is an ion, an electrolyte, the situation is somewhat different. Because they carry electric charge, ions can be moved both by physical forces dependent on concentration and by electric forces. If, for instance, a cell interior is electrically negative relative to the outside solution there will be a tendency for ions carrying positive charges, cations, to move down the gradient of electrical potential into the cell. The greater the potential difference across the membrane the greater this tendency will become and, as we shall see, it may be strong enough to counteract a concentration gradient so that an ion may move from a dilute external solution into the concentrated cell sap without increasing its free energy; consequently, no work need be performed to accomplish the movement.

Both chemical and electric potentials contribute to the *electrochemical potential* of the ion, $\bar{\mu}$, which is defined by eq. (3.2)

$$\bar{\mu}_j = \bar{\mu}_j^* + RT \ln C_j + z_j F \psi \tag{3.2}$$

$\bar{\mu}_j^*$ is the electrochemical potential of the j ion in its standard state and in subsequent equations it will be left out, but should be remembered as a constant added to the sum of the other two terms; $\ln C_j$ is the natural log of the concentration of the ion j (again we make the assumption that the activity of the ion j is approximately equal to its concentration); z_j is the algebraic valency of the ion j; F is the *Faraday* constant which defines the amount of charge carried by 1 gram equivalent of the ion; and finally ψ refers to the electric potential of the system which contains the ion—note that it is not a specific property of the ion but is determined by the total electrolyte composition of the compartment. The units of electrochemical potential are joules per mole.

The driving force on an ion is the gradient of electrochemical potential, $d\bar{\mu}_j/dx$, which is simply obtained by differentiating eq. (3.2)

$$-\frac{d\bar{\mu}_j}{dx} = -\frac{RT}{C_j}\cdot\frac{dC_j}{dx} - z_j F\frac{d\psi}{dx} \tag{3.3}$$

The negative signs in front of the terms are to indicate that the driving force is in a direction which is thermodynamically downhill, i.e., a negative gradient. As an ion moves downhill it loses some of its free energy; exactly how much is determined by the difference in the electrochemical potential between the cell interior and the outside solution. Thus, $\bar{\mu}_j^o - \bar{\mu}_j^i$ defines the decline in free energy of 1 gram-mole of the ion if it is moving passively downhill into the cell. If, on the other hand, it moves into the cell against a gradient of electrochemical potential (i.e., $\bar{\mu}_j^o < \bar{\mu}_j^i$) then the difference in electrochemical potential defines the minimum amount of energy which must be supplied to the system to 'drive' 1 gram-mole of an ion up the gradient.

The Nernst potential and the detection of active transport

We have already seen in our case history of *Hydrodictyon* (see page 12) that the ions in the cell interior are more highly concentrated than in the external medium. The following procedure can be used to decide whether it is necessary to do any work, or supply any energy, to maintain this steady state.

First of all let us begin by considering a cell which is in equilibrium where an ion j is at the same electrochemical potential on either side of the plasmalemma, i.e.,

$$\bar{\mu}_j^i = \bar{\mu}_j^o \tag{3.4}$$

Since there is no electrochemical potential gradient there is no driving force on the ion. Ion movements across the membrane will occur by random thermal motion of the ions but inward and outward movements are balanced. Such a system is said to be in *passive flux equilibrium*. If we expand eq. (3.4) using (3.2), and cancelling $\bar{\mu}_j$ and $\bar{\mu}_j^*$ on each side of the equation we can write

$$RT \ln C_j^o + z_j F \psi^o = RT \ln C_j^i + z_j F \psi^i \tag{3.5}$$

When we measure the electric potential difference across the cell membrane (see page 83) we measure $\psi^i - \psi^o$ which is, for convenience, referred to as E. We can gather the electrical terms in eq. (3.5) to the left hand side of the equation and define E as a function of concentration, e.g.,

$$\psi^i - \psi^o = \frac{RT}{z_j F} \ln \frac{C_j^o}{C_j^i} = E_j^{\mathrm{N}} \tag{3.6}$$

When a system is in passive flux equilibrium E will be given a special superscript E_j^{N} which will indicate the Nernst potential of the ion j. Thus, the Nernst potential is the electric potential difference necessary to maintain an asymmetric distribution of an ion across a membrane at equilibrium.

One very important conclusion can be reached if we inspect the worked example below. If we give the constants R and F their numerical values (0·31 and 96·5, respectively), consider a temperature of 20 °C (293 K) and convert from natural logs to logarithm base 10 (multiply by 2·303), the left-hand side of eq. (3.6) can be much simplified:

$$E_j^{\mathrm{N}} = \frac{58}{z_j} \log \frac{C_j^o}{C_j^i} \text{ millivolts (mV)} \tag{3.7}$$

If we consider for a moment a monovalent cation (e.g., potassium) $z_j = 1$ then the equation is further simplified. Suppose we find by direct measurement that $C_j^i = 100C_j^o$, the equation will tell us how large E has to be in order to establish equilibrium and stabilize the asymmetric distribution of j, e.g.,

$$
\begin{aligned}
E_j^N &= (58/1)\log(1/100) \\
&= -58 \log 100 \\
&= -58 \times 2 \\
&= -116 \text{ mV}
\end{aligned}
$$

(Figure 3.1 can be used to estimate the Nernst potential of an ion if its concentration ratio across the membrane is known.) We can see here that an electrical gradient of −116 mV counteracts the 100-fold concentration gradient. The important conclusion is, therefore, that simply because an ion is more concentrated in the cell interior there is no reason to assume that work is necessarily performed in moving that ion into the cell; the process may be passive. We shall see later in this chapter (page 61) that electric potential differences of the

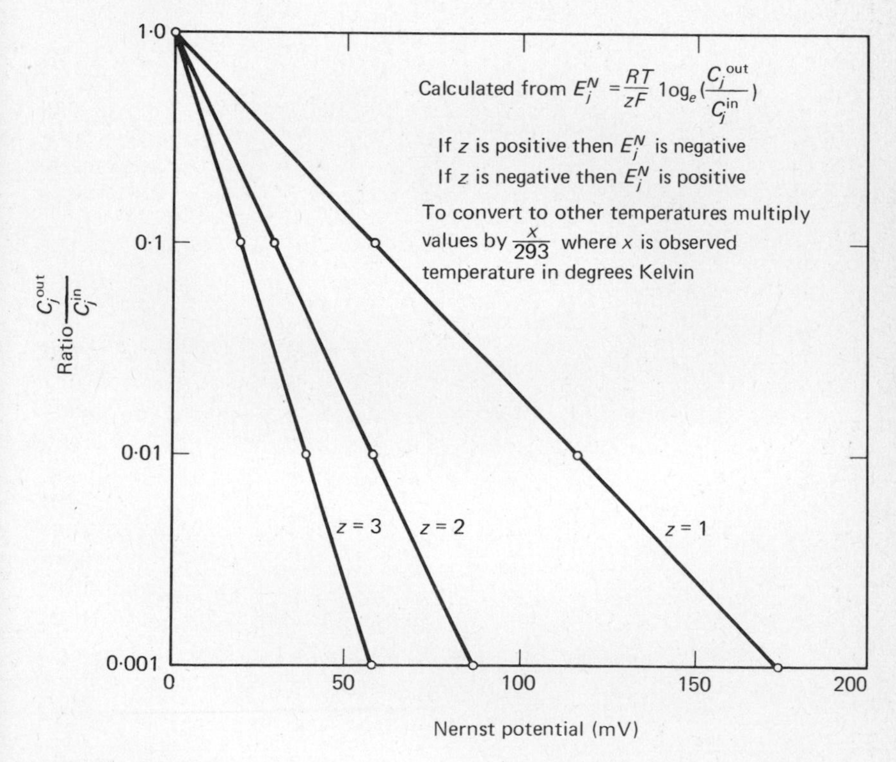

FIG. 3.1 Nernst potential, E_j^N, as a function of the concentration ratio between two compartments at 20 °C (293 K).

order of -116 mV are found frequently in a variety of cells, indeed, the interior of most cells is electrically negative relative to the surrounding medium. This poses a problem in the transfer of anions from the external solution into the cell since, in most cases, they move up not only a concentration gradient, see Table 1.1, but an electrical potential gradient also. If for instance a 100-fold concentration difference of a monovalent anion, e.g., chloride, were in passive flux equilibrium the membrane potential would have to be $+116$ mV. If we find by measurement that the observed electrical potential, E_{obs}, differs from the Nernst potential as calculated from the steady state concentrations, we must conclude that energy is being expended and the ion is not in passive flux equilibrium. Thus, the calculation of Nernst potentials provides a simple test for detecting the ionic species whose transport into the cell is active.

In practice, it is unusual to find ions which are in passive flux equilibrium across the membrane so that in most cases $\bar{\mu}_j^i \neq \bar{\mu}_j^o$. Thus, there usually is an electrochemical potential difference, and expanding eq. (3.2) we can write

$$\Delta\bar{\mu}_j = \bar{\mu}_j^i - \bar{\mu}_j^o = (RT \ln C_j^i + z_jF\psi^i) - (RT \ln C_j^o + z_jF\psi^o) \tag{3.8}$$

$$= z_jF(\psi^i - \psi^o) - RT \ln\left(\frac{C_j^o}{C_j^i}\right) \tag{3.9}$$

We can simplify eq. (3.9) if we inspect eq. (3.6) and see that

$$RT \ln \frac{C_j^o}{C_j^i} = z_jFE_j^{\mathrm{N}} \tag{3.10}$$

Substituting this in eq. (3.9) we get

$$\bar{\mu}_j^i - \bar{\mu}_j^o = z_jF(\psi^i - \psi^o) - z_jFE_j^{\mathrm{N}} \tag{3.11}$$

Now $\psi^i - \psi^o$ is the electric potential difference E across the membrane, and E_j^{N} is the Nernst potential at passive flux equilibrium, thus

$$\bar{\mu}_j^i - \bar{\mu}_j^o = z_jF(E - E_j^{\mathrm{N}}) \tag{3.12}$$

Both E and E_j^{N} are expressed in millivolts, thus we can express the difference in electrochemical potential as a millivolt value, ΔE_j, which is the difference between the value of E and the Nernst potential of the ion.

For cations, a positive value of ΔE_j indicates that the net physical driving force is in an outward direction so that transport of the ion into the cell is likely to be active. If ΔE_j is negative, conversely this indicates that passive movement of the ion into the cell could occur since the net driving force is inwards. For anions, the reverse of these two statements is true.

Driving forces for ion movements in *Hydrodictyon africanum*

Reconsidering the data for *Hydrodictyon africanum*, we can redraft Table 1.1 to include information (Table 3.1) on the electric potentials between the external solution and the cytoplasm (plasmalemma potential) and between the cytoplasm and the vacuole (tonoplast potential). The concentration of the ions in the two cellular compartments is also given. The potential across the plasmalemma is -116 mV (cytoplasm negative) and the tonoplast potential is $+26$ mV (vacuole positive). The resultant potential between the vacuole and the external solution is the sum of these two potentials, i.e., $-116 + 26 = -90$ mV. It can be seen that none of the calculated Nernst potentials for the ions in the cytoplasm matches the observed value of the membrane potential E_{co}. For the observed K^+ concentration gradient to be stabilized without active transport the electric potential difference, E, across the plasmalemma would have to be -173 mV. Similarly at passive flux equilibrium the observed Cl^- concentration would require a potential of $+94$ mV to stabilize it.

Table 3.1 *Calculation of Nernst potentials and the direction of passive driving forces on ions in the cytoplasm and vacuole of* **Hydrodictyon africanum**

(a) Distribution of ions between outside and cytoplasm

$E_{co} = -116$ mV, $T = 287$ °K

Ion	C_o	C_c	E_j^N	ΔE_j	Direction of driving force
K^+	0·1	93	−173	+58	Cytoplasm—outside
Na^+	1·0	51	−99	−17	Outside—cytoplasm
Cl^-	1·3	58	+94	−210	Cytoplasm—outside (very strong)

(b) Distribution of ions between the cytoplasm and the vacuole

$E_{vc} = +26$ mV

Ion	C_c	C_v	E_j^N	ΔE_j	Direction of driving force
K^+	93	40	+21	−5	Close to equilibrium
Na^+	51	17	+27	+1	At equilibrium
Cl^-	58	38	−10	+36	Cytoplasm—vacuole

Concentrations C_o, C_c, C_v refer to outside solution, cytoplasm and vacuole respectively and are in millimoles per litre. Potentials are expressed in millivolts. Activity coefficients of unity are assumed.

By contrast the internal concentration of Na^+ is less than could be supported at equilibrium by the observed potential across the plasmalemma. This result implies that Na^+ is either excluded from the cell or actively pumped out into the medium surrounding it. These data for the cytoplasm relative to the outside contrast with those for the cytoplasm relative to the vacuole, where it can be seen that both K^+ and Na^+ are close to their equilibrium concentrations (i.e., $E_{cv} \simeq E^N_K$ and E^N_{Na}) and that the driving force on Cl^- is into the vacuole.

Large differences between E_{obs} and E^N_j indicate that most of the ion j in a compartment must be maintained there by active transport. This point is perhaps made a little clearer in Table 3.2 in which the

Table 3.2 ***Predicted concentration of ions (in millimoles per litre) in the cytoplasm if $E_{co} = E^N_j$ for each ion***

$E_{co} = -116$ mV

Ion	C_o	C_c observed	C_c predicted
K^+	0·1	93	10
Na^+	1·0	51	95
Cl^-	1·3	58	0·018

measured potential difference across the plasmalemma is taken as the Nernst potential for each ion and concentration inside the cell predicted from this value (i.e., $E_{co} = E^N_j$). We can see that the predicted value for potassium is about 10 per cent of the observed concentration, so that 90 per cent of K^+ and almost all of the chloride must be transported actively across the plasmalemma. By contrast with the other two ions, the concentration of sodium observed in the cytoplasm is about half that predicted if $E_{co} = E^N_{Na}$, indicating active transport in an outward direction. Since the distributions of the ions between cytoplasm and vacuole are much more nearly in electrochemical equilibrium in *Hydrodictyon*, it is evident that most active transport occurs at the plasmalemma and that this membrane represents the principal permeability barrier of the cell. For some purposes, it is convenient to consider the movement of ions from the outside solution into the vacuole. When this is done the relevant potential to use is the value E_{vo} which as we have seen is -90 mV.

Activity of ions in cytoplasm

While the vacuolar sap is a relatively simple solution of mainly inorganic electrolytes the cytoplasm contains ions both free in solution and associated with charged sites on macromolecules, principally protein. In addition, there are numerous organelles and vesicles which may contain higher concentrations of certain ions than the cytoplasm. We may well ask whether the average concentration of an ion in the cytoplasm is a sound basis on which to calculate equilibrium potentials across the plasmalemma. The activity of an ion will be greatly reduced if an appreciable fraction of it is not in free solution. In calculating the electrochemical potential difference of an ion (eqs. 3.8 to 3.12), we should have used the ratio of its activity in the cell and the outside solution rather than its concentration ratio. We have assumed, however, that for practical purposes the activity and concentration are more or less equivalent. We shall now see how good this assumption is.

There are some authors (a minority) who believe that most of the cations in the cytoplasm are bound electrostatically by association with negatively charged groups on the protein in the cytoplasm. Ling (1962, 1965) has attacked the membrane theories of selective ion transport by purporting to show that the binding sites in the cytoplasm are highly selective for potassium and discriminate against sodium. Bound potassium will have a low mobility, hence the activity of potassium in the cytoplasm would be much lower than the measured concentration. There would, therefore, be no electrochemical potential gradient and the whole system would be in equilibrium. Similar ideas were developed in the sorption theory of Troshin (1961).

This matter can be resolved if direct measurements are made of the activity of potassium and sodium in the cytoplasm. Ion-selective electrodes have been developed for this purpose (see page 81). Although most observations have been made on muscle from various animals and on toad oocytes, the activity of potassium in the cytoplasm of two algae (*Chara australis* and a species of *Griffithsia*) was measured by Vorobiev (1967). This author recalculated Nernst potentials for potassium based on his activity ratios and found that the passive driving forces were all in the same direction as those calculated earlier from concentration measurements. Spanswick (1968) pointed out that such conclusions gave a strong indication that potassium was not appreciably bound in the cytoplasm of these algae. A more comprehensive survey of potassium and sodium activity in the cytoplasm is described by Dick and McLaughlin (1969) for toad oocytes. Using K-selective and Na-selective electrodes these authors compared the activities of potassium, a_K, and sodium, a_{Na}, with the values of C_K and C_{Na} from the same cells. Concentration and activity are related in eq. (3.13).

$$C_j = a_j/f_j \tag{3.13}$$

where f_j is the activity coefficient of the ion j. The activity coefficient is determined by the total ionic strength of the solution and decreases as the total electrolyte concentration increases. The appropriate activity coefficient to apply to measurements of the ionic activity of the cytoplasm can only be calculated precisely from the total electrolyte composition of the cytoplasm, but an approximate value can be inferred from an isotonic Ringer solution. This value was 0·75 for f_K and f_{Na}. If all the potassium or sodium found in concentration measurements were in solution eq. (3.13) should hold. If a certain amount is bound the expression must be rewritten as

$$C_j = a_j/f_j + C_j^s \tag{3.14}$$

where C_j^s indicates the concentration of bound or sequestered ion. It was found that C_K^s was less than 3 per cent of the value of C_K but that about half of the intracellular sodium was located in C_{Na}^s. Thus, the cytoplasm would seem to preferentially bind sodium rather than potassium in direct contradiction to the hypothesis of Ling and Troshin. If a_{Na} is appreciably smaller than C_{Na} in plant cells also, it would have the effect of steepening the inward electrochemical potential gradient, increasing the energy used up in pumping Na^+ out of the cell. In *Hydrodictyon africanum* the value of $E_{co} - E_{Na}^N$, would be −34 mV rather than −17 mV as calculated in Table 3.1.

Ling (1969) pointed out that the insertion of the electrode may disrupt the delicately balanced structure of protein micelles whose binding properties would be altered releasing free potassium in the vicinity of the electrode. If this were the case the activity recorded by the electrode should decrease exponentially with time as potassium, accumulated locally, diffuses away from the electrode into the cytoplasm. In practice, the values of a_K remain constant for some hours after the insertion of the electrode. There can, therefore, be no major gradient of a_K near the electrode tip. Other problems in interpreting measurements made by ion selective electrodes in the cytoplasm are discussed on page 82.

Consumption of ions in metabolism

Three of the most important anions in cellular physiology have received surprisingly little attention from workers in algal cells. Nitrate, sulphate and phosphate share the common characteristic that after entering the cell as ions they become incorporated into organic compounds some of which are electrically neutral. Much of the incoming nitrate will be rapidly reduced by the enzyme nitrate reductase and become incorporated into amino acids; this is especially likely to be the case in photosynthesizing cells, e.g., *Hydrodictyon*. In a very early study by Hoagland and Davis (1929) on *Nitella*, the authors were unable to detect any NO_3^- in the vacuolar sap although the pond water in which the alga grew contained measurable amounts of NO_3^-. Since there can be no doubt whatever that *Nitella* had absorbed very large quantities of NO_3^- during its life we are left to conclude that all of it had been converted into other nitrogenous compounds. What, then, can we say about the driving force on nitrate movement across the plasmalemma and tonoplast? It seems entirely possible that the concentration of NO_3^- inside the cell is smaller than that outside, but precise estimates have not been made. It is partly due to the lack of a suitable isotope with which to study NO_3^- fluxes and partly due to lack of endeavour that factors controlling the active transport of NO_3^- (if, indeed, it is actively transported) are almost entirely unknown.

The fluxes of sulphate across membranes are generally much slower than nitrate (see chapter 4 for a partial explanation), but the major portion of sulphate transferred to the cytoplasm will be found in the −SH groups of sulphur-containing proteins. Again it is difficult to calculate a precise value for the Nernst potential for sulphate in cytoplasm. Using data from Hoagland and Davis (1929) and assuming a value for the membrane potential of −140 mV, the Nernst potential for SO_4^{2+} in the vacuole would be +38 mV. Thus SO_4' would seem to be a long way from electrochemical equilibrium in *Nitella* and must be actively transported inwards.

Phosphate ions, H_2PO_4', appear to be actively transported and incorporated quickly into phosphorylated compounds. The utilization of these essential anions in metabolism and growth effectively reduces the electrochemical potential gradient up which they must be transported on their entry into the cell.

Flux ratio as a test for active transport

Using suitable radioactive tracers the fluxes of ions in and out of cells can be measured (see page 88). The results of tracer experiments of this kind can be used as an additional test of active transport. Since the net flux into a cell is composed of two terms

$$\text{Net flux} = \text{flux}_{in} - \text{flux}_{out} \qquad (3.15)$$

it can be appreciated that in a cell which has established a steady state with its environment and which is not growing, the net flux is zero since $\text{flux}_{in} = \text{flux}_{out}$. In a growing cell there is usually a net flux of ions inward. The fact that a cell may have reached a steady state does not, of course, indicate that it is at passive equilibrium from a thermodynamic point of view. If there is a tendency for an ion to diffuse out of the cell, work must be performed to counter this driving force so that the cell is kept 'topped-up' with the ion in question. The ratio of the two ion fluxes will be unity if the electrochemical potential difference across the membrane is zero. In general terms, the flux ratios and driving forces have been related by an equation derived by Ussing and by Teorell and known as the Ussing–Teorell equation

$$\frac{\phi_j^{\text{in}}}{\phi_j^{\text{out}}} = \frac{C_j^{\text{out}}}{C_j^{\text{in}}} \exp\left(z_j \frac{FE}{RT}\right) \tag{3.16}$$

where ϕ_j^{in} and ϕ_j^{out} are the partial fluxes across the membrane and E is the observed electric potential difference.

By inspection of eqs. (3.8) and (3.9) we can rewrite eq. (3.16) so as to emphasize that the driving force across the membrane is the difference in electrochemical potential $\bar{\mu}_j$ ($\bar{\mu}_j = z_jFE - RT \ln [C_j^{\text{out}}/C_j^{\text{in}}]$)

$$RT \ln\left(\frac{\phi_j^{\text{in}}}{\phi_j^{\text{out}}}\right) = \bar{\mu}_j^o - \bar{\mu}_j^i = \Delta\bar{\mu}_j \tag{3.17}$$

At flux equilibrium, $\Delta\bar{\mu}_j = 0$ and the flux ratio is unity ($\log_e 1 = 0$).

In practice, most cells in a constant environment are found to be in a steady state if they are not growing. Thus, for most ions $\phi^{\text{in}}/\phi^{\text{out}}$ will be unity. If there is a difference in the electrochemical potential of ion j between the cell and the surroundings (i.e., $\Delta\bar{\mu}_j \neq 0$) then we find, by experiment, that the flux ratio is still unity, this means that either influx or efflux is faster than can be accounted for by passive driving forces alone and we might conclude that active transport is taking place.

We recall that $\Delta\bar{\mu}_j$ can be expressed as the difference between the Nernst potential and the observed electric potential across the membrane (eq. 3.12). Substitution of this in eq. (3.17) gives

$$RT \ln\left(\frac{\phi^{\text{in}}}{\phi^{\text{out}}}\right) = z_jF(E_j^{\text{N}} - E_{\text{obs}}) \tag{3.18}$$

thus

$$\ln\left(\frac{\phi^{\text{in}}}{\phi^{\text{out}}}\right) = \frac{z_jF(E_j^{\text{N}} - E_{\text{obs}})}{RT} \tag{3.19}$$

therefore

$$\frac{\phi_j^{\text{in}}}{\phi_j^{\text{out}}} = \exp\left(\frac{z_jF(E_j^{\text{N}} - E_{\text{obs}})}{RT}\right) \tag{3.20}$$

We can apply this equation to the electrical data obtained from *Hydrodictyon* and predict the flux ratio for each ion if only passive driving forces were involved in its movement. It is apparent that Table 3.3 is the flux equivalent of Table 3.2 in which equilibrium concentrations were predicted for given values of the membrane potential.

In each case the observed flux ratio differs greatly from the value

***Table 3.3 Predicted and observed flux ratios between the outside and the vacuole of* Hydrodictyon africanum**

Value of $E_{vo} = -90$ mV at 14 °C

Ion	$E_j^N - E_{vo}$ (mV)	ϕ^{in}/ϕ^{out}		Observed partial fluxes (pmoles cm^{-2} sec^{-1})	
		Predicted	Observed	In	Out
K^+	−63	0·095	0·94	1·4	1·6
Na^+	+20	2·250	0·70	0·7	1·0
Cl^-	+173	0·001	2·30	1·4	0·6

predicted on the assumption that the ions are at electrochemical equilibrium. If, for instance, active Cl^- transport were to cease, the efflux would greatly exceed the influx until a new steady state was reached.

Membrane fluxes and permeability coefficients

The flux ratio is a dimensionless number which tells us nothing about the actual rates of movement of ions across the membrane. The membrane is a barrier to diffusion composed of lipid material and we might suspect that the behaviour of an ion in the lipid phase will be an important factor in determining how quickly it crosses the barrier. The driving forces determined by the gradient of electrochemical potential $d\bar{\mu}/dx$ will determine what the equilibrium concentrations of ions would be at passive flux equilibrium, but they only partly determine the rate at which the equilibrium can be reached. Thus, the rate of transfer of an ion is determined by several factors which may be equated in the general form

$$\text{Flux} = \text{Driving force} \times \text{Activity}^{(\text{mem})} \times \text{Mobility}^{(\text{mem})}$$

The superscripts (mem) denote that activity of the solute and its mobility in the membrane itself are important parameters. These two terms are important in determining the permeability of the membrane to a given ion, e.g.,

$$P_j = \frac{U_j^* K_j^* RT}{L} \qquad (3.21)$$

where U_j^* = mobility of the ion j in the membrane,

K_j^* = the partition coefficient of the ion between water and the membrane phases. It thus defines the C_j^* (the concentration of ion j in the membrane),

L = the length of the pathway followed by the ion.

Although it defines P_j, this equation is not very helpful because it contains two quantities which are unmeasurable and to which it is very difficult to assign sensible arbitrary values. These difficulties stem from the non-homogenous nature of the membrane itself. We have seen that the membrane has both a matrix and an infrastructure which differ in chemical composition and physical properties. It is probable that ions, during their movement across the membrane, may follow several pathways in which their mobilities may be different.

Since there is a high probability that at least one of the pathways is hydrophilic, the concentration of the ion in the membrane cannot be reliably inferred from its partition coefficient between water and lipid.

Notwithstanding these difficulties, it is relatively easy to get practical measurements of P_j from tracer experiments in which the initial influx or efflux of ions is measured. A valid measurement of P_j can be made only if the flux is driven passively. Inspection of Nernst potential calculations usually indicates which of the partial fluxes is passive (e.g., Table 3.1). Using eqs. (3.22) and (3.23) which were derived by Goldman (1943) and by Hodgkin and Katz (1949), P_j can be calculated from one of the partial fluxes. Since all terms other than P_j can be measured or are constants, the solution for P_j is simple

$$\phi_j^{\text{in}} = -\frac{z_j FEP_j}{RT} \times \frac{C_j^o}{1 - \exp(z_j FE/RT)} \tag{3.22}$$

$$\phi_j^{\text{out}} = -\frac{z_j FEP_j}{RT} \times \frac{C_j^i \exp(z_j FE/RT)}{1 - \exp(z_j FE/RT)} \tag{3.23}$$

The form of the equations is made simpler if $z_j FE/RT$ is designated as x.

We can now write equations for P_K, P_{Na} and P_{Cl}

$$P_K = \frac{\phi_K^{\text{out}}}{C_K^i} \cdot \frac{1}{x} (e^{-x} - 1) \tag{3.24}$$

$$P_{Na} = \frac{\phi_{Na}^{\text{in}}}{C_{Na}^o} \cdot \frac{1}{x} (e^{x} - 1) \tag{3.25}$$

$$P_{Cl} = \frac{\phi_{Cl}^{\text{out}}}{C_{Cl}^i} \cdot \frac{1}{x} (e^{x} - 1) \tag{3.26}$$

ϕ_j has the units mole cm^{-2} s^{-1} and P_j is expressed as cm s^{-1}. Substituting the values for the membrane potential given in Table 3.1 and the partial fluxes from Table 3.3, we find the following values for the permeability coefficients at 14 °C; $P_K = 9{\cdot}2 \times 10^{-7}$ cm s^{-1}, $P_{Na} = 6{\cdot}8 \times 10^{-8}$ cm s^{-1}, $P_{Cl} = 7 \times 10^{-9}$ cm s^{-1}.

There is evidence that the practical value of P_j is in fact the sum of the permeabilities of the several pathways which an ion may follow in its passage across the membrane. Since the environment experienced by the plasmalemma may modify the properties of these pathways differentially, we should not necessarily expect that P_j will remain constant. There is a growing body of evidence that the value of P_K, for instance, can be modified by temperature, hydrogen ion concentration and the concentration of potassium in the external medium. In some cases, however, the ratio of the permeability coefficients of mobile ions is useful in understanding how the membrane potential is controlled.

Factors controlling the electrical potential across membranes

The electrical potential difference is clearly of cardinal importance in determining the magnitude and the direction of driving forces on ions. It is essential that the nature of this potential and its origin are clearly understood. It may be helpful to consider a very simple model (Fig. 3.2).

(a) Consider a system of two compartments containing differing concentrations of KCl and separated by a membrane.
(b) Suppose that the permeability of the membrane to K^+ is very large relative to Cl^-.
(c) There will be a tendency for both K^+ and Cl^- to flow from the more concentrated compartment, which we shall designate as *i*, to the less concentrated compartment, *o*. The rate at which K^+ can leave compartment *i* will be faster than the rate for Cl^- because of the properties of the membrane.
(d) What happens? For a brief moment in time compartment *i* loses K^+ faster than Cl^- and an electric potential of negative sign is developed because of this slight separation of charge.
(e) The negative potential slows down the movement of K^+, attracting it back towards compartment *i*. When the potential has developed a steady value the net movement of K^+ from *i* to *o* will cease even though $[K_i] > [K_o]$.
(f) It is clear from this model that the membrane potential is strongly determined by its permeability to K^+ and Cl^-.

We may define the potential as the tendency of electric charges of different sign to diffuse across the membrane at unequal rates. In most living cells, the cytoplasm has a small excess of negative charge relative to the external medium. In practice, the amount of charge which must be separated to give rise to electric potentials of the

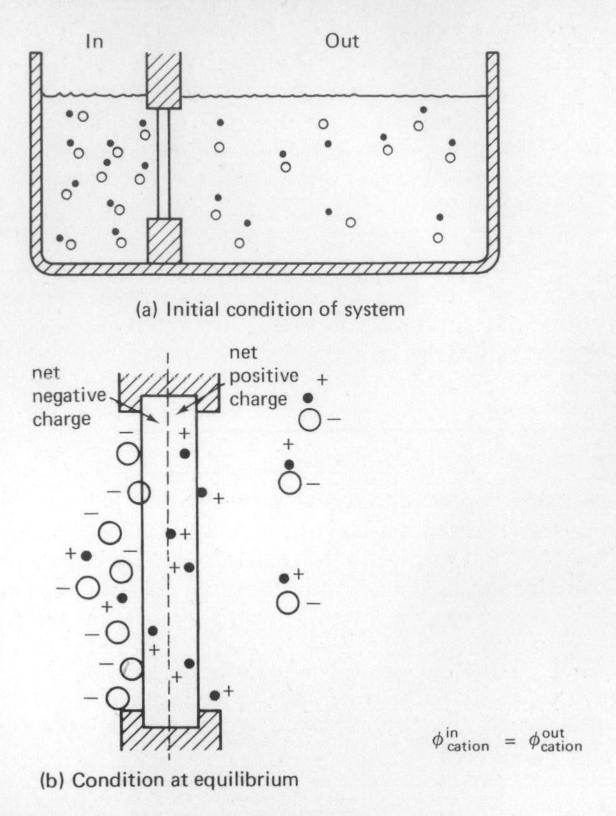

FIG. 3.2 ***Development of a diffusion potential in a model system. The two compartments are separated by a membrane which is cation-selective. The accumulation of indiffusible anion at the membrane surface sets up an electrical gradient across the membrane which slows down the diffusion of the outgoing cations by charge repulsion so that the cation influx and efflux become balanced.***

order of -100 mV is minute in relation to the sum of positive and negative charges in the cytoplasm. The quantity of charge can be calculated from eq. (3.27)

$$Q = \frac{C \times E}{F} \tag{3.27}$$

where Q = the quantity of charge in equivalents cm^{-2}; C = the capacitance of the membrane (usually taken as 1×10^{-6} Farad cm^{-2}); E = the electric potential difference in volts and F = the Faraday (10^5 coulombs $equivalent^{-1}$).

If we consider a large spherical coenocyte of *Hydrodictyon africanum* with radius 0·5 cm and with a layer of cytoplasm 20 μm thick the volume of the cytoplasm will be 0·06 cm^3 and its surface area 3·14 cm^2. The potential difference across the plasmalemma of -116 mV or $1{\cdot}16 \times 10^{-1}$ volts would result from the interior possessing a net negative charge of 6×10^{-7} equivalents per litre in a system containing 2×10^{-1} equivalents of positive and negative charge per litre, i.e., less than a millionth of the charge is unbalanced.

From the Nernst eq. (3.6) we know that an electric potential of 58 mV of opposite sign to the ion under consideration is required to stabilize each tenfold difference in concentration of a monovalent ion in two compartments at 20 °C. In the simple physical system described above a steady state would be achieved only if the membrane were completely impermeable to Cl^-. If the membrane had an appreciable chloride leak then both the membrane potential and the steady state differential [K] would decline.

With our simple system we know how the concentration difference between *i* and *o* arose; we filled the compartments with KCl solutions of different strength. Obviously, the higher concentration of ions inside cells relative to their surroundings does not arise in this way and we also know that the plasmalemma does have an appreciable passive permeability to chloride. In the simple model, the principal role played by the indiffusible chloride ions was to provide negative charge. Their place could be taken by a negative current electrode or some other source of negative charge. As it happens, living cells synthesize many protein molecules carrying a net negative charge to which the membrane is impermeable. We shall see in the next section how potent this fixed charge system can be in sustaining the membrane potential. The presence of indiffusible protein anions does not, of course, explain how the chloride gets into the cell. In fact, almost nothing we have learnt so far in this chapter helps very much in this respect.

In our simple model, the electric potential difference was determined solely by the passive diffusion of the permeable ion, K^+. If the membrane has been permeable to Cl^- we would have found that the electric potential difference would no longer be a simple function of the concentration of K^+ in the two compartments but would depend on Cl^- also. The relative influences of the two ions on the membrane potential would depend on their relative mobilities in the membrane as well as on their concentrations in the two compartments. What we need is some expression which will equate the membrane potential with these two sets of properties. In the Goldman equation we can see how this is done; assuming that ions are diffusing passively, then,

$$E = \frac{RT}{F} \ln \frac{P_{\mathrm{K}}[\mathrm{K}]^o + P_{\mathrm{Na}}[\mathrm{Na}]^o + P_{\mathrm{Cl}}[\mathrm{Cl}]^i}{P_{\mathrm{K}}[\mathrm{K}]^i + P_{\mathrm{Na}}[\mathrm{Na}]^i + P_{\mathrm{Cl}}[\mathrm{Cl}]^o} \tag{3.28}$$

where E is the electric potential difference across the membrane. Notice that the ratio of the chloride terms is inverted relative to the cationic terms. This is explained because the chloride concentration gradient will tend to reduce any negative potential set up by the asymmetric distribution of the cations. Terms for other diffusible ions can be added to the equation.

Let us state again what eq. (3.28) predicts: the electric potential across a membrane is determined by the steady state concentration of ions across the membrane and their permeabilities. The equation predicts what is often found in practice, viz., that E is most sensitive to the ion with the largest permeability coefficient and subject to the largest passive driving force.

There are many cases in which the value of E can be seen to vary as predicted by the Goldman equation although in recent years the interpretation of the membrane potential solely in terms of the passive diffusion of ions has come increasingly under fire. It has been found by Thorhaug (1970) that changes in E in *Valonia* spp with temperature do not fit the Goldman equation. The data might be explained by variations in P_K, P_{Na} and P_{Cl} with temperature (see page 125) or they might be related to the activity of ion pumps which contribute directly to the potential by pumping ions of one sign in one direction only. Such pumps are described as *electrogenic* and we will discuss them in greater detail in a later section. There is a growing number of reports of electrogenic pumps in a variety of plant and animal species and particularly in the membranes of chloroplasts and mitochondria.

The role of calcium in controlling membrane permeability

In some experiments where the membrane potential of *Chara australis* was predictably determined by the diffusion potentials of K^+ and Na^+, divalent calcium ions were omitted from the bathing solution (Hope and Walker, 1961). Since calcium is ubiquitous in natural waters this is an improbable growth medium in which the permeability properties of the membrane might be expected to be abnormal. Small amounts of calcium added to the outside solution seem to desensitize the membrane potential of *Nitella translucens* to changes in the K^+ concentration in the range 0·1–1·0 mM. (Spanswick, Stolareck and Williams, 1967). Similar results were found by Kitasato (1968) working with *Nitella clavata* where the membrane potential was insensitive to K^+ in the concentration range 0·1–10·0 mM when 1·0 mM $CaCl_2$ was present in the solution bathing the cell. This author showed that Ca^{2+} had a major effect in reducing the value of P_K so that the membrane resistance to the passive movement of K^+ increased. It seems probable that the permeability coefficient of sodium will be similarly affected. Since the membrane potential in these circumstances does not appear to be primarily determined by the Goldman equation it seems that the diffusion of some other ion, e.g., H^+, or electrogenic ion pumps are of significance in its control.

Because Ca^{2+} ions reduce the leakage of ions from the cytoplasm to the outside solution they can be said to have an energy-conserving function. Less leakage of ions results in less wastage of the free energy consumed in their active transport across the membrane.

Origin of the membrane potential

The ionic asymmetries determining E are created and maintained by the active transport of ions across the plasmalemma and through the synthesis of negatively charged proteins. The properties of the membrane are dependent on metabolism either directly, as in the case of ion pumps, or indirectly through the control of phospholipid synthesis and membrane assembly. Before we proceed, we should be quite clear in our minds that the passive transport of one ion is often dependent on the active transport of another so that by one means or another both are controlled by metabolism.

Donnan distribution of indiffusible anions. The cytoplasm contains a large number of protein molecules, some with enzymic activity, which have charged sites associated with the α-carboxyl groups of amino acid residues of the polypeptides. Many of these proteins carry a net negative charge at the pH of the cytoplasm. The protein molecules are too large to diffuse across the plasmalemma by normal processes and so they operate in much the same way as Cl^- in the simple model.

Consider a situation (Fig. 3.3) in which there are two compartments separated by a membrane which is impermeable to protein. Initially in the compartment marked i there is a quantity of negatively charged protein and an equivalent amount of K^+. In compartment o there is a solution of KCl. The membrane is freely permeable to both K^+ and Cl^-. The system is now left to reach equilibrium. Some Cl^- will diffuse into i from compartment o accompanied by K^+, thus the concentration of K^+ will increase. The tendency for K^+ to flow back to o and the indiffusible nature of the protein will give rise to an electric potential of negative sign which effectively repels the influx of Cl^- into the compartment i. Thus, at equilibrium there will be an imbalance of mobile ions across the membrane. This is known as the Gibbs–Donnan effect. The concentration gradients of the asym-

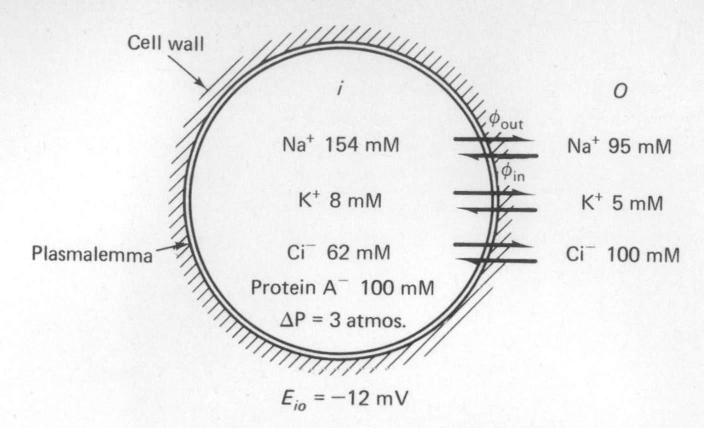

***FIG. 3.3** Equilibrium distribution of ions and the membrane potential, E_{io}, in a cell containing 100 m eqv indiffusible anion associated with protein. The model makes the assumption that the plasmalemma is permeable to K^+, Na^+ and Cl^-. Because the internal solute concentration is greater than that in the surroundings, water will move into the cell causing an osmotic pressure of three atmospheres above ambient.*

metric ions give rise to an electric potential, known as the Donnan potential, E_D, hence

$$E_D = \frac{RT}{F} \ln \frac{[K^+]^o}{[K^+]^i} = \frac{RT}{F} \ln \frac{[A^-]^i}{[A^-]^o} \tag{3.29}$$

(Notice that this is identical with the expression for the Nernst potential, eq. (3.6), for univalent ions.) The greater the indiffusible protein content of compartment *i* the greater the discrepancy at equilibrium between the ionic concentrations in the two compartments.

Figure 3.3 illustrates a set of conditions in which the fixed anion charge on protein within a cell is equivalent to 100 m eqv. The membrane is permeable to all of the mobile ions. At equilibrium, the ionic asymmetry of the mobile ions is such that a potential of −12 mV is produced across the membrane (inside negative). This value is small in relation to the size of the potentials found in most plant cells but approximates to the equilibrium potentials found in some bacteria.

The Donnan potential depends ultimately on the metabolic energy which is expended on the synthesis of proteins and other charged macromolecules.

Neutral ion pumps and the asymmetric distribution of ions across the plasmalemma. The term 'pump' has wide usage in the description of active transport processes but readers should not pursue the analogy of a mechanical pump with valves too far. The mechanisms responsible are pumps merely in the sense that they move ions against a gradient of electrochemical potential. They, therefore, must perform work and consume free energy. The mechanism by which one ion pump operates will be outlined in chapter 5, but some of its general features can be conveniently mentioned at this stage.

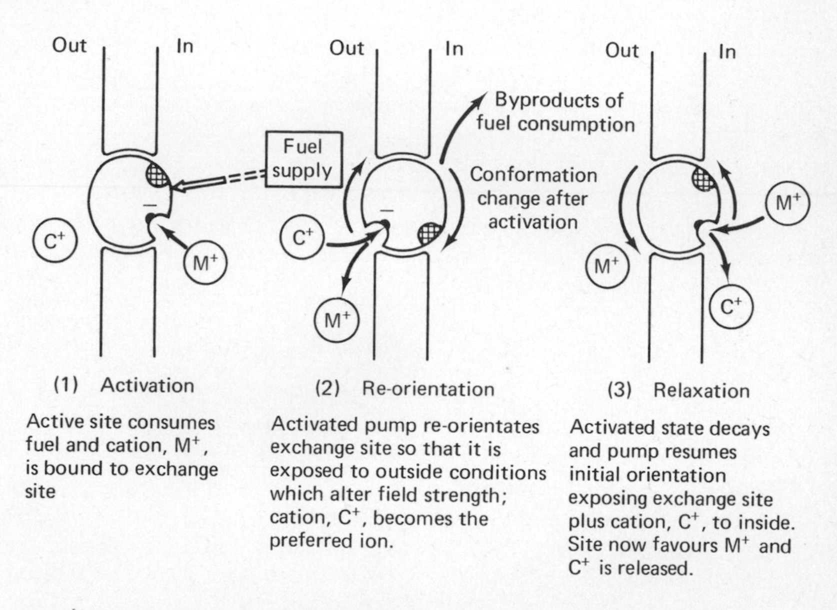

FIG. 3.4 Some features of a hypothetical 'neutral' ion exchange pump.

Figure 3.4 shows a pumping mechanism which might bind potassium at the outer surface of the plasmalemma. The presence of the bound ion causes some change in the orientation of the pump so that K^+ is carried to the inner surface of the membrane. At this point, the local chemical environment is such that K^+ is released from the pump and its place on the binding site is taken by another cation M^+ which has a higher affinity for the binding site than K^+ (in chapter 4, page 121 I shall indicate how the chemical environment of a molecule affects the field strength and the ion-selectivity of its binding sites). With M^+ attached to the binding site the pump again alters its position exposing M^+ to the outside solution where conditions are such that the binding of K^+ is favoured and M^+ is released. By continuously recycling, this pump will enrich the interior with K^+ but because it carries M^+ on the return stroke it will not contribute directly to the electrical potential. It contributes to the potential by increasing the asymmetry of K^+ across the membrane so that in the Goldman equation the term

$$\ln \frac{P_K[K]^o}{P_K[K]^i}$$

will become increasingly negative.

The calculations in Tables 3.1 and 3.3 indicate that at least some of the K^+ in *Hydrodictyon* must arrive at the cell interior by such a pumping mechanism. Since all but a trivial amount of the anion Cl^- must be transported actively some kind of pump must carry this ion; in *Hydrodictyon*, the characteristics of this pump are such that it is probably electrically neutral so that some other anion, possibly OH^-, moves in the opposite direction across the membrane.

What will be the immediate consequences of blocking a neutral ion pump by some inhibitor molecule? We see that the concentration of K^+ in *Hydrodictyon* is not at electrochemical equilibrium. The pump keeps the cell topped up with K^+ which continually leaks back into the outer solution. If the pump stops, the concentration difference of K^+ across the membrane will diminish and, since the membrane is appreciably more permeable to K^+ than the other ions, the value of E will fall and eventually a new equilibrium will be established.

If the treatment applied does not alter the permeability coefficients, stopping an electrically neutral ion pump will bring about a gradual change in the value of E as K^+ diffuses out of the cell. There is another type of pump which contributes directly to E so that when it is stopped abrupt changes in the membrane potential are recorded.

Electrogenic ion pumps. An electrogenic ion pump differs from the neutral pump illustrated in Fig. 3.4 only in that on the return stroke the pump does not carry an ion and thus remains charged. The effect of this pump will be to enrich the interior in either negative or positive charge depending on whether a cation or anion is carried. An electrogenic pump could also pump an ion out of the cell leaving a charge disbalance in its wake. A mechanism of this kind separates charge directly and, of course, will quickly produce a suitable electrical gradient which will promote the diffusion of ions of opposite sign in the same direction as the pump, or of similarly charged ions in the opposite direction.

There is increasing evidence that electrogenic pumps of this kind are widely distributed. In the giant algal coenocyte *Acetabularia mediterranea* there is an inwardly directed electrogenic chloride pump which contributes about −90 mV to the observed potential of −170 mV between the cytoplasm and the outside. An electrogenic H^+ extrusion pump has been described in *Nitella clavata*, and Smith (1970) has suggested that such a pump may be a controlling factor in Cl^- uptake by *Chara* and other coenocytes. Other examples of electrogenic H^+ pumps are found in the fungal hyphae of *Neurospora*, the roots of barley and in chloroplasts and mitochondria which have been examined in great detail (see chapter 5).

An electrogenic pump contributes directly to the membrane potential and its existence may be inferred if blocking the pump produces an immediate change in its value. In *Acetabularia* a variety of treatments inhibiting metabolism result in an almost immediate depolarization of the membrane potential from −170 mV to −80 mV (Saddler, 1970). In *Neurospora* (Fig. 3.5) an even greater depolarization occurs when the H^+ extrusion pump is inhibited by sodium azide, a respiratory poison. Notice that the original value of the potential is restored when the pump is unblocked (Slayman, 1970).

The composite nature of the membrane potential. The membrane potential may be composed of several components:

(a) The Donnan potential resulting from indiffusible anions.
(b) The ionic asymmetry created by neutral ion pumps and the consequent potential resulting from differential rates of diffusion of anions and cations back into the outside medium.
(c) The separation of charge caused by electrogenic ion pumps.

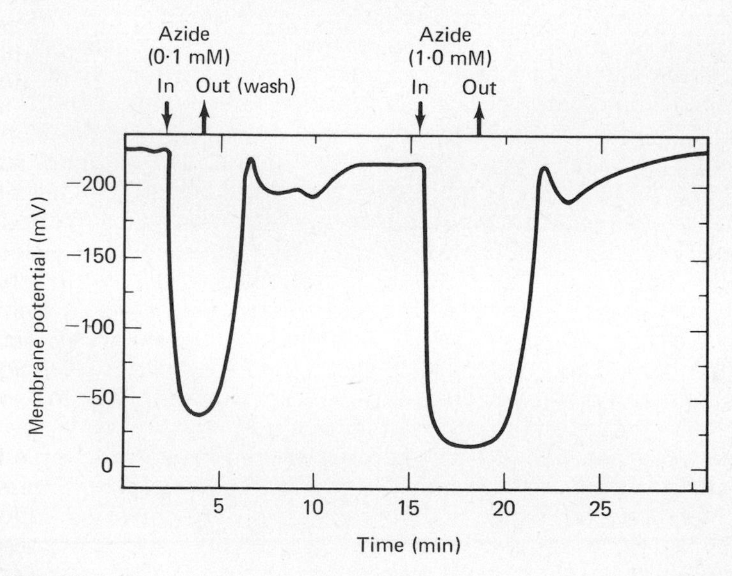

FIG. 3.5 The effect of sodium azide on the membrane potential of Neurospora. *Metabolic inhibition by azide stops an electrogenic pump and the cell depolarizes very rapidly (data from Slayman, 1970).*

Each of these three components is dependent on the metabolism of the cell.

Conductance and electrical resistance of membranes

The conductance of the plasmalemma can be estimated from the fluxes of the ions crossing it in response to a known driving force. It has been assumed that, in certain conditions, most of the passive ion current across the plasmalemma is carried by sodium and potassium. The conductance, G^+_{pl}, to cations is then given by

$$G^+_{\text{pl}} = \frac{F^2}{RT}\,(\phi^{\text{in}}_{\text{K}} + \phi^{\text{in}}_{\text{Na}}) \tag{3.30}$$

where G^+_{pl} has the units $\text{ohm}^{-1}\ \text{cm}^{-2}$ and $\phi^{\text{in}}_{\text{K}}$ and $\phi^{\text{in}}_{\text{Na}}$ are the measured influxes of K^+ and Na^+ in pmoles $\text{cm}^{-2}\ \text{s}^{-1}$. The other symbols have their usual meanings.

The resistance of the plasmalemma R_{pl} is the reciprocal of the conductance and using eq. (3.30) Williams, Johnston and Dainty (1964) arrived at a value of 300 kilohms (kΩ) cm^2. Estimates made by other workers using similar calculations lie in the range of 150–300 kΩ cm^2.

Estimates of membrane resistance made as above can be compared with direct electrical measurements in which a known current is driven across the plasmalemma (see methods page 91) and the change in E_{co} recorded. Surface resistance can be calculated from Ohm's law. The reported results in the literature are quite variable and even in a single organism *Nitella translucens* have been found to be from 15 kΩ cm^2 in solutions containing no calcium to 120 kΩ cm^2 in solutions containing approximately 1 mM $CaCl_2$. Although variable it is clear that the actual resistance of the plasmalemma is considerably less, by factors ranging from $\frac{1}{200} - \frac{1}{3}$, than the estimates produced on the assumption that K^+ and Na^+ carry all the membrane current (eq. 3.30). If these ions do not carry all the current, what others are available to fulfil this role? Measured fluxes of calcium show that it is too impermeable to be the missing term in eq. (3.30).

The research of Kitasato (1968) on *Nitella clavata* indicates that the essential flux missing from eq. (3.29) is that of hydrogen ions. If electrodes are inserted into the cell a current can be made to flow across the plasmalemma. The total current will be the sum of that carried by the electrolytes moving through the membrane. The ratio of current carrying capacity of the charged species will be approximately the same as the ratio of their permeability coefficients. Figure 3.6 shows that only a minor portion of the total current crossing the membrane was carried by K^+. The straight line drawn on this figure is the relationship which would have been found if the current

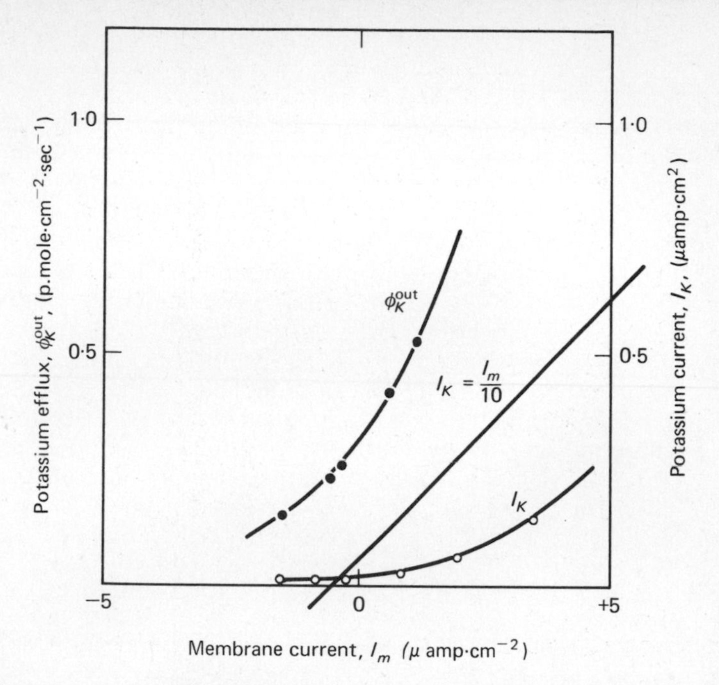

FIG. 3.6 The relationship between K^+ efflux, the electric current carried by K^+ (I_K) and the total current (I_m) across the plasmalemma of Nitella clavata. *The cells were bathed in a dilute nutrient solution containing 1·0 mM Ca^{2+}. The theoretical curve (straight line) shows the situation where one-tenth of the total current would be carried by K^+; in fact I_K is much less than one-tenth of the total current. (Data from Kitasato, 1968.)*

carried by K^+, I_K, had been as much as one tenth of the total membrane current, I_m. If K^+ is not carrying the current some other ion must be doing so. Kitasato points out H^+ is the only other ion in the system which could do this and goes on to show that the membrane potential is very sensitive to the asymmetric distribution of H^+ across the plasmalemma. If H^+ carries the current then its permeability coefficient, P_H, must be very much higher than that of K^+. According to this author, P_H would be in the order of 10^{-3} cm s^{-1} whereas the value P_K was measured as approximately 10^{-7} cm s^{-1}. These findings have recently caused other workers to re-evaluate their earlier experiments. Lannoye, Tarr and Dainty (1970) found that as the pH of the bathing solution was lowered from 6·0 to 4·5 the membrane potential depolarized by 75 mV, but on measuring fluxes of sodium they found that the depolarization was due to marked increases in passive Na^+ influx and in the value of P_{Na}. With appropriate adjustments to P_{Na} the change in membrane potential could be predicted from the Goldman equation (eq. 3.29). Re-examining their data for *Chara australis* Walker and Hope (1969) concluded that although the sum of the partial conductances of K^+, Na^+ and Cl^- through the plasmalemma account for only one half of the membrane conductance, the discrepancy is more likely to be due to errors associated with the measurement of unidirectional fluxes across the membrane rather than to the conductance of H^+. They cite some unpublished data by Coster and Hope where it was found that in this species the conductance of the membrane to H^+ was as low as $5 \times 10^{-7}\ \Omega^{-1}\ cm^{-2}$; the total conductance was approximately $5 \times 10^{-3}\ \Omega^{-1}\ cm^{-2}$.

Clearly, these conflicting views about the importance of H^+ fluxes in controlling the membrane potential cannot be reconciled as they stand. In all the studies, adequate care seems to have been taken to check the sources of error and there seems to be no reason to doubt the validity of the results. Kitasato's results have been confirmed by other workers using *Nitella clavata* who found a large efflux of H^+ associated with localized zones of the cell surface (Spear, Barr, and Barr, 1969). The only suggestion is that the differences are genuine and that the two species differ vastly in the values of P_H.

Practical measurements and their interpretation

Only a limited number of measurements are necessary to determine the electrochemical potential difference of an ion between a cellular compartment and the outside solution. The internal and external concentration of the ion (or more correctly, the activities of the ion) and the electric potential difference are all we need to know for the calculations. If, however, the results of such a calculation tell us, as they frequently do, that the ion is not in electrochemical equilibrium across the plasma membrane we may wish to measure the magnitude of active and passive fluxes.

Internal concentration of ions

In chapter 1 we pointed out that giant algal cells have enormous advantages over the cells of higher plants since it is possible to sample the vacuolar sap and cytoplasm directly and to measure the concentration of ions in them. With the long cylindrical internode cells of members of the family *Characeae* (which includes the genera *Chara, Nitella, Nitellopsis*) the vacuolar sap can be sampled without significant contamination from the cytoplasm after the cells have been centrifuged at a low speed (300 × gravity) for a few minutes. About half of the cytoplasm in these cells is in a gel state and remains closely associated with the plasmalemma and the cell wall. A second, inner, layer of cytoplasm is in a more fluid condition and centrifugation causes this to move to one end of the cell; the sap in the vacuole moves to the other. The sap remains surrounded by the intact tonoplast. After centrifugation, the sap and cytoplasm remain in separate parts of the cell for some time. The junction of the phases, which can be seen quite sharply when the cells are examined under a dissecting microscope, is then pinched off with a pair of fine forceps and the cell severed on the vacuole side of the constriction. A measured volume of the sap can then be withdrawn with a calibrated capillary, diluted to a convenient volume, and the concentration of ions in it measured by a number of standard techniques. Major cations, e.g., K^+, Na^+, Ca^{2+} can be estimated using flame emission or atomic absorption spectrophotometry, while Cl^- is usually measured by an electrometric titration described by Ramsay, Brown and Croghan (1955). The flowing cytoplasm may be sampled in a similar manner. In some of the larger algal coenocytes, e.g., *Valonia* and *Hydrodictyon africanum* vacuolar sap may be obtained either by inserting a microcapillary directly into the vacuole and withdrawing the sap, or by cutting off a small portion of the cell and withdrawing the contents rather like cutting open an egg to expose the yolk. Direct estimates of the concentration of ions in the cytoplasm and vacuole of smaller cells and in particular the cells of higher plants cannot be made by direct measurements and a variety of indirect techniques may be applied, such as extracting tissue with hot water (Higinbotham, Etherton and Foster, 1964) or compartment analysis from tracer efflux experiments (see page 90).

Activity measurements of ions in cytoplasm

Ion-selective electrodes function in a manner analogous to the more familiar glass pH electrode; their selectivity depends on the relative proportions of SiO_2, Al_2O_3 and Na_2O in the glass (Hinke, 1959). Figure 3.7 illustrates an electrode selective for potassium, the tip of which is small enough to be thrust into the cytoplasm of a cell. The electrode measures a number of electric potentials in series; these are indicated in the figure by the symbol E. A small potential E_1 will be developed between the silver–silver chloride electrode, a second boundary potential E_2 will exist between the solution filling the electrode (0·1 M KCl) and the inner surface of the glass; the potential between the outer surface of the glass, E_3, and the outer solution is the one we wish to measure and is the only potential which varies when the electrode is inserted into a series of solutions of varying potassium activity. The electrode is calibrated by recording millivolt

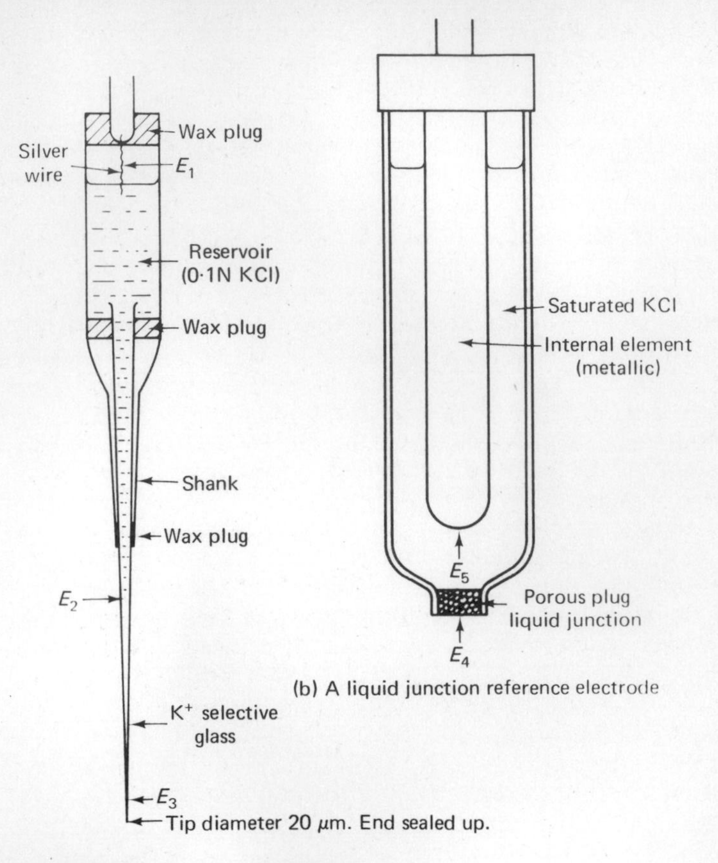

FIG. 3.7 A simple K-selective glass electrode based on the design of Hinke (1959) and a standard type of liquid junction reference electrode.

readings obtained when it is immersed in solutions of known potassium concentration.

The electrode is pushed through the wall of the cell and inserted in the cytoplasm using a micromanipulator (this procedure and the effect it has on the cell are described in more detail in the following section).

There are a number of problems in the interpretation of the results of activity measurements in the cytoplasm. A selective ion electrode usually gives its most unequivocal results when it is immersed in a simple salt solution of known hydrogen ion concentration. There are numerous phases in the cytoplasm and the electrode may make contact with macromolecules or cellular structures which have ionic gradients surrounding them in the form of electrical double layers. The nature of the difficulty is analogous to that of measuring the hydrogen ion activity (pH) in a suspension of soil particles and in the clear supernatant above them. The particles carry fixed charges which give rise to diffuse electrical double layers in which hydrogen ions may be more, but are usually less, abundant than in the true solution depending on the ion-selectivity of the site. Hence, the activity of hydrogen ions recorded in the vicinity of the double layer differs from that in the true solution. In the cell, the relative volume occupied by electrical double layers and the true solution is unknown and the influence of the former on activity measurements cannot be assessed accurately.

In spite of these difficulties, the activity measurements which were described earlier (page 63) have been useful in establishing the fact that much, if not most, of the potassium in the cytoplasm is in free solution.

Electrical potential measurements using micro-electrodes

The electrodes used to measure the electric potential across the cell membrane and the tonoplast are very fine, open ended capillaries filled with an electrolyte solution (usually 3N KCl) and connected through a silver–silver chloride or calomel electrode to a millivolt meter of high input impedance. The second half of the circuit is completed through a KCl electrode and its associated silver–silver chloride or calomel electrode immersed in the external solution. The completed circuit is as illustrated in Fig. 3.8.

In the circuit there are several potentials but those between the silver–silver chloride electrode and the salt bridge, E_1 and E'_1, and between the electrode tips and the liquid junctions, E_1 and E'_2 are equal or nearly so on the two sides of the plasmalemma and hence cancel one another out. The potential measured between the two KCl electrodes is, therefore, the sum of two potentials, i.e., E_3 across the plasmalemma and E_4 across the cell wall (a Donnan system containing indiffusible anions). In practice, E_4 is quite small in relation to E_3 and variations in the magnitude of E_3 are frequently accompanied by proportional variations in the magnitude of E_4.

The electrodes are prepared by stretching semi-molten glass tubing in a special microforge to produce a very fine tip, tapering quite abruptly from a stout shank and reservoir. The electrode used by Walker (1955) in his admirable study is illustrated in Fig. 3.9. The electrode shown is equipped with a fine needle-like plunger which can be pushed down the bore of the capillary to free it from any cell wall debris or any cytoplasmic material which may come to

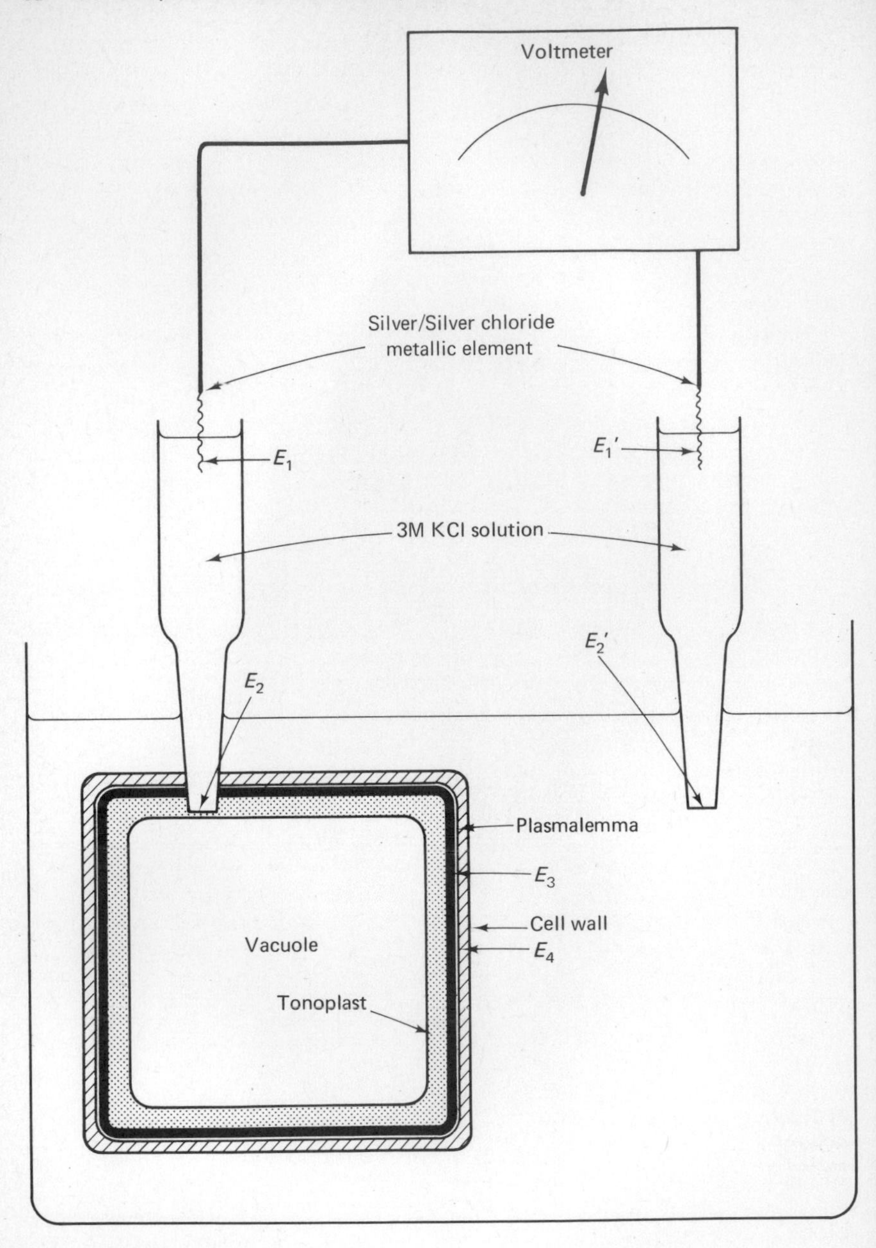

FIG. 3.8 Electrode system and potentials measured in the estimation of transmembrane potentials in plant cells. For explanation, see page 83.

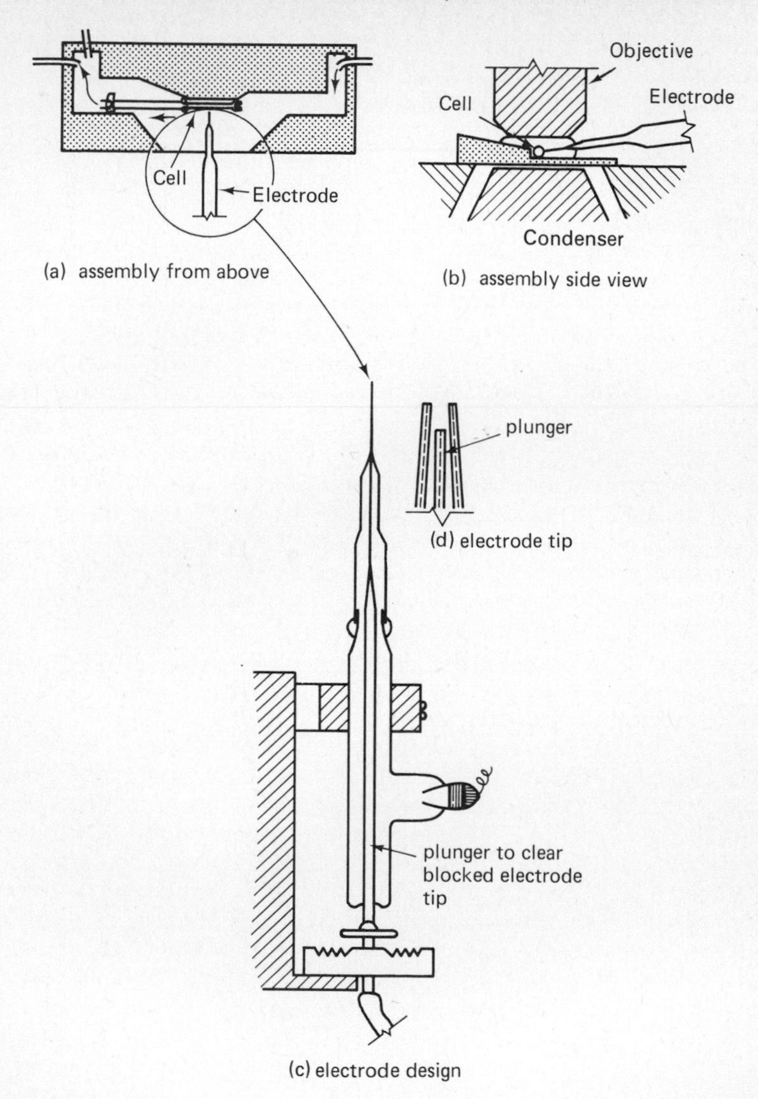

FIG. 3.9 Microelectrode assembly used for measuring transmembrane potential differences in* Nitella. *(Redrawn from Walker, 1955, by kind permission of the author.)

block it when the electrode is pushed into the cell. In practice, such a device can be used only if the tip of the micro-electrode is greater than 10 μm. Tip diameters smaller than this (1 μm) are more commonly used.

Micro-electrodes are inserted into cells using a micromanipulator, the process being observed by the operator through a microscope. Figure 3.9 illustrates the system used by Walker (1955) in his study on *Nitella* in which the insertion of the electrode is observed through the water immersion objective of a microscope. It is particularly important to be able to do this if the tip of the electrode is to be inserted in the very thin layer of cytoplasm between the cell wall and the vacuole. Electrodes inserted more deeply impale the tonoplast and hence measure the potential difference between the vacuole and the outside solution. If two electrodes are inserted the potential difference may be measured between the cytoplasm and the vacuole directly. If only one is used, the value of the potential between the cytoplasm and the vacuole can be obtained by successive measurements of the potentials between the cytoplasm and the outside solution and the vacuole and the outside solution; the value E_{cv} is then determined by difference. Although large algal cells provide the most convenient subjects for micro-electrode studies, in recent years progressively smaller cells have been impaled with micro-electrodes, root hairs by Etherton and Higinbotham (1960), cells in the oat coleoptile and pea root by Higinbotham, Etherton and Foster (1964, 1967), cells from storage tissue of beetroot by Poole (1966) and a *tour de force* by Barber (1968) who reproducibly measured the potential across the plasma membrane of *Chlorella*, a unicellular alga only 8–12 μm in diameter.

How does the cell react to being impaled in this way? The sequence of events reported by Walker (1955) and Spanswick and Williams (1964) for *Nitella* have not been observed in many other tissues but in this organism the following events are observed. The point at which the cell is ruptured heals quickly and there appears to be very little loss of cytoplasmic material through the wound. At the moment of entry the cyclosis of the flowing cytoplasm stops, but begins again 2 to 4 minutes later. Shortly after this a refractile material forms around the shaft of the micro-electrode where it enters the cell, and this material flows up the shaft towards the tip which it ultimately covers. When this cap is finally formed the measured potential difference between the cytoplasm and the outside solution falls abruptly by +30 to +40 mV. Walker (1955) found that this process occurred even when the tip of the electrode was deeply inserted in the vacuole. The refractile material pushed up the tonoplast in front of it so that with time the electrode was covered by a cap of cytoplasm. This sequence of events is illustrated in Fig. 3.10. Spanswick and Williams (1964) observed that flowing cytoplasm follows a helical path in one direction along the cell and returns along an adjacent helix. If the electrode used for measuring the vacuolar potential difference is inserted about 100 μm into the cell at the junction of these opposed

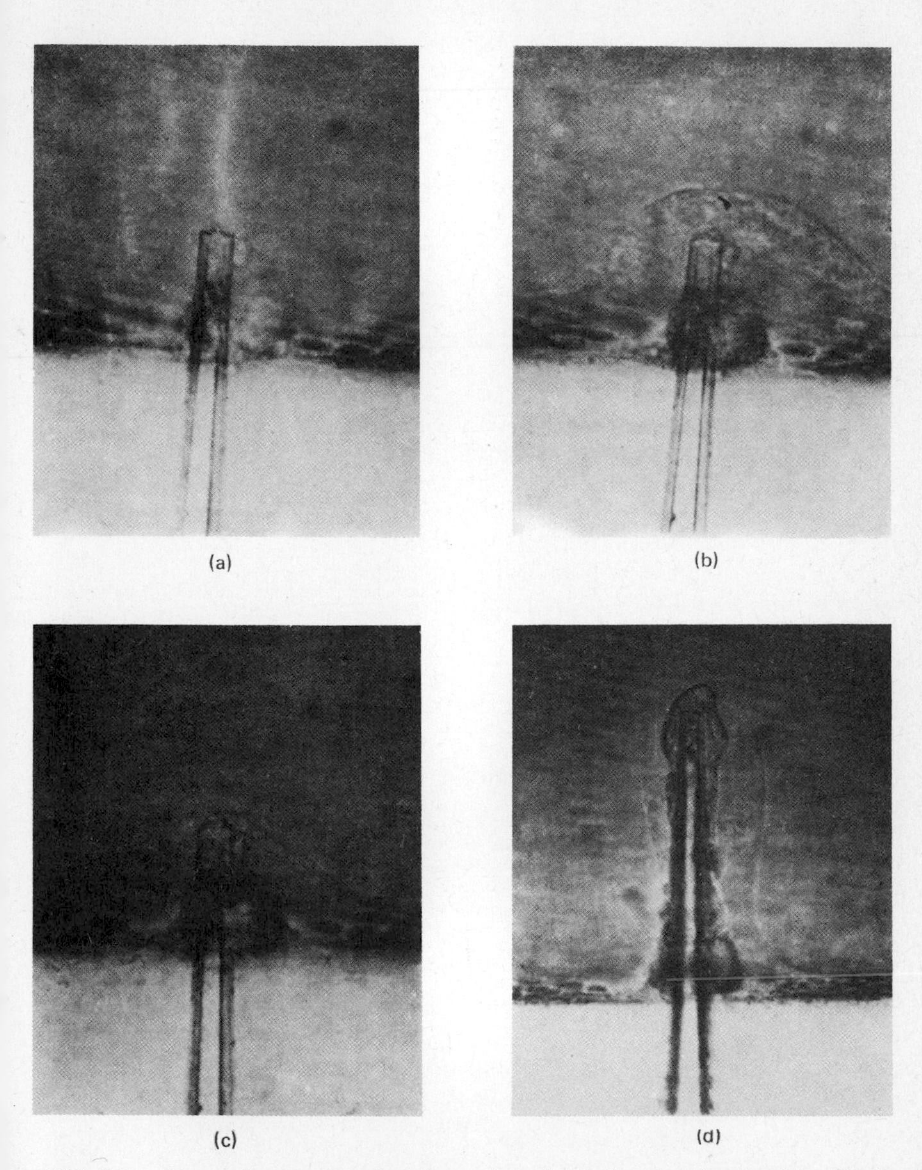

FIG. 3.10 Photomicrographs of a glass microelectrode in an internodal cell of Nitella. *(a) in the vacuole; (b) with cytoplasm covering the electrode; (c) with a seal forming on the electrode; (d) another electrode with the tip fully sealed. The tip diameter of the electrode shown here was almost 10 μm. (From Walker, 1955, by kind permission.)*

streams, the cytoplasmic cap does not build up and measurements can be made over many hours. Spanswick (1970) has found that in *Nitella translucens* the value of the vacuolar potential increased from about −120 mV to −130 mV in a 4 to 6 hour period following placement of the electrode. This change in potential was accompanied, and greatly exceeded, by changes in the electrical resistance of the plasma membrane.

If the tip of the electrode becomes clogged with cell wall material, protein or other substances with ion exchange properties, then the potentials between the tips of the electrodes and the solution in which they are placed (viz., E_2 and E_2'' may no longer be equal). Ion exchange will modify the mobility of ions diffusing from the electrode tip. If two electrodes, one of them which has become blocked or partially blocked in this way during its insertion into the cell, are placed in a common electrolyte solution without an intervening barrier a potential difference will be recorded between them. This potential will, of course, be additive to the actual membrane potential. As a standard test for tip potentials most workers do compare the tip potentials of electrodes after physiological measurements have been made. Potential readings from electrodes where the tip potential is appreciable are usually rejected.

Flux experiments

Healthy, mature cells which are no longer growing come to a state of flux equilibrium if they are left in a solution whose concentration is kept constant. In this condition, the net flux of a given ion is zero since the partial fluxes, ϕ_j^i and ϕ_j^o, are equal and opposite. It is possible to measure the influx and efflux in this condition if cells are labelled with a small number of radioactive atoms which serve as tracers. For each of the major nutrient elements, except nitrogen, a convenient radioactive isotope is available.

We will denote the concentration of radioactive tracer in the following equations by an asterisk, C_j^*, to distinguish it from the concentration of unlabelled ions C_j. The most important assumption made in tracer experiments is that the cell membrane does not distinguish between radioactive and non-radioactive ions of a given element so that the tracer and stable ions enter the cell in the same ratio as in the outside solution. This ratio, C_j^*/C_j is known as the specific activity. The rate of accumulation of tracer is given by eq. (3.31)

$$V\frac{dC_j^*}{dt} = A\left[\phi_j^{\text{in}}\frac{C_j^{o*}}{C_j^o} - \phi_j^{\text{out}}\frac{C_j^{i*}}{C_j^i}\right] \tag{3.31}$$

where V is the volume of the cell in cubic centimetres and A = the surface area of the cell in square centimetres.

Influx can be measured from the initial rate of increase in cellular activity by placing an unlabelled cell in tracer solution of known specific activity. During the course of such an experiment C_j^{i*} is negligibly small in comparison with C_j^i and hence the efflux tracer is also negligible. Equation (3.31) can, therefore, be rearranged to calculate influx.

$$\phi_j^{\text{in}} = \frac{V\left(\dfrac{dC_j^{i*}}{dt}\right)}{A\left(\dfrac{C_j^{o*}}{C_j^o}\right)} \tag{3.32}$$

Efflux is measured by loading up the cell for some hours with tracer from a labelled culture solution, washing the cell briefly with water and placing it in a solution of similar ionic concentration but lacking radioactive tracer. The external solution is changed repeatedly so that, in practice, C_j^{o*} remains zero. In such circumstances, the radioactivity of the cell will decrease since the influx of tracer will also be zero. Since the internal concentration of the ion, i.e., $C_j^i + C_j^{i*}$, remains constant during the efflux the internal specific activity decreases and so does the probability of a radioactive ion moving across the plasma membrane. The loss of radioactivity from the cell is not, therefore, constant but exponential and depends on the natural logarithm ($\log_e$) of C_j^{i*}, thus

$$\ln (VC_j^{i*}) = \frac{A\phi_j^{\text{out}}}{VC_j^i}\cdot t \tag{3.33}$$

If $\ln (VC_j^{i*})$ is plotted against time, a straight line should result and the value of efflux is determined from its slope.

The description above is somewhat simplified and applies only if the efflux and influx occur across a single boundary between two compartments. In plant cells, the principal complication is caused by the cell wall which has a relatively high ion exchange capacity. The first process to occur when an unlabelled cell is placed into a tracer solution is that radioactive ions exchange for unlabelled ones pre-existing on the cell wall. A new equilibrium is, however, quickly established with the external solution. If we plot the total radioactivity of the cell with time without correcting for this initial adsorption we get the curvilinear relationship (Fig. 3.11). If the linear portion is extrapolated back to the ordinate we may measure the quantity of material accumulated by the exchange sites on the cell wall. Since the tracer on these sites remains exchangeable it can be removed by washing the cell briefly in unlabelled culture solution. When this

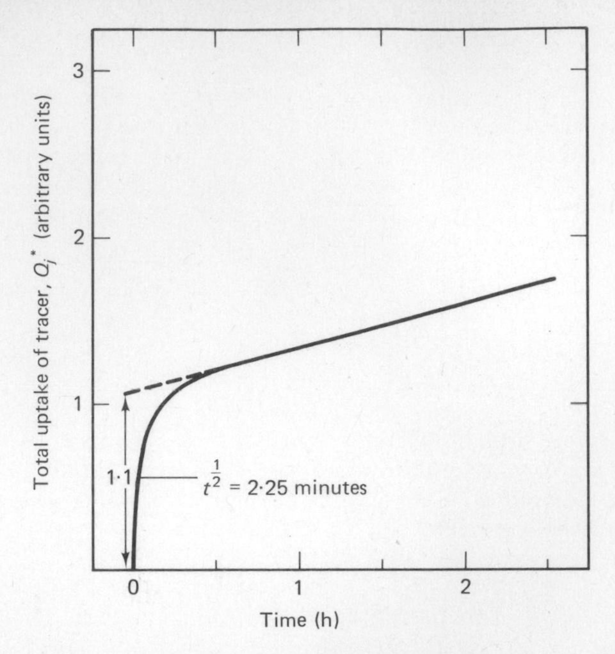

FIG. 3.11 Contrived data to illustrate the uptake of a labelled ion by a plant cell. Extrapolation of the linear portion to t_0 gives the quantity of tracer in the cell wall. The half-time, $t^{1/2}$, for the build-up of tracer in the wall is 2·25 min.

is done in *Hydrodictyon africanum* the value VC_j^{i*} increases linearly with time and a steady flux across the plasmalemma may be measured for some hours. In most cells treated with dilute solutions, the plasmalemma flux controls the rate of accumulation of radioactivity and only one rate for the increase of internal activity is observed. If, however, the flux of a given ion across the tonoplast is slower than that across the plasmalemma, the influx with time will have two phases: the first measuring the filling up of the cytoplasm, and hence the flux across the plasmalemma, the second the filling up of the vacuole and hence the flux across the tonoplast.

By similar arguments, it may be seen that the efflux could be resolved into two or three components (Fig. 3.12). Extrapolation of the linear portions of these curves to the ordinate estimates the amount of tracer in each phase and the slopes give the efflux across the plasmalemma and the tonoplast. This particular technique enables the measurement of the total ionic concentration in these compartments to be made by an indirect method. This matter will be dealt with in more detail in chapter 7. By applying a specific activity correction factor to the data in the histogram in Fig. 3.12 it is possible

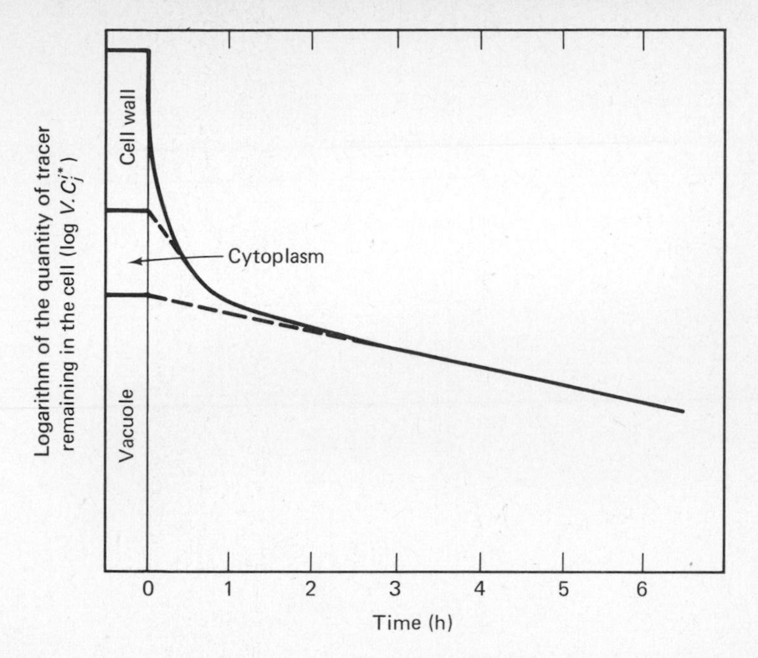

***FIG. 3.12** Contrived data to illustrate the various phases in the elution of a tracer from a labelled cell into a tracer-free medium. Extrapolation of linear portions to t_0 gives the quantity of tracer in three compartments.*

to calculate the actual concentration of ion in each of the three compartments. Examples of this kind of analysis on tissues from higher plants will be found in chapter 7.

Electrical resistance measurements

Potassium chloride micro-electrodes of the type described for potential difference determinations can be used in a slightly different way to measure the electrical resistance of cell membranes. The principle involved in these measurements is quite simple. Two electrodes are inserted into the cell vacuole, one of them is connected through a circuit of the kind illustrated in Fig. 3.13 and is used in an exactly similar way to measure the electric potential difference between the vacuole and the outside solution. The second electrode is connected to an electric pulse generator which passes a current across the cell and the plasmalemma to a silver wire which is placed parallel to the long axis of the cell in the external solution. The plasmalemma and the tonoplast can be regarded as two resistors in series. During the passage of current across them Ohm's law predicts that if R remains

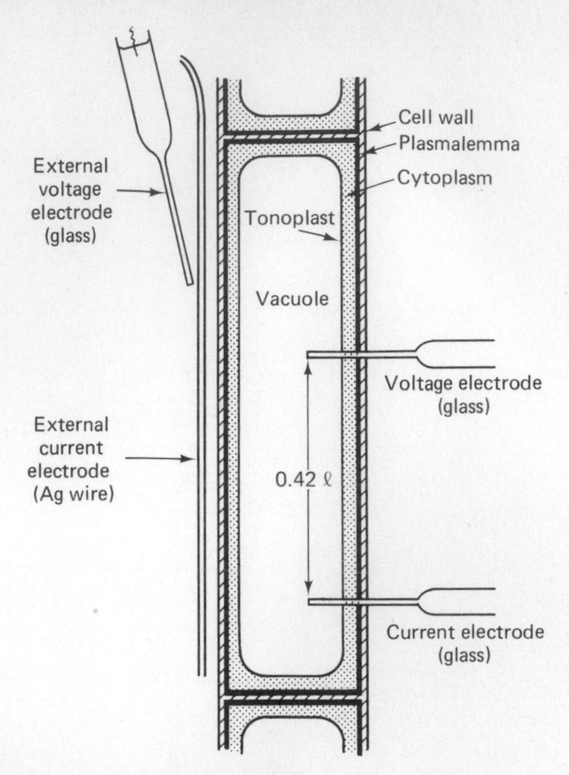

FIG. 3.13 Electrode assembly for measuring electrical resistance of cell membranes. The internal current and voltage electrodes are separated by 0·42l, where l = the length of the cell. See text for further explanation.

constant then the potential difference will decrease. Thus, if the current and the membrane potential are both measured, the resistance of the two membranes in series is simply calculated. The insertion of a third micro-electrode into the cell with its tip in the cytoplasm will permit simultaneous measurements of the change in p.d. across the plasmalemma and the tonoplast during the current pulse.

This simple procedure was outlined by Hogg, Williams and Johnston (1968) and little more needs to be said about it save that for geometrical reasons the internal voltage electrodes should be at a distance of 0·42*L* from the centrally located current electrode, where *L* is the length of the cell (see Fig. 3.13). Recently, Spanswick (1970) has shown that the resistance between the vacuole and the outside solution changes considerably over a period of 4 to 6 hours after the insertion of the electrodes. It is not clear what causes this change, which is proportionally much larger than that of the resting potential

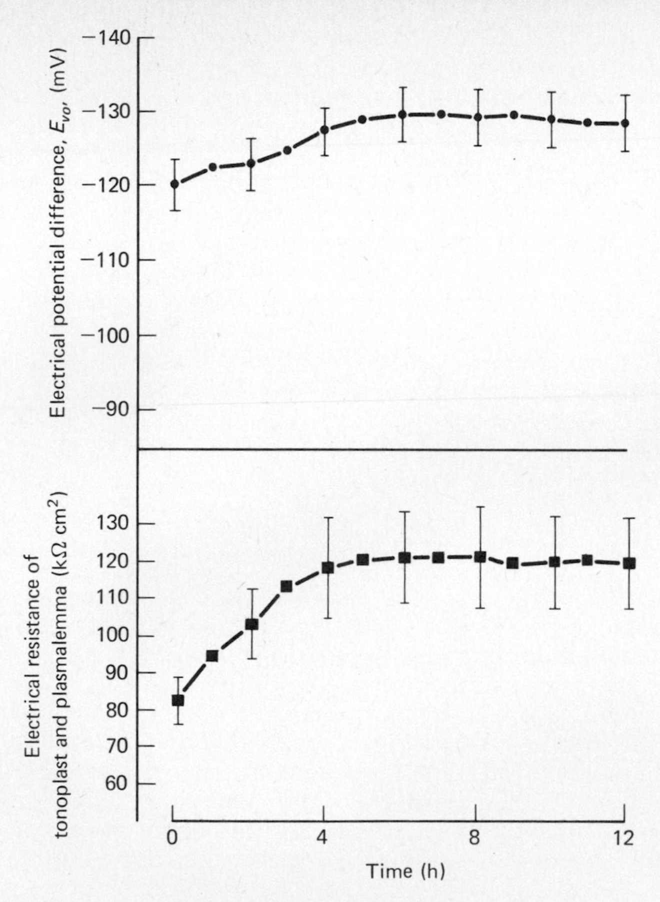

FIG. 3.14 Variation in the membrane potential, E_{vo}, *and membrane resistance with time after the insertion of two micro-electrodes into the vacuole of* Nitella. *The standard error of the mean is shown for alternate observations; each point is the mean value for twenty cells. (Redrawn from Spanswick, 1970.)*

of the cell (Fig. 3.14). Earlier reports of the membrane resistance of cells were usually made a short time after the insertion of the electrodes and Spanswick believes that these results may underestimate considerably the actual values. His results also show that the resistance of the plasmalemma is much greater than that of the tonoplast and that it is the plasmalemma resistance which increases with time.

Voltage clamp technique

Several authors have used a system similar to the one above to manipulate potential difference across the plasmalemma and to hold it at a certain value while the external concentration of an ion, which usually affects the value of the potential, is raised or lowered. Electronic circuitry of a negative feedback loop is employed for this purpose. We have already described an experiment by Kitasato (1968) where it was found that an increasing amount of current had to be passed through the cell to maintain a constant potential when the hydrogen ion concentration in the medium was raised. In this technique the cell is prevented from coming into equilibrium with the new set of conditions which were imposed by the concentration change in the external solution so that a flux of the ion in question can be measured as it flows down a controlled electrochemical potential gradient. Such experiments provide information on the current-carrying capacity of the membrane and of its passive permeability to the ion being examined. The voltage clamp technique has been used in studies on various algal cells by Kitasato (1968) and Gutknecht (1967).

Vacuolar perfusion technique

In some of the very large algal coenocytes, e.g., *Valonia*, *Halicystis*, it is possible to insert quite wide capillaries into the vacuole without causing more than transient effects on the electrical properties and permeability characteristics of the cell. The insertion of two such capillaries provides an inlet and an outlet through which the vacuole may be perfused with solutions of predetermined composition. The attractive feature of this technique is that it permits the experimenter to control the ionic composition of the medium on *both* sides of the cytoplasm. The most usual procedure is to make the solution perfusing the vacuole the same as the external solution and in such a situation the electrochemical potentials of the ions in the vacuole are equal to those in the outside solution; passive driving forces are, therefore, eliminated. The persistence of the net flux of an ion either into or out of the vacuole in such circumstances provides an excellent test for active transport and potentially the system is very useful in determining effects of inhibitors and environmental factors such as light and temperatures on the operation of active ion pumps. Gutknecht (1967) has made very interesting use of this technique with the alga *Valonia* and has incorporated current and voltage electrodes so that any contribution to the electric potential made by active ion pumps of an electrogenic kind can be short circuited.

Because of the physical problems involved in the insertion of capillaries of sufficiently large bore this technique is applicable only to the larger algal cells but even with this restriction it remains one of the most interesting of the newer methods of observing active transport in a living cell.

Bibliography

a Further reading

1. The following three review articles all adopt a similar approach and perhaps only one of them need be read. They are all excellent for reference material because not only do they collect and interpret current research but they outline with great clarity the theoretical basis of modern thinking about ion transport; particular emphasis is placed on the nature of passive driving forces.

Dainty, J. (1962) Ion transport and electric potentials in plant cells. *Ann. Rev. Pl. Physiol.* **13**, 379–402.

Gutknecht, J. and Dainty, J. (1968) Ionic relations of marine algae. *Oceanogr. Mar. Biol. Ann. Rev.* **6**, 163–200.

Gutknecht, J. (1970) The origin of bioelectric potentials in plant and animal cells. *Am. Zoologist* **10**, 347–354.

2. Slayman, C. L. (1970) Movements of ions and electrogenesis in microorganisms. *Am. Zoologist* **10**, 377–392.

This review advances admirably the case for regarding all ion transport as being dependent on metabolism. It tends to correct the impression, which might be gained from superficial readings of the three reviews above, that passive driving forces can be considered in isolation from metabolism.

3. Ling, G. N. (1962) *A Physical Theory of the Living State: the Association–Induction Hypothesis*. Blaisdell, N.Y.

Not every worker on ion transport accepts that the ion-selectivity of cells is determined by the properties of the plasmalemma. This author has consistently interpreted his experimental data in a way which indicates that the K-selectivity of the cell is determined by the delicately balanced structure of protein micelles in the cytoplasm and not by ion selective membrane transport. One of his basic tenets is that the most cellular K is bound or adsorbed in the cytoplasm so that chemical estimations of cytoplasmic K give no hint of its real activity. He believes that the activity gradient for K is small and that active pumping of K across the plasmalemma is an entirely unnecessary concept. The serious student should read this book because its arguments against membrane theories of ion transport are very persuasive at times; it is a classic work by an 'outsider'.

b References in the text

Barber, J. (1968) Measurement of the membrane potential and evidence for active transport in *Chlorella pyrenoidosa*. *Biochim. Biophys. Acta* **150**, 618–625.

Dick, D. A. T. and McLaughlin, S. G. A. (1969) The activities and concentrations of sodium and potassium in toad oocytes. *J. Physiol.* **105**, 61–78.

Etherton, B. and Higinbotham, N. (1960) Transmembrane potential measurements of cells of higher plants as related to salt uptake. *Science* **131**, 409–410.

Goldman, D. E. (1943) Potential, impedance and rectification in membranes. *J. Gen. Physiol.* **27**, 37.

Gutknecht, J. (1967) Ion fluxes and short circuit current in internally perfused cells of *Valonia ventricosa*. *J. Gen. Physiol.* **50**, 1821–1834.

Higinbotham, N., Etherton, B. and Foster, R. J. (1964) The effect of external K, NH_4, Na, Ca, Mg and H ions on the cell transmembrane potential of *Avena* coleoptile. *Plant Physiol.* **39**, 196–203.

Hinke, J. A. M. (1959) Glass microelectrodes for measuring intracellular activities of sodium and potassium. *Nature* **184**, 1257–1258.

Hoagland, D. R. and Davis, A. R. (1929) The uptake and accumulation of electrolytes by plant cells. *Protoplasma* **6**, 610–626.

Hodgkin, A. L. and Katz, B. (1949) The effect of sodium ions on the electrical activity of the giant axon of the squid. *J. Physiol.* **108**, 37–77.

Hogg, J., Williams, E. J. and Johnston, R. J. (1968) A simplified method for measuring membrane resistances in *Nitella translucens*. *Biochim. Biophys. Acta* **150**, 518–520.

Hope, A. B. and Walker, N. A. (1961) Ionic relations of *Chara australis* R.BR. IV. Membrane potential differences and resistances. *Aust. J. Biol. Sci.* **14**, 26–44.

Kitasato, H. (1968) The influence of H^+ on the membrane potential and ion fluxes of *Nitella*. *J. Gen. Physiol.* **52**, 60–87.

Lannoye, R. J., Tarr, S. E. and Dainty, J. (1970) The effects of pH on the ionic and electrical properties of the internodal cells of *Chara australis*. *J. Exp. Bot.* **21**, 543–551.

Ling, G. N. (1962) *A Physical Theory of the Living State: the Association–Induction Hypothesis*, chapter 4. Blaisdell, N.Y.

Ling, G. N. (1965) The membrane theory and other views for solute permeability, distribution and transport in living cells. *Perspect. Biol. Med.* **9**, 87–106.

Ling, G. N. (1969) Measurements of potassium ion activity in the cytoplasm of living cells. *Nature* **221**, 386–387.

Poole, R. J. (1966) The influence of the intracellular potential on potassium uptake by beetroot tissue. *J. Gen. Physiol.* **49**, 551–563.

Ramsay, J. A., Brown, R. H. J. and Croghan, P. C. (1955) Electrometric titration of chloride in small volumes. *J. Exp. Biol.* **32**, 822–829.

Saddler, H. D. N. (1970) The ionic relations of *Acetabularia mediterranea*. *J. Exp. Bot.* **21**, 345–359.

Slayman, C. L. (1970) Movements of ions and electrogenesis in microorganisms. *Am. Zoologist* **10**, 377–392.

Smith, F. A. (1970) The mechanism of chloride transport in *Characean* cells. *New Phytol.* **69**, 903–917.

Spanswick, R. M. (1968) Measurements of potassium ion activity in the cytoplasm of the *Characeae* as a test of the sorption theory. *Nature* **218**, 357.

Spanswick, R. M. (1970) Electrophysiological techniques and magnitude of membrane potentials and resistances in *Nitella translucens*. *J. Exp. Bot.* **21**, 617–627.

Spanswick, R. M. and Williams, E. J. (1964) Electrical potentials and Na, K and Cl concentrations in the vacuole and cytoplasm of *Nitella translucens*. *J. Exp. Bot.* **15**, 193–200.

Spanswick, R. M., Stolarek, J. and Williams, E. J. (1967) The membrane potential of *Nitella translucens*. *J. Exp. Bot.* **18**, 1–16.

Spear, D. G., Barr, J. K. and Barr, C. E. (1969) Localization of hydrogen ion and chloride ion fluxes in *Nitella*. *J. Gen. Physiol.* **54**, 397–414.

Thorhaug, A. (1971) Temperature effects on *Valonia* bioelectric potential. *Biochim. Biophys. Acta* **225**, 151–158.

Troshin, A. S. (1961) Sorption properties of protoplasm and their role in cell permeability. In: *Membrane Transport and Metabolism* (A. Kleinzeller and A. Kotyk eds.), 45–53. Academic Press, N.Y.

Vorobiev, L. N. (1967) Potassium ion activity in the cytoplasm and vacuole of cells of *Chara* and *Griffithsia*. *Nature* **216**, 1325–1327.

Walker, N. A. (1955) Microelectrode experiments on *Nitella*. *Aust. J. Biol. Sci.* **8**, 476–489.

Walker, N. A. and Hope, A. B. (1969) Membrane fluxes and electric conductance in Characean cells. *Aust. J. Biol. Sci.* **22**, 1179–1195.

Williams, E. J., Johnston, R. J. and Dainty, J. (1964) The electrical resistance and capacitance of the membranes of *Nitella translucens*. *J. Exp. Bot.* **15**, 1–14.

four

Passive movements of ions across membranes and ion-selectivity

Some of the ion fluxes across cell membranes are driven passively down gradients of electrochemical potential; such movements are brought about, therefore, by diffusion. If an ion is at electrochemical equilibrium in two compartments separated by a membrane an interchange of ions between the two compartments may still occur; this can be readily observed if radioactive tracer is added to one compartment and its appearance in the other is monitored. Whether an ion diffuses down a free-energy gradient or moves across the membrane by random thermal motion certain common difficulties present themselves if the ion is obliged to leave the aqueous phase and enter the lipid phase during its passage across the membrane. The first of these difficulties concerns the water of hydration which, in all but special circumstances, surrounds the ion and adds very considerably to its bulk.

Hydrated and non-hydrated ions

Monovalent cations, Li^+, Na^+, K^+, Rb^+ and Cs^+ form a series in which the mass and charge on the atomic nucleus increase, with lithium the lightest nucleus and caesium the heaviest. The number of electrons surrounding the nucleus also increases with mass number and so does the volume in space occupied by the electron orbitals. By analogy, the atomic nucleus of lithium may be thought of as being more accessible to the outside world because the cloud of electrons covering it is very thin. By similar reasoning, the nucleus of caesium is much less accessible since the cloud of electrons is multilayered and dense. These features are reflected in the radii of the non-hydrated ions which are as follows: Li^+, 0·06 nm; Na^+, 0·095 nm; K^+, 0·133 nm; Rb^+, 0·148 nm; Cs^+, 0·169 nm. In aqueous solutions water molecules are bound to ions by electrostatic forces emanating from the charged particles in the atomic nucleus. The more closely water molecules can approach the charged centre of the nucleus the more firmly they will be bound and the greater the free-energy change involved in this hydration process (see Table 4.1). As we have seen, the electron shell around lithium is thinner than that surrounding caesium, thus water molecules come closer to the centre of charge in the former case; in practice, this means that more water molecules are associated with the primary hydration shell of lithium. In Table 4.1

Table 4.1 ***Numbers of water molecules in the primary hydration shell of some ions in aqueous solution and the heats of hydration****

Cation	Hydration number (n)	Heat of hydration (kcal mole^{-1})	Anion	Hydration number (n)	Heat of hydration (kcal mole^{-1})
H^+	5	−276	F^-	4	−91
Li^+	5	−133	Cl^-	1	−59
Na^+	5	−115	Br^-	1	−52
K^+	4	−90	I^-	1	−45
Rb^+	3	−81	SO_4^{2-}	8	?
Cs^+	—	−73			
Mg^{2+}	13	−501			
Ca^{2+}	10	−428			
Al^{3+}	21	−1010			

* *Data derived from Tables XI, XVI and XIX in chapter 2,* **Modern Aspects of Electrochemistry**, *Ed. J. O'M. Bockris (Butterworth, London).*

the number of water molecules associated with the primary hydration shell of a number of common ions is given together with the heats of hydration. Notice that the amount of water held by polyvalent cations and, consequently, the energy necessary to dehydrate them is much greater than for monovalent cations. The hydrated radii of ions are very difficult to estimate and there does not appear to be general agreement about the correct values. The water molecules associated with the ions slow down their movement through liquid water, so that observed values for ionic mobilities give a relative measurement of the relative size of the ions. Although Li^+ is undoubtedly the smallest non-hydrated ion it has the lowest mobility (Table 4.2) when an electric field is applied to the solution.

Table 4.2 Mobilities of ions in aqueous solutions at 25 °C

Cations	Mobility $cm^2\ s^{-1}\ V^{-1}$	Anions	Mobility $cm^2\ s^{-1}\ V^{-1}$
K^+	$7{\cdot}62 \times 10^{-4}$	SO_4^{2-}	$8{\cdot}27 \times 10^{-4}$
Na^+	$5{\cdot}19 \times 10^{-4}$	Cl^-	$7{\cdot}91 \times 10^{-4}$
Li^+	$4{\cdot}01 \times 10^{-4}$	NO_3^-	$7{\cdot}40 \times 10^{-4}$

The water of the hydration shell effectively increases the size of the ions and the hydrated ionic radii of alkali cations are in the reverse order to the non-hydrated radii, so that the solvated ion of lithium is the largest.

At distances greater than the first hydration shell the electric field of the central ion is still strong enough to interfere with the normal orientation of water molecules and results in a secondary shell (Fig. 4.1).

When an ion crosses a lipid membrane it must either be stripped of its associated water or pass into some hydrophilic pathway sufficiently large to accommodate the rather bulky hydrated ion. In practice, the activation energies for permeation of ions through bio-

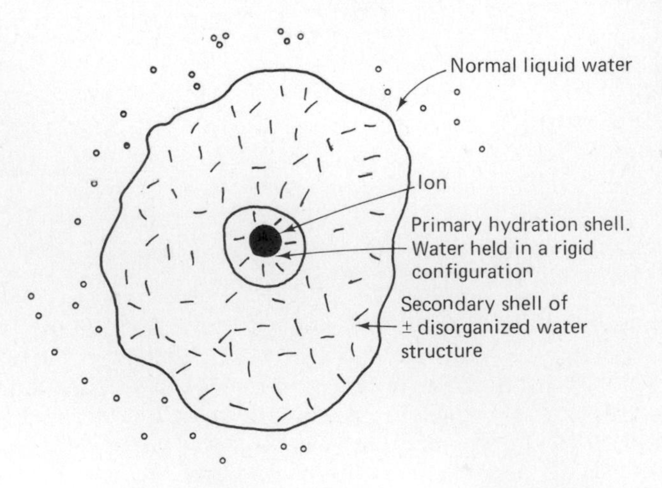

FIG. 4.1 The hydration shells surrounding an ion in an aqueous solution.

logical membranes are much lower than would be found if the ions were dehydrated during their movement through a lipid phase of low dielectric (see page 126). There are grounds, therefore, for preferring some interpretation based on a hydrophilic pathway through the membrane. The nature of this pathway in membranes is not known but most evidence suggests that no single part of it is a permanent structural entity. In synthetic lipid bilayers, it is possible to add molecules which serve as ion 'carriers' and this discovery lends support to those hypotheses which require carrier molecules to account for the rapid permeation of biological membranes by ions. (A careful distinction should be drawn between ion pumps and ion carriers; although both carry ions, pumps are linked directly to some metabolic reaction, carriers are not.) Ion carriers are usually envisaged as being more soluble in the lipid phase of the membrane than the ions which they carry, but having some essential structural feature which allows the ion to be retained in a polar environment. In other words, carriers operate a kind of shuttle service between the aqueous phases on either side of the membrane and in so doing they allow ions to bypass the hydrophobic interior of the membrane. In what follows we will consider how this may be brought about in more detail and try to explain how carrier molecules are able to discriminate between various species of closely related ions. For this purpose, research studies of the properties of purified and modified phospholipid bilayers will be extremely useful.

Permeation of synthetic lipid bilayers by ions and water

Formation and properties of synthetic lipid bilayers. Phospholipid bilayers can be formed spontaneously when phospholipids are dispersed in water (see chapter 2, page 29). Perhaps the simplest technique was used by Bangham, Standish and Watkins (1965) who found that dried lecithin (phosphatidyl choline) shaken vigorously in warm water and then cooled produced small vesicles surrounded by concentric arrays of phospholipid which resembled the myelin form of the plasmalemma surrounding nerves. Electron microscopic investigation revealed that the membrane had the familiar trilamellar structure of two electron-dense bands sandwiching an inner clear zone. Some properties of this system will be discussed below.

Another useful system for experimental observations is one in which a bilayer is formed across a perforation in a plate separating two compartments of aqueous solution. This is done by taking a solution of membrane-forming lipids in a suitable solvent and painting a small amount of the mixture over the hole in the plate while it is immersed in water. At first the hole becomes plugged with the lipid mixture but subsequently it thins down by draining (see Fig. 4.2) until it reaches the thermodynamically stable configuration of the bilayer. The direction of drainage is, of course, upwards because the lipids have a lower specific gravity than water. Other more elaborate techniques for special applications are described by Goldup, Ohki and Danielli (1970).

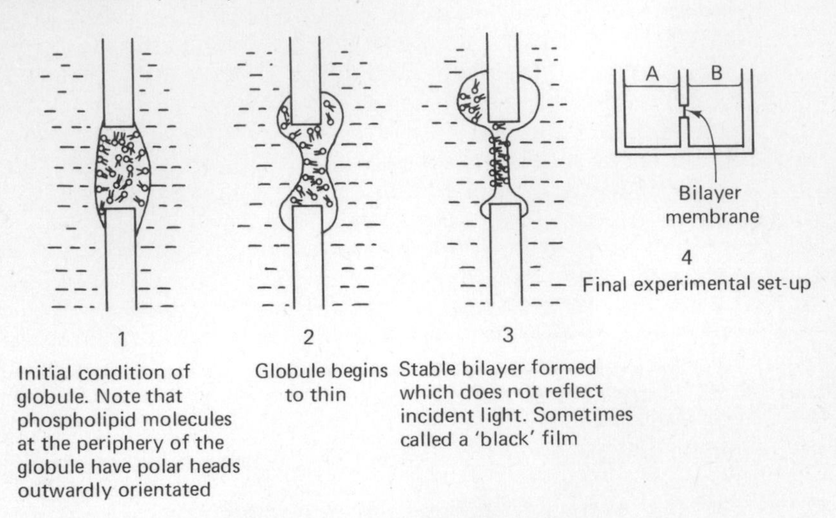
A
B
Bilayer membrane
4
Final experimental set-up
1
2
3
Initial condition of globule. Note that phospholipid molecules at the periphery of the globule have polar heads outwardly orientated
Globule begins to thin
Stable bilayer formed which does not reflect incident light. Sometimes called a 'black' film

FIG. 4.2 Formation of a phospholipid bilayer across a perforation in a plastic plate separating two aqueous compartments.

Phospholipids both purified and of mixed biological origin can be used to form membranes and they may be mixed with synthetic or naturally occurring neutral lipids, e.g., dodecane or cholesterol, in a chloroform–methanol solvent.

Membranes prepared by this technique from lecithin and n-tetradecane (neutral lipid) have a thickness of 7·2 nm; another system using 1:1 mixture of phosphatidylethanolamine and cholesterol in hexane/decane solvent were found to be 7·34 nm. Thus, the bilayers produced are somewhat thinner than the unit membranes seen in organisms. They are generally thinner because they lack the peripheral addition of protein to the membrane surfaces and any lipid/protein structures in the membrane matrix.

When precautions are taken to purify the lipid and hydrocarbon components of the membrane-forming mixture, the electrical resistance of bilayers is extremely high, being usually greater than $10^7 \Omega\ cm^2$. The conductance of the membranes to ions and electric current is, therefore, very low and their permeability to most ions, particularly to cations, is very much lower than is found in biological membranes. By contrast the permeability of bilayers to water is frequently very comparable with values from biological membranes (Table 4.3).

Table 4.3 Water permeability of bilayer and biological membranes

Membrane type	Permeability to water (cm s^{-1}) × 10^4	
	Isotopic flux measurement	Flux under influence of osmotic pressure
Bilayer:		
Lecithin-tetradecane	4·4	17·4–104
Lecithin-decane	2·2	19·0
Biological membranes (range)	0·23–53·0	0·37–270

Note that in both synthetic and natural membranes the rate of water permeation under the influence of an osmotic pressure difference (this measures the hydraulic conductivity, L_p) is considerably larger than the observed diffusion of labelled water across the membrane (this measures the diffusional coefficient, P_d). In highly purified lipid bilayers, which usually lack pores, this difference is most likely to be due to the influence of unstirred layers.

Unstirred layers, in which there will be incomplete mixing of tracer and unlabelled water, occur near the surface of the membrane. The actual concentration profiles of tracer (i.e., its specific activity) in these layers are extremely complex. The layers themselves can vary in thickness from 30–300 μm and are, therefore, very much broader than the membrane itself. In effect, these layers extend the length of the diffusion pathway and increase its resistance. The influence of unstirred layers on the value of P_d can be calculated from eq. (4.1) if their thickness can be reasonably estimated

$$P_d \text{ (observed)} = P_d \cdot \frac{1}{1 + [P_d/(D/\delta)]} \qquad (4.1)$$

where P_d (observed) and P_d are the observed and actual values of the diffusion coefficient; D is the self-diffusion coefficient of water and δ is the thickness of the unstirred layer.

When appropriate corrections of this kind are applied, the ratio L_p:L_d in bilayers prepared from purified phospholipid components is usually close to unity (see page 116). In biological membranes, however, L_p:P_d is usually much greater than unity even after appropriate corrections are made for unstirred layers. This difference in the two permeability coefficients can be used to calculate the dimensions of hypothetical pores in biological membranes (see Stein, 1967, chapter 3, for a full discussion).

A remarkable number of other physical properties are comparable in bilayer and biological membranes (Table 4.4) which encourages the view that the lipid bilayer is the key structure determining many of the characteristics of natural membranes.

Table 4.4 Comparison of some properties of bilayers and biological membranes*

Property	Biological membranes at 20–25 °C	Bilayers at 36 °C
Electron microscope image	Trilaminar	Trilaminar
Thickness (nm)	6·0–10·0	6·0–7·5
Capacitance (μmf cm^{-2})	0·5–1·3	0·38–1·0
Resistance (Ω cm^{2})	**10^{2}–10^{5}**	**10^{6}–10^{9}**
Surface tension (dyn cm^{-1})	0·03–1·00	0·5–2·5
Hydraulic conductivity (μm s^{-1} atm^{-1})	0·37–270	30
Activation energy for water permeation (kcal mole^{-1})	9·6	12·7

* *Data taken from Thompson, T. E. and Henn, F. A. (1968),* **Structure and Function of Membranes in Mitochondria and Chloroplasts.** *Ed. E. Racker (Rienhold, N.Y.).*

The most important respect in which the two types of membrane differ is in their electrical resistance which may be as much as seven orders of magnitude less in biological membranes. Natural membranes are much more permeable to ions than bilayers and, more than this, they are a good deal more ion-selective. We shall now see how a simple bilayer can be modified so that its properties come to resemble biological membranes more closely still.

Modification of bilayer permeability and ion-selectivity

As we learnt in chapter 2, the phospholipid molecules which determine the structure of the bilayer are divisible into a polar head region which comes into direct contact with the aqueous environment and a non-polar hydrocarbon tail which is hydrophobic. Appropriate additions of a number of substances can be shown to selectively affect the chemical environment of either of these two regions and from such studies we gain some clues on that part of the bilayer which has the maximum influence in determining its permeability characteristics.

Surface charge on bilayer. When phospholipid liposomes are prepared by shaking lecithin in water the membranes carry almost no net charge. If the liposomes are formed in a solution containing salt labelled with radioactive tracer it is possible to follow the rate at which the tracer is eluted from the liposomes when they are suspended in tracer-free solution. Observations of this kind by Bangham, Standish and Watkins (1965) showed that liposomes were selectively anion permeable and that cations were eluted from them very slowly, i.e., the reverse of the situation in most biological membranes. If a small amount of a long-chain anion, dicetyl phosphoric acid, was added to the lecithin the bilayer surface carried a net negative charge. Conversely, the addition of a long-chain cation, stearyl amine, produced a net positive charge. Both of these molecules have non-polar regions which solubilize in the lipid interior and polar regions which become ionized in the same layer as the phospholipid 'heads Figure 4.3 shows that increasing the net negative charge on the surface greatly increased the potassium efflux, while the addition of the cationic detergent, which made the surface positively charged, virtually eliminated the cation permeability.

In natural membranes, the component phospholipids can carry either positive, negative or no net charge. If these phospholipids are extracted and separated and then incorporated in synthetic membranes the surface charge on the phospholipid appears to be the main

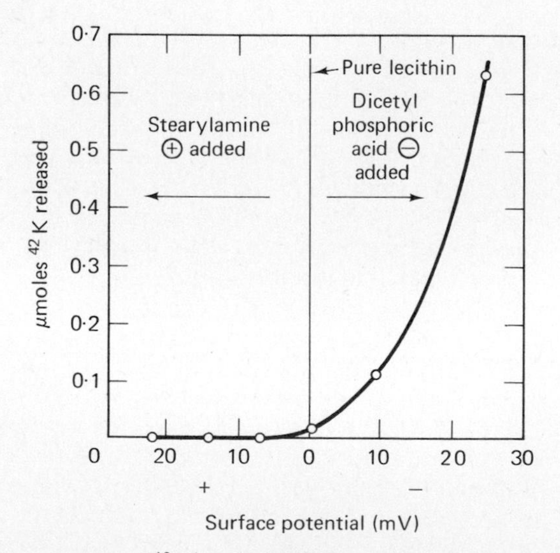

***FIG. 4.3** The amount of $^{42}K^+$ released from liposomes made from lecithin as a function of the surface potential and the incorporation of long chain ions. The measurements were made at 37 °C after 30 minutes dialysis against 0·145 mM KCl (adapted from Bangham, Standish and Watkins, 1965).*

property determining the selectivity characteristics of the membrane (Hopfer, Lehninger and Lennarz, 1970). Phospholipids of all three types can be extracted from a single tissue, and it seems possible that in the natural membrane there may be a mosaic of charge distribution pattern so that some areas of the membrane could be selective for cations and other areas for anions. This purely speculative suggestion has some precedence in studies of liposomes. Bangham *et al.* (1965) found that the anion fluxes out of liposomes were largely unaffected by the presence of detergent molecules even though these radically altered the cation effluxes. They suggested that anions and cations may enter the bilayers at different places on the surface and that only the cation entry points were influenced by the treatments.

A recent publication by Papahadjopoulos (1971) further supports the view that in both synthetic and biological membranes there may be a mosaic of phospholipid areas which have differing properties. Using highly purified phospholipids to produce vesicles, he found that those prepared from phosphatidylserine and phosphatidylglycerol had a very marked selectivity for K^+, as opposed to Na^+, permeation. The K/Na elution ratio varied from 5·1 to 9·0 depending on the pH of the eluting solution. Vesicles prepared from lecithin (phosphatidylcholine) or from phosphatidylethanolamine did not discriminate between the two ions at all, thus confirming the earlier observation of Bangham *et al.* (1965). Mixtures of phosphatidylserine and lecithin exhibited a reduced but still significant discrimination in favour of K^+. Phosphatidylserine is found in most cell membranes where it comprises from 10 to 20 per cent of the total phospholipid; in some microorganisms and in mitochondria and chloroplasts where it is not found, its place is taken by phosphatidylglycerol or related glycerol-containing phospholipids which also are K-selective. In the biological membrane, the author suggests that the K-selective phospholipids might be segregated into special areas which could represent sites concerned with specific permeability functions. In this connection, it is interesting to note that the two K-selective phospholipids can activate ($Na^+ - K^+$) ATPase preparations; this enzyme probably acts as the Na/K pump in the plasmalemma of a huge range of organisms and cells (see page 147). Neither lecithin nor phosphatidylethanolamine, which do not show K-selectivity, can activate the enzyme.

The surface charge on synthetic membranes can be modified by the adsorption of protein by electrostatic binding to the phospholipid head regions. Since at neutral or physiological pH values most proteins carry a slight negative charge their presence on the bilayer tends to make it more cation selective but it seems that, only when there is some interaction between the protein and the hydrocarbon interior, is there any marked influence on ion-selectivity. Tsofina *et al.* (1966) showed that lipid–protein bonding greatly increased the electrical conductance and K-selectivity of a lecithin bilayer containing bovine serum albumin. A similar result was obtained by Kimelberg and Papahadjopoulos (1971) when they incorporated cytochrome *c* into phospholipid vesicles formed from phosphatidylserine. This association increased both the K^+ and Na^+ permeability of the vesicles 100- to 1000-fold, but even the most permeable vesicles discriminated in favour of K^+ so that the K/Na efflux ratio was approximately 2.

The cation permeability of lipid bilayers can be substantially reduced in the presence of polyvalent cations. Both thorium, Th^{4+}, and iron, Fe^{3+}, slowed down the transfer of K^+ and Na^+ relative to Cl^- in bilayer membranes (Gutknecht and Tosteson, 1970).

We can summarize this section by saying that the surface charge on the membrane can determine (1) the relative permeability of anions and cations, largely by modifying the cation permeability, and (2) the selectivity of the membrane for various cations. Without further modification, however, the total ionic conductance of the membrane does not increase to the point where it becomes comparable with biological membranes.

The effects of alcohols. A variety of small molecules can be incorporated into phospholipid bilayers greatly increasing their ionic conductance; among these the effects of alcohols are particularly well documented.

Gutknecht and Tosteson (1970) prepared bilayers by a procedure similar to that illustrated in Fig. 4.2; their source of phospholipid was erythrocytes from the blood of sheep. Various amounts of a number of alcohols were added to a 100 mM solution of potassium chloride in the two compartments separated by the membrane. In every case, the electrical resistance of the bilayer (R_m) was substantially reduced (Fig. 4.4). Straight-chain alcohols with 2–8 carbon atoms were used and it is clear from Table 4.5 that the alcohols increased in effectiveness as the chain length increased. Alcohols of chain lengths greater than 8 were not sufficiently soluble in the aqueous phase to be used in the experiments. The addition of the alcohols did not reduce the mechanical stability or the thickness of the bilayer until the resistance R_m fell to below $10^5\ \Omega\ cm^2$ when the films ruptured. The effect of the alcohols was reversible and, after replacement of the solution by alcohol-free KCl, the resistance of the bilayer returned to its original value in 10–20 minutes.

A more extensive examination of the effect of heptanol (7·8 mM) indicated that it increased the K^+-selectivity of the bilayer bathed in mixed solutions of KCl and NaCl. The transfer of $K^+ + Na^+$ (T_{K+Na}) across the membrane increased relative to chloride, T_{Cl}, on the addition of heptanol; $T_{K+Na}:T_{Cl}$ increased from 6 to 21 and the ratio $T_K:T_{Na}$ increased from 3 to 21. In the absence of alcohols, the relative permeability of the bilayer to ions of the alkali cation series was as follows:

$$Rb^+ > K^+ > Cs^+ > Na^+ > Li^+$$

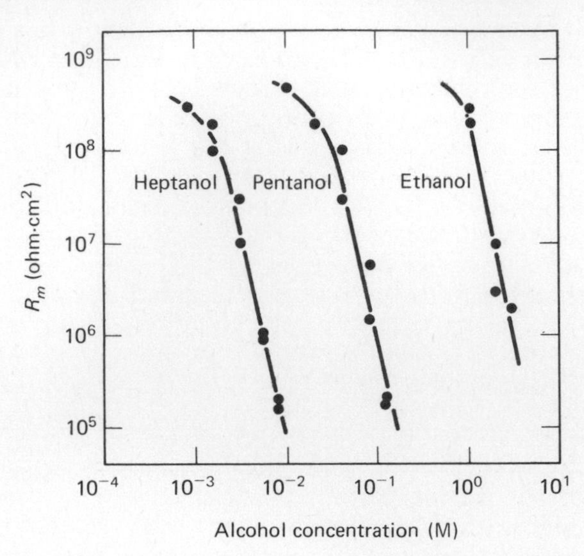

FIG. 4.4 The effects of certain n-alkyl alcohols on the electrical resistance, R_m, of lipid bilayer membranes. The aqueous phases separated by the membrane contained 100 mM KCl and the alcohols at the concentrations indicated. (From Gutknecht and Tosteson, 1970, by kind permission.)

Table 4.5 Potency of alcohols in reducing the electrical resistance of phospholipid bilayer membranes

Alcohol	Molar concentration in aqueous phase required to give R_m of 10^6 Ω cm²	Molar concentration of a saturated aqueous solution
Ethanol	3·0	Freely miscible
n-Propanol	1·4	Freely miscible
n-Butanol	0·30	1·07
n-Pentanol	0·10	0·306
n-Hexanol	0·028	0·058
n-Heptanol	0·0054	0·0078
n-Octanol	>0·00038	0·00038

In the presence of heptanol, the bilayer became most permeable to potassium although the transfer numbers of both K^+ and Rb^+ were increased very considerably relative to other ions.

The addition of the alcohol apparently had no effect on the osmotic permeability coefficient (L_p) of the bilayer. With or without heptanol (7·8 mM) in the aqueous phase, L_p was 20–30 $\times 10^{-4}$ cm s^{-1} (compare this value with those in Table 4.2).

It would seem that the addition of alcohol makes the bilayer behave much more like a biological membrane both with respect to its conductance and its selectivity. This does not mean that alcohols are an essential feature of the structure of biological membranes; other substances can have similar effects. Alcohols may exert their influence by modifying the structure of hydrocarbon tails. It has been found that alcohols inserted into lipid monolayers disrupt the hydrophobic bonds between the fatty acid chains of the lipids thus making the lipid phase more liquid by reducing its structural order. We can see from this that the physical conditions in the lipid phase may be very important in determining membrane conductance or resistance.

Alcohol treatment did not increase the hydraulic conductivity of the membrane even though its electrical conductance was greatly increased. Treatments which increase bilayer conductance by forming pores usually increase the hydraulic conductivity (see page 117) and from this it seems unlikely that alcohol produces its effect by making the membrane more porous. Since the alcohol treatment may make the lipid phase more fluid, it is possible that the increased electrical conductance might be related to the rapidity with which structural defects, or ion-binding sites, can be moved across the lipid phase, as predicted in the theory of Träuble outlined in chapter 2 (page 34). The increased fluidity of the membrane might be expected to speed up the movements of 'kinks' in the hydrocarbon tails, but if this were the complete explanation the diffusional water permeability (P_d) of the bilayer should also increase.

Other molecules, mostly short-chain, surface active agents, which break up the hydrophobic bonding in the lipid phase, increase membrane conductance; it has also been suggested that the presence of protein in the lipid phase of membranes might have the same effect (Cherry, Dodd and Chapman, 1970).

Increased conductance caused by uncouplers of phosphorylation. In chapter 5 we shall describe the relationships which exist between electron flow and the phosphorylation of ADP to form ATP. The two processes are coupled suggesting that some common intermediate step or compound is shared by the two processes. According to one theory the common step is the creation of a hydrogen ion activity gradient across the membrane at which phosphorylation is occurring.

Certain inhibitor molecules have the capacity to uncouple the two processes; when they are applied to photosynthesizing or respiring cells and organelles electron-flow continues while phosphorylation is inhibited. Some uncouplers are effective in very low concentrations.

The mechanism by which uncouplers disrupt phosphorylation has been clarified by studies using bilayer membranes. Three main classes of uncouplers have been examined in this connection, viz., substituted phenols, e.g., dinitrophenol (DNP), benzamidazoles and substituted carbonylcyanide phenylhydrazones, e.g., carbonylcyanide *m*-chlorophenyl hydrazine (CCCP) all of which are weak acids with appreciable lipid solubility. In each case, these compounds increase greatly the ionic conductance of bilayers, their effect being maximal at the pH value at which the uncoupler dissociates in aqueous solution. In the region of the maximum (see Fig. 4.5), the uncoupler-

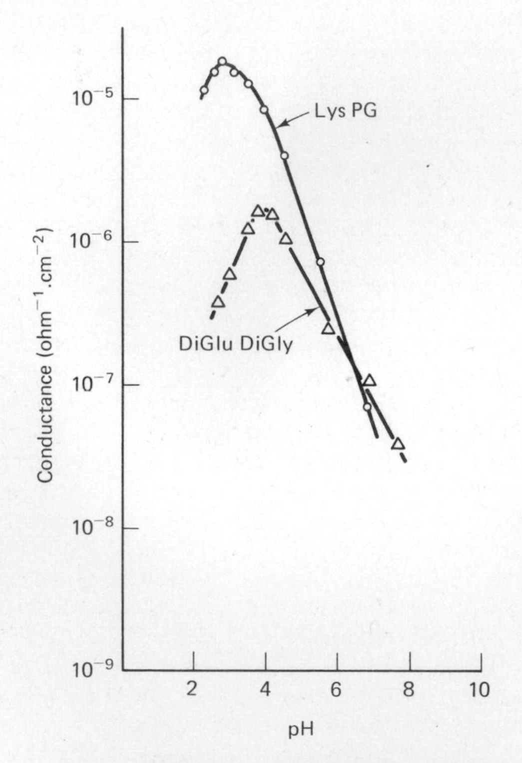

***FIG. 4.5** The influence of pH on the electrolytic conductance of bilayers in the presence of 93 μM dinitrophenol, an uncoupler. The bilayers were formed from phospholipids extracted from the bacterium* Staphylococcus aureus. *Lys PG = Lysyl phosphatidyl glycerol (+vely charged); Di Glu Di Gly = diglucosyl diglyceride. (From Hopfer et al., 1970, by kind permission.)*

induced conductance is almost entirely due to a high proton permeability (Hopfer, Lehninger and Thompson, 1968). These authors found that the uncoupler, carbonyl cyanide *p*-trifluoromethoxyphenylhydrazone, induced very little change in K^+ or Cl^- transfer rates across the bilayer.

These findings are consistent with the role of uncouplers as 'carriers' for hydrogen ions (protons) across the hydrocarbon interior of the membrane and explain how it is that, in their presence, organelles are not able to build up a sufficiently energetic proton potential to provide the free energy for phosphorylation. This matter will be discussed from another viewpoint in chapter 5 (page 160).

Increased conductance caused by antibiotics and macrocyclic compounds. A range of antibiotic molecules produced by bacteria and fungi has been found to greatly increase the permeability of lipid bilayers and biological membranes to ions and in some instances the increased permeability makes the membrane more ion-selective. Because these molecules are synthesized by organisms, and because they are frequently active at very low concentrations (monactin is effective over the range 10^{-11} to 10^{-6} M) there is particular interest concerning the mechanism by which these 'ionophores' effect their changes in the properties of the membrane. This interest stems, of course, from the speculation that compounds with similar properties may be present in biological membranes. Table 4.6 lists the origins, chemical nature and properties of some of the substances used in research on lipid bilayers and membranes.

There are apparently two principal ways in which these compounds exert their influence. The first type, which will be exemplified by monactin and valinomycin, appears to act as a true ion-carrier, forming a complex with the transported ion which diffuses across the lipid phase of the membrane several orders of magnitude faster than the ion alone. The second type of induced conductance appears to be due to the formation of transient pores in the membrane which admit ions and small non-electrolytes.

a Ion-carrying antibiotics. Valinomycin and monactin are very effective in increasing the conductance of lipid bilayers at very low concentrations (10^{-11} to 10^{-6} M). In single salt solutions of alkali cation chlorides, the increased conductance caused by monactin is strongly dependent on the species of alkali cation present. Figure 4.6 shows that when lithium chloride (100 mM) was present marked increases in the electrical conductance of the bilayer were found only when the antibiotic concentration exceeded 10^{-8} M. In the presence of 100 mM KCl, this concentration of monactin increased the electrical conductance of the bilayer by nearly five orders of magnitude. Other

***Table 4.6* Some macrocyclic antibiotics which affect the conductance and selectivity of bilayers and biological membranes**

Molecule	Biological origin	Molecular weight	Chemical nature
Valinomycin	Bacterial: *Streptomyces fulvissimus*	1111	Cyclic polypeptide forming 36 membered ring. Highly selective for K^+ and H^+. Non-ionic. Sparingly water soluble.
Nonactin	Bacterial: *Streptomyces* spp. (several)	736	Macrotetralide ring structure. Selective for K^+. Non-ionic. Sparingly soluble in water and lipid.
Enniatin A and B	Fungal: *Fusarium orthoceras* var. Enniatum	821 and 779	Cyclic polypeptide. Smaller ring. Selective for K^+ but less so than valinomycin.
Gramicidin A, B and C	Bacterial: *Tyrothrix* spp. *Bacillus brevis*	1400	Linear polypeptides. Dispersible in polar lipids. Strongly cation selective but specificity among alkali cations not great.
Nystatin	Bacterial: *Streptomyces noursei* *S. aureus*	932	A polyene: a cyclic molecule incorporating conjugated double bonds. Appreciably anion selective.

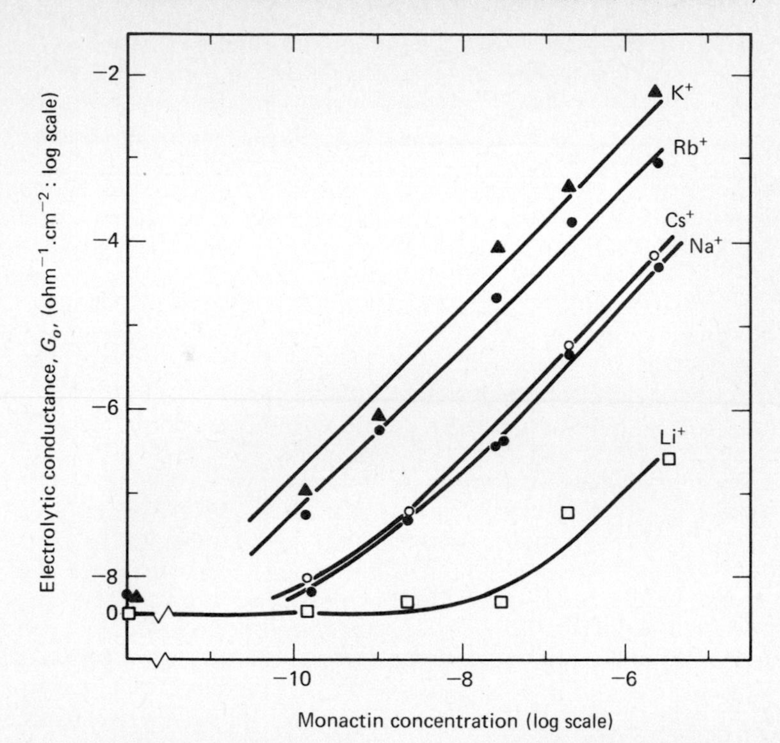

FIG. 4.6 The influence of the 'ionophore', Monactin, on the electrolytic conductance of a phospholipid bilayer in the presence of single salt solutions (100 mM) of alkali cations. (Redrawn from Eisenman, Ciani and Szabo, 1968.)

ions were intermediate in their effects. From Fig. 4.6 we can determine an ion selectivity series for monactin:

$$K^+ > Rb^+ \gg Cs^+ > Na^+ \gg Li^+$$

Eisenman, Ciani and Szabo (1968) found that the almost linear dependence of the membrane conductance on the concentration of the antibiotic, and the close agreement between the conductance and the measured permeability coefficients of the various ions tested, were consistent with a model in which monactin and a cation combine to form a complex. Hence, the permeability of the membrane for a given ion will depend on the number of carrier complexes crossing the membrane in unit time and the affinity of the carrier for the ion.

An examination of the structure of the monactin molecule immediately gives a clue to the nature of its association with ions. It has a ring structure in which all of the chemical groups situated on the outside of the ring are non-polar; this ensures that the molecule is soluble in the non-polar hydrocarbon environment at the centre of the bilayer. At the inner face of the ring there are four oxygen atoms on furane rings and two keto groups on opposite sides of the ring. These sites are potentially available for binding cations electrostatically; thus, the interior of the ring provides a polar environment. In the atomic model constructed by Eisenman *et al.* (1968), the space in the centre of the monactin molecule was about 0·7 nm in diameter. We can imagine an ion attached to the exchange site in the centre of the molecule surrounded by a hydrophobic coating. Thus, during its journey across the bilayer, the ion never leaves a polar environment.

Valinomycin is also ring-structured and most of its effects on membrane conductance are closely similar to those of monactin. The demonstration by Pressman and co-workers (1967) that valinomycin can carry appreciable quantities of potassium and rubidium into a non-polar solvent and the linear dependence of conductance on valinomycin concentration indicates that this antibiotic also acts essentially by complexing and diffusion through the membrane lipid. Further indirect evidence supports this view. The rate of diffusion of an ion-carrier through lipid will be influenced by the viscosity of the medium, thus any treatment increasing the viscosity or structure of the hydrocarbon tails should reduce the conductance of the membrane. In chapter 2 we saw that cholesterol stabilizes the hydrocarbon environment of membranes and makes the structure less fluid; low temperatures should have a similar effect of increasing the viscosity of the lipid phase. Mueller and Rudin (1967) found that cholesterol-containing bilayers had a lower conductance in the presence of valinomycin than cholesterol-free bilayers. It has also been found that the conductance in the presence of valinomycin has a strong dependence on temperature, the Q_{10} being approximately 8.

It seems probable that the action of other cyclic polypeptides, e.g., Enniatin A and B, will also involve the transport of the ion as part of an antibiotic-ion complex.

b 'Pore' forming antibiotics. The second type of antibiotic-induced conductance is found in a group of substances known as polyenes of which Nystatin is an example (see Fig. 4.7b). The most striking difference between this molecule and monactin is that nystatin is amphipathic, i.e., hydrophilic groups project from both the outer and inner faces of the ring. For this reason, the molecule is practically insoluble in hydrocarbons. This does not appear to be a very promising characteristic for a molecule which assists ions to cross a lipid barrier and yet, when nystatin is applied to solutions bathing bilayer membranes, very large increases in conductance

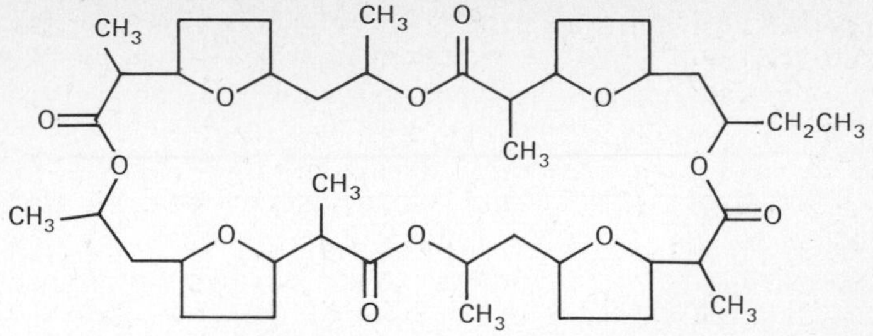

(a) Structural formula of Monactin

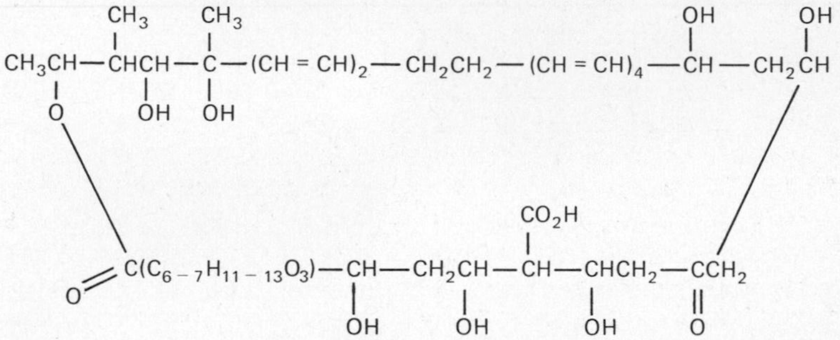

(b) Structural formula of Nystatin

FIG. 4.7 Structural formulae of two macrocyclic antibiotics used in studies on membrane permeability. (a) The 'carrier' molecule, Monactin. (b) The 'pore-former', Nystatin.

result. In a solution of 100 mM NaCl, the addition of 4 μg ml^{-1} of nystatin (approximately 5×10^{-6} M) to both of the compartments separated by a membrane increases the initial conductance from $5 \times 10^{-9}\ \Omega^{-1}\ cm^{-2}$ to approximately $10^{-3}\ \Omega^{-1}\ cm^{-2}$. Conductance varied as a high power of the molar concentration of nystatin (Cass, Finkelstein and Krespi, 1970). In the presence of the antibiotic, the bilayer was selectively permeated by chloride so that an electric potential of 49 mV developed across a bilayer separating compartments having a tenfold difference in sodium chloride concentration (more highly concentrated compartment, +ve).

Nystatin and other polyene antibiotics are unlikely to be operating by the formation of ion-carrier complexes; results of several kinds give important clues to their mode of action. Unlike monactin or valinomycin the presence of nystatin has a marked effect on the permeation of the bilayer by water. Figure 4.8 shows that there was a linear dependence of both the water diffusion coefficient, $(P_d)_w$, and the osmotic water permeability coefficient (hydraulic conductivity), L_p, on the conductance of the membrane. The increases in the value of these coefficients in response to the nystatin treatment are, of course, relatively small in comparison with the increased permeability of the membrane to anions, e.g., 1000-fold increase in conductance was matched with a 15-fold increase in P_d and 35-fold increase in L_p. Before the application of nystatin the ratio of the two permeability coefficients L_p/P_d was approximately 1 when P_d was corrected for the effect of unstirred layers. When the conductance

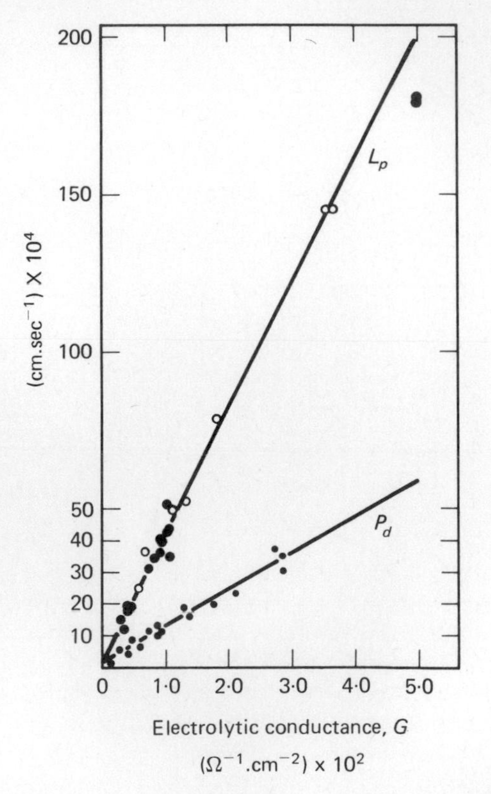

FIG. 4.8 The hydraulic conductivity, L_p, and diffusion coefficient, P_d, for water as a function of the conductance of nystatin-treated phospholipid bilayers. The membranes were examined at 25 °C in 100 mM NaCl, 5 mM sodium phosphate buffer, pH 7 and 1–5 μg ml^{-1} nystatin. The osmotic pressure gradient across the membrane was produced by dissolving either additional NaCl or glucose in one of the aqueous compartments. (Adapted from Holz and Finkelstein, 1970.)

was increased L_p/P_d became 3·3. Such findings are usually taken to mean that, in the membrane with high conductance, there are channels through which columns of water molecules can be moved cooperatively when an osmotic or hydrostatic pressure is applied to the system. In the absence of nystatin, a ratio of 1 implies that water at one side of the membrane cannot 'see' water on the other side and that, either in self-diffusion or under the influence of a driving force, water crosses the membrane in individual packets of one or several molecules at a time.

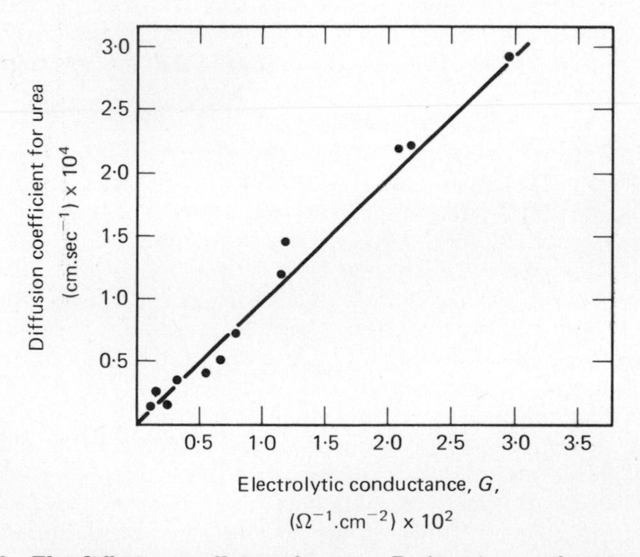

FIG. 4.9 ***The diffusion coefficient for urea, P_d (urea), as a function of the conductance of nystatin-treated phospholipid bilayers. (From Holz and Finkelstein, 1970, by kind permission.)***

The suggestion that nystatin opens up pores in the membrane is even more persuasive when we find that the addition of the antibiotic promotes the permeation of a number of small non-electrolyte molecules in addition to water. In the absence of nystatin, the membrane is essentially impermeable to urea (radius 0·227 nm), glycerol (radius 0·274 nm) and glucose (radius 0·400 nm). Addition of nystatin to bring the conductance of the bilayer up to $10^{-2}\ \Omega^{-1}\ \text{cm}^{-2}$ makes the membrane appreciably permeable to urea (Fig. 4.9) and glycerol but still not permeable to glucose. The permeability of a range of small hydrophilic solutes with radii less than 0·400 nm was inversely correlated with their radius. The results are consistent with nystatin creating a pore in the membrane of approximately 0·4 nm radius. This value is very close to the value, encountered in chapter 2, which is necessary to explain the anomalously high flux of water and small non-electrolytes into the cells of the coenocytic alga *Chara ceratophylla*. The linear relation between conductance and the diffusion coefficients of water implies that the pores also serve to conduct ions through the membrane. If this is the case, it would be interesting to see if the membrane conductance induced by nystatin is much less for a bulky anion, e.g., sulphate which has perhaps as many as eight water molecules in its hydration shell (Table 4.1) than for the relatively small halide ions. Holz and Finkelstein estimate that even in a high conductance membrane the area occupied by the pores is relatively small, e.g., for a bilayer of conductance $10^{-2}\ \Omega^{-1}\ cm^{-2}$ less than 0·01 per cent of the membrane surface area would be occupied by pores.

It remains to speculate on how the pores are created. The size of the hole in the ring structure of nystatin is consistent with a pore with a radius of approximately 0·4 nm. One molecule of nystatin, however, is not sufficiently large to span the width of the lipid barrier and it seems most unlikely that pores are produced by stacks of molecules. Cass *et al.* (1970) pointed out that the related molecule, filipin, appears to make much larger pores in the membrane than nystatin even though the hole at the centre of its ring is much smaller. It seems more probable that the polyene antibiotics interact in some way with the cholesterol, which was present in the bilayer, so as to induce the formation of lipid micelles. In chapter 2 (page 33) we learnt that close packing of micelles in the membrane would leave gaps of approximately 0·4 nm radius. The pores are believed to be unstable physical entities which are created and destroyed at different locations continuously. Temperature greatly affects the rate at which these processes occur and Cass *et al.* found that conductance at low temperature, when the lipid phase was less fluid, was much higher than at elevated temperatures. By increasing the viscosity of the lipid, low temperature would tend to stabilize pores, extending their transient existence at any point on the membrane surface. This dynamic view of pore formation is very similar to the ideas of a dynamic equilibrium in a biological membrane between micellar and bilayer orientation of the phospholipid (Lucy, 1964, 1969).

The properties of a bilayer membrane treated with nystatin with respect to its permeation by water, small non-electrolytes and anions resemble so closely the membrane of a red blood cell that Holz and Finkelstein conclude: 'It is possible that these similarities are more than phenomenological, and it therefore might not be completely fruitless to attempt to extract a polyene or polyene-like molecule from red cell membranes.'

A bilayer membrane incorporating both nystatin and monactin or valinomycin would approximate quite closely the properties of a wide range of biological membranes since it would consist of anion selective water filled pores and mobile cation carriers of high selectivity.

Ion-selectivity of binding sites

In chemical and physical properties the ions of the alkali cations are closely related; as far as their interactions with water and simple anions are concerned: 'The five alkali cations can be described as rigid, nonpolarizable, monopolar spheres differing only in radius' (Diamond and Wright, 1969).

The ability of biological membranes or the carriers therein to recognize and discriminate between the various members of the series poses a problem of basic physico-chemical interest as well as being crucial for the proper functioning of cells and membrane-bound organelles. There are equally interesting selection mechanisms in the divalent alkaline earth series (Mg^{2+}, Ca^{2+}, Sr^{2+}, Ba^{2+}, Ra^{2+}) and among the halide anions (F^-, Cl^-, Br^-, I^-). There are many instances of discrimination among these various ions in non-biological systems, such as clay minerals, glasses and ion exchange resins which are strikingly similar to discrimination by organisms and have the advantage that they can be more easily studied. The characteristics of these systems have been reviewed by Diamond and Wright (1969) in an excellent paper which describes the principles governing ion-selectivity in biological systems.

Selective binding of alkali cations by non-biological materials. The binding of alkali cations to minerals at high temperature is highly selective for lithium and the firmness with which other ions are bound is inversely related to their non-hydrated radii. The binding selectivity is therefore:

$$Li^+ > Na^+ > K^+ > Rb^+ > Cs^+$$

Lithium is the most firmly bound because its positively charged centre can make the closest approach to the electronegative binding sites on the mineral. As we have seen, hydration reverses the order of the ions with respect to size. It has been found that the binding of cations from aqueous solutions onto aluminosilicates ($AlOSi^-$) can produce a selectivity series in which caesium being the smallest hydrated ion is most firmly bound. Such sequences are readily explained in terms of the hydrated or non-hydrated ionic radii. In other classes of aluminosilicate, e.g., Zeolites, sequences which are less readily explained have been found:

$$Na^+ > K^+ > Rb^+ > Li^+ > Cs^+$$

and

$$K^+ > Cs^+ > Rb^+ > Na^+ > Li^+$$

With five elements there are, by permutation, 120 possible permeability sequences but only 11 of these have ever been observed in physical systems. It is most interesting that the same 11 sequences have been found in biological systems of one kind or another. None of the other 109 has been reported and this encourages the belief that the same factors determine the selectivity of clays, minerals, glasses, organic polymers (e.g., DNA) and organisms for alkali cations and that some general principles can be described which will be applicable to each case.

Origin of alkali cation selectivity. The general principles for binding site specificity have been described by Eisenman (1962) after a long series of empirical observations with electrodes made from glasses of different composition. He found that the selectivity of the electrodes depended on the pH and the chemical composition of the glass in such a way as to suggest that binding affinity was controlled by the strength of the negatively charged sites. Weak sites had their fixed negative charges ($AlOSi^-$) screened either by protons, if the solution was acid, or by fixed positive charges in the glass if its composition had been contrived to place them there. Weak sites selected ions on the basis of their hydrated radius; thus, caesium or rubidium would be preferentially bound. If the screening effect of protons was reduced and the electrode allowed to re-establish equilibrium in solutions of decreasing hydrogen ion concentration the binding sequences changes:

at pH 2·0

$$Rb^+ > Cs^+ > K^+ > Na^+ > Li^+$$

pH 3·5–5·0

$$Na^+ > K^+ > Rb^+ > Cs^+ > Li^+$$

pH 6·0

$$Na^+ > K^+ > Rb^+ > Li^+ > Cs^+$$

pH 7·0

$$Na^+ > K^+ > Li^+ > Rb^+ > Cs^+$$

Notice as the pH increases, and with it the electronegativity of the binding site in the electrode glass, lithium moves progressively to the left (its affinity increasing) and rubidium and caesium to the right.

To explain these results a simple theory was advanced by Eisenman which was based on three facts.

(a) Where a bare ion becomes either hydrated or bound to a fixed electronegative site there is a decrease in free energy in the system. The equilibrium cation specificity of a site depends on the free energy differences between ion:site and ion:water interactions.

(b) The free energies of interaction involving alkali cations depend largely on electrostatic forces.
(c) The observed selectivity patterns indicate that the attraction between positively charged ions and electronegative binding sites varies as the inverse square of the distance between them. Such conditions are characteristic of *Coulombic* forces and it is readily apparent how the strength of the site: ion interaction is determined by the effective radius of the ion.

Consider a surface bearing fixed negative charges immersed in an aqueous solution of two cations *a* and *b*. The cation which will be preferred by the binding site will be the one which experiences the larger decrease in free energy when it transfers from water to the site. In general terms, then, binding strength is given by eq. (4.2):

$$\text{Affinity} \propto \Delta F_s - \Delta F_w \qquad (4.2)$$

where ΔF_s is the change in free energy of interaction of the site and the ion and ΔF_w is the free energy of hydration of the ion. Considering again the case where there are two ions present, *a* and *b*, the relative affinity of the site for *a* and *b* will be governed by eq. (4.3)

$$\text{Relative affinity} \propto \Delta F_s^a - \Delta F_s^b - (\Delta F_w^a - \Delta F_w^b) \qquad (4.3)$$

If we are considering a negative site of high field strength so that site: ion interactions involve a very large change in free energy, then the terms ΔF_w^a and ΔF_w^b become negligible and the relative affinity will be determined by which ion can get closest to the site. In such situations, the ion becomes effectively dehydrated so that in a mixture of alkali cations lithium would be the preferred ion. The most complete contrast is one where the field strength of the negative site is very low so that ΔF_w^a and ΔF_w^b become the controlling factors. As we saw in Table 4.1, the free energy of hydration of lithium is greatest and thus it would have the most unfavourable ratio of $\Delta F_s - \Delta F_w$. In intermediate situations, the balance between ΔF_s and ΔF_w is more subtle and it is not difficult to envisage how intermediate situations arise. Take the following series: Na $>$ K $>$ Rb $>$ Cs $>$ Li; the field strength of the binding site is just strong enough to dehydrate the sodium ion, the position of K^+, Rb^+, Cs^+ then follows a logical sequence but Li^+ is misplaced because in this instance the ΔF_s interaction was not sufficiently strong to overcome the large value ΔF_w for this ion.

Binding sites in biological materials and factors affecting their field strength. The choice of chemical groups in biological materials which could act as ion-binding sites is very restricted, e.g., for cations most available sites will be either carboxyl groups (COO^-) or phosphoric acid groups ($H_2PO_4^-$) and for anions, amino groups (NH_3^+). It is, however, possible to produce, from this restricted choice, a range of binding sites differing in field strength and ion-selectivity. The presence of strongly electronegative groups or atoms in the vicinity of the charge site can have a considerable effect on its field strength. This effect is brought about by the tendency of an electronegative group to draw electrons towards itself, thus distorting the electron orbitals of other atoms in the molecule, see Fig. 4.10. In the figure it can be seen that the electron-withdrawing effect of group X allows the cation M^+ to approach closer to the centre of charge of the negative site than would be the case with a similar site in a molecule lacking group X. The chemical nature of X and its distance from the

(a) In a case where there is no induction *rf* is the closest approach the ion can make to O^-

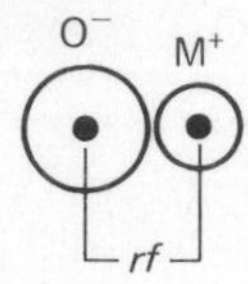

(b) Effect of induction on the field strength of a negatively charged site

Strongly electronegative atom, X, tends to withdraw electrons from the shell of atom A_1 making distribution of electron about the nucleus asymmetric

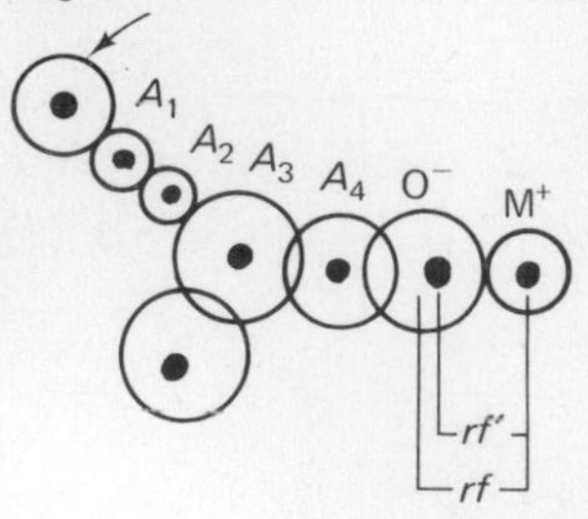

The effect is communicated through the other atoms $A_2 - A_4$ and hence to the negatively charged site.

1. The electron density of O^- will decrease
2. M^+ and O^- can approach one another more closely. The distance *rf* decreased to *rf'*
3. The field strength of O^- increases in proportion to Δrf

***FIG. 4.10** Scheme to illustrate the inductive effect of an electronegative atom on the field strength of a remote ion-building site. For further explanation see text.*

negatively charged site will vary the intensity of the inductive effect and produce sites with the required range of field strength, whose ion-selectivity will consequently vary.

Selectivity of binding sites for anions and divalent cations. Similar analogies have been found between the ion selectivity of non-biological materials with respect to anions of the halogens and divalent cations of the alkaline earth series, viz., Mg^{2+}, Ca^{2+}, Sr^{2+}, Ba^{2+}. The same general principles describing selectivity in terms of ΔF_s and ΔF_w can be shown to be applicable to these groups of ions also.

One interesting note should be made of the competition for charged sites between monovalent and divalent cations. The relative affinities of sites for monovalent or divalent cations depend critically on inter-site spacing; widely separated sites will tend to select monovalent cations preferentially, while closely spaced sites favour divalent cations.

Passive permeation of algal cells by ions

By this point some readers may have formed the opinion that we have strayed rather too far from *Hydrodictyon* and other plants. It is reasonable to ask how useful synthetic model systems are in explaining what is going on in a living cell.

The plasmalemma of *Hydrodictyon africanum* and of all other plant cells has a matrix which almost certainly consists of a bilayer of phospholipid and neutral lipid molecules. The electrical resistance of the plasmalemma is usually in the range 10^3–10^4 Ω cm^2 which is probably very much lower than the bilayer which could be produced from its extracted lipid components (on the basis of other studies we would expect values of 10^7–10^9 Ω cm^2 for such a bilayer). Thus, the algal membrane has a much higher conductance than can be accounted for by its unmodified phospholipid components. The permeability coefficients of monovalent cations are higher than for chloride and in *Hydrodictyon* there is considerable K^+-selectivity. From what we have learnt we might suggest that the basic cation selectivity of the membrane is accounted for by a net negative charge at the membrane surface. This negative charge might originate from the phospholipid heads or from the protein layer bonded to them. We have seen that the conductance of bilayer membranes can be made comparable to biological membranes by including a small number of molecules which can either transport ions across the hydrocarbon interior as a complex, e.g., monactin, or induce the formation of transient pores in the hydrocarbon phase, e.g., nystatin.

It is relevant to inquire whether there is any evidence that such carriers or pores play a part in controlling passive ion movements across the plasmalemma.

Water movement through pores

In chapter 2 we saw that the alga *Chara ceratophylla* has anomalously high permeability to water and some other small electrolytes. The diffusive movement of water across the membrane might be accounted for by the voids formed in the 'kinks' in the hydrocarbon chains (Träuble, 1971; see chapter 2) but evidence from Dainty and Ginzberg (1964) indicates that the hydraulic conductivity L_p, of *Nitella translucens*, a related alga, is greater than the diffusion coefficient for water, P_d. As we indicated earlier, this departure of the ratio L_p/P_d from unity is taken as evidence for the existence of water-filled pores. On this basis and from data which indicate that molecules with a radius greater than 0·4 nm obey the 'rules' and cross the membrane by dissolving in the hydrocarbon layer, it has been proposed that the plasmalemma has in it pores of radius no more than 0·4 nm. There seems to be no reason why such pores should not provide a pathway for at least some of the smaller hydrated ions.

Influence of temperature on permeability coefficients

In a most interesting paper Hope and Aschberger (1970) examined the effects of temperature on the permeability of *Chara corallina* and *Griffithsia pulvinata*, a red alga, to K^+ and Na^+. Lowering the temperature from 25 °C progressively depolarized the membrane potential. By measuring tracer fluxes of Na^+ and K^+ in cells stabilized at temperatures in the range 0–25 °C these authors were able to calculate the permeability coefficients, P_K and P_{Na}, at different temperatures (Fig. 4.11). They found that the ratio $P_{Na}:P_K$ remained constant in *Chara corallina* over the range 25–13 °C. Below 13 °C the ratio increased substantially indicating that the permeability to Na^+ was increasing relative to K^+. If the logarithms of the permeability coefficients P_K and P_{Na} are plotted against the reciprocal of the absolute temperature the slopes of the curves obtained give the enthalpy change (ΔH^*) for the permeation of the membrane. In Table 4.7 the results of the experiments are summarized; some of them are rather surprising, especially those for sodium.

The potential energy barrier for K^+ permeation, ΔH^*_K, decreased with increasing temperature; this result accords with the fact that the membrane potential became more negative as the temperature increased. In *Chara corallina* the membrane potential is largely determined by the diffusion of K^+, therefore a decrease in the potential energy barrier to K^+ diffusion will cause K^+ to diffuse faster. In the same species, the effect of increasing temperature was to *increase* the potential energy barrier to Na^+ permeation. This paradoxical situation, in which the enthalpy of activation of the more

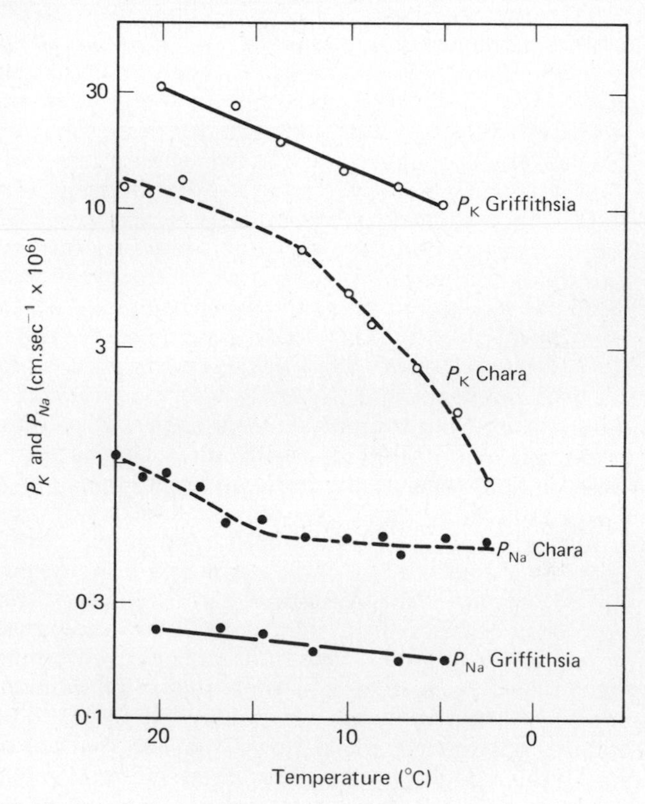

FIG. 4.11 Permeability coefficients for potassium and sodium at various temperatures. (Adapted from Hope and Aschberger, 1970.)

Table 4.7 Mean enthalpies of activation for permeation of the plasmalemma of two algae at various temperatures

Species	ΔH^*_K (kcal mole^{-1})	ΔH^*_{Na} (kcal mole^{-1})	Temperature range (°C)
Chara corallina	28·0	3·5	3–13
	10·1	14·0	13–25
Griffithsia pulvinata	14·0	3·4	3–20

Data from Hope and Aschberger (1970).

permeable ion (K^+) is greater at low temperatures than the less permeable ion (Na^+) has also been noted in *Nitella translucens* by Hogg, Williams and Johnston (1968). In *Griffithsia pulvinata* the enthalpies showed no marked changes at 13 °C, but in this species also ΔH^*_K is higher than ΔH^*_{Na} even though $P_K > P_{Na}$.

Some of the things we have learnt from the study of bilayers treated with either monactin or nystatin are helpful in explaining this paradox.

The values of ΔH^*_K and ΔH^*_{Na} in either temperature range are appreciably lower than the heats of hydration for these ions as given in Table 4.1 where the enthalpies for the ion: water interactions are -90 and -115 kcal mole^{-1} for K^+ and Na^+ respectively. It seems most improbable that either ion crosses the hydrocarbon in a non-hydrated form. If they do not then we know that they might move *via* carriers or through pores. The effect of temperature on P_K is consistent with the movement of a K^+-carrier complex across the membrane, the diffusion of which would be slower at lower temperatures, 13 °C being a critical point in *Chara corallina*. Hope and Aschberger (1970) point out that the change in the enthalpy values at this point may result from a phase change or partial solidification of some component in the lipid phase. Such a suggestion seems very reasonable; in lipid bilayers the induced conductance of K^+ by valinomycin was reduced both by low temperature or by increasing the viscosity of the hydrocarbon phase by incorporating cholesterol into it. On the other hand, both cholesterol and low temperature enhanced the conductance of nystatin-treated bilayers. This enhancement probably resulted from the stabilization of water-filled pores in the more viscous lipid phase and it might well be that the lowered value of ΔH^*_{Na} is a reflection of a similar phenomenon in *Chara corallina*. In *Griffithsia pulvinata* the ΔH^*_{Na} is lower than ΔH^*_K at all temperatures suggesting, by analogy, that the composition of the membrane in this species is such that pores are relatively stable over the whole temperature range studied.

If there are, indeed, pores in the plasmalemma it seems probable that both ions can travel through them. We can say nothing about the ion selectivity of such pores except that, since K^+ has the smallest hydrated radius, it might be expected to diffuse quicker than Na^+. Since the flux of K^+ is very much higher than Na^+ it seems probable that there is a specific pathway for K^+ movement in addition to its movement through the pore. The observed permeability coefficient at 25 °C would then be composed of two terms

$$P_K = P_K^{carrier} + P_K^{pore} \tag{4.4}$$

while P_{Na} is determined very largely by P_K^{pore}. By analogy with the bilayer model, decreased temperature should increase ΔH^*_K for the carrier pathway but decrease ΔH^*_K for movement *via* the pores. Instability in the hydrocarbon interior will disrupt the pores and will lead to increased ΔH^*_K and ΔH^*_{Na} for the pore pathway.

Membrane resistance and K^+ selectivity

A model containing ion-selective carriers and pores might also be used to explain some other well documented phenomena. Spanswick (1970) has found that the electrical resistance of the plasmalemma, R_{pl}, in *Nitella translucens* decreases sharply when the K^+ concentration in the external solution is increased from 0·1 to 10·1 mM, indicating an increased K^+ conductance, G_K. In nystatin-treated bilayers, it was found that conductance through pores increased with increasing electrolyte concentration, but in monactin-treated bilayers conductance was determined by the concentration of the carrier. By analogy, then, the maximum conductance of the carrier pathway will be reached when the carrier is saturated with K^+; beyond this K^+ concentration conductance *via* the pores will be of increasing importance in determining the value of G_K and hence R_{pl}. One of the consequences we might predict from this is that the selectivity of the membrane for K^+ will diminish as the pore conductance becomes greater, since the pore is a common pathway for K and Na while the carrier is relatively specific for K.

It has been noted frequently, especially in the tissues of higher plants, that a high degree of selectivity in the absorption of K relative to Na is found when these ions are mixed at low concentrations in the external solution; this specificity markedly declines when the concentrations exceed 10 mM.

The movement of hydrated ions through a narrow pore will be subject to a number of restraints; in addition to electrostatic interactions between charged groups in the pore and the ion there will be electrokinetic reactions between the water molecules and the pore lining. This interaction produces a surface potential which organizes water molecules at the pore surface into a layer of soft ice (Davies and Rideal, 1961). The thickness of this layer is known to decline with increasing electrolyte concentration and consequently the effective resistance of the pore will decrease.

The speculative nature of much of this discussion should be heavily stressed and, although certain of the conclusions may well prove to be wrong, the simplified system of a bilayer with mobile carriers and transient pores can provide a way of thinking about the more complex situation in biological membranes, and in some cases suggests experiments which may give further insight into the way membranes control the diffusion of substances in and out of organisms.

Bibliography

a Further reading

Goldup, A., Ohki, S. and Danielli, J. F. (1970) Black lipid films. *Rec. Prog. Surface Sci.* **3**, 193–260.

This is a comprehensive and readable account of the 'state of the art' up to 1970. It includes recipes and techniques for making bilayers which are clearly illustrated so that the student might be encouraged to develop this technology for himself.

Pressman, B. C. (1970) Energy linked transport in mitochondria. In *Membranes of Mitochondria and Chloroplasts*. (Ed. E. Racker), 213–250. Van Nostrand–Reinhold, N.Y.

Although most of this review is more relevant reading for chapter 5, there is an excellent account of ionophorous antibiotics with which the author has worked for many years.

Cass, A., Finkelstein, A. and Krespi, V. (1970) The ion permeability induced in thin lipid membranes by the polyene antibiotics Nystatin and Amphotericin B. *J. Gen. Physiol.* **56**, 100–124.

This reference is given in the general bibliography but attention is drawn to this explicit and interestingly written account of 'pore' formation in a lipid bilayer membrane.

Diamond, J. M. and Wright, E. M. (1969) Biological membranes: the physical basis of ion and non-electrolyte selectivity. *Ann. Rev. Physiol.* **31**, 581–646.

This long and comprehensive review explores every aspect of solute-selectivity by membranes. The brief account on ion-selectivity in the present work was largely summarized from this paper which should be consulted where further details are required.

b References in the text

Bangham, A. D., Standish, M. M. and Watkins, J. C. (1965) Diffusion of univalent ions across the lamellae of swollen phospholipids. *J. Mol. Biol.* **13**, 238–252.

Bockris, J. O'M. (1954) *Modern Aspects of Electrochemistry*. Butterworths, London.

Cass, A., Finkelstein, A. and Krespi, V. (1970) The ion permeability induced in thin lipid membranes by the polyene antibiotics Nystatin and Amphotericin B. *J. Gen. Physiol.* **56**, 100–124.

Cherry, R. J., Dodd, G. D. and Chapman, D. (1970) Small molecule-lipid membrane interactions and the puncturing theory of olfaction. *Biochim. Biophys. Acta* **211**, 409–416.

Dainty, J. and Ginzburg, B. Z. (1964) The measurement of hydraulic conductivity (osmotic permeability to water) of internodal Characean cells by means of transcellular osmosis. *Biochim. Biophys. Acta* **79**, 102–111.

Davies, J. T. and Rideal, E. K. (1961) *Interfacial Phenomena*. Academic Press, N.Y. and London.

Diamond, J. M. and Wright, E. M. (1969) Biological membranes: the physical basis of ion and non-electrolyte selectivity. *Ann. Rev. Physiol.* **31**, 581–646.

Eisenman, G. (1962) Cation selective glass electrodes and their mode of operation. *Biophys. J.* **2**, 259–323.

Eisenman, G., Ciani, S. M. and Szabo, G. (1968) Some theoretically expected and experimentally observed properties of lipid bilayer membranes containing neutral molecular carriers of ions. *Fed. Proc.* **27**, 1289–1304.

Gutknecht, J. and Tosteson, O. C. (1970) Ionic permeability of thin lipid membranes. Effects of n-alkyl alcohols, polyvalent cations and a secondary amine. *J. Gen. Physiol.* **55**, 359–374.

Hogg, J., Williams, E. J. and Johnston, R. J. (1968) The temperature dependence of the membrane potential and resistance in *Nitella translucens*. *Biochim. Biophys. Acta* **150**, 640–648.

Holz, R. and Finkelstein, A. (1970) Water and non-electrolyte permeability induced in thin lipid membranes by the polyene antibiotics Nystatin and Amphotericin B. *J. Gen. Physiol.* **56**, 125–145.

Hope, A. B. and Aschberger, P. A. (1970) Effects of temperature on membrane permeability to ions. *Aust. J. Biol. Sci.* **23**, 1047–1060.

Hopfer, U., Lehninger, A. L. and Lennarz, W. J. (1970) The effect of polar moeity of lipids on bilayer conductance induced by uncouplers of oxidative phosphorylation. *J. Membrane Biol.* **3**, 142–155.

Hopfer, U., Lehninger, A. L. and Thompson, T. E. (1968) Protonic conductance across phospholipid bilayer membranes induced by uncoupling agents for oxidative phosphorylation. *Proc. Natl. Acad. Sci* (U.S.) **59**, 484–490.

Kimelberg, H. K. and Papahadjopoulos, D. (1971) Interactions of basic proteins with phospholipid membranes. Binding and changes in the sodium permeability of phosphatidylserine vesicles. *J. Biol. Chem.* **246**, 1142–1167.

Lucy, J. A. (1964) Globular lipid micelles and cell membranes. *J. Theoret. Biol.* **7**, 360–368.

Lucy, J. A. (1969) Lysosomal membranes. Chapter 11 in *Lysosomes in Biology and Pathology*. North Holland Publishing Company, Amsterdam.

Mueller, P. and Rudin, D. O. (1967) Development of $K^+ - Na^+$ discrimination in experimental bimolecular lipid membranes by macrocyclic antibiotics. *Biochim. Biophys. Res. Commun.* **26**, 398–403.

Papahadjopoulus, D. (1971) $Na^+ - K^+$ discrimination by 'pure' phospholipid membranes. *Biochim. Biophys. Acta* **241**, 254–259.

Pressman, B. C., Harris, E. J., Jagger, W. S. and Johnson, L. H. (1967) Antibiotic-mediated transport of alkali ions across lipid barriers. *Proc. Natl. Acad. Sci.* (U.S.) **58**, 1949–1952.

Spanswick, R. M. (1970) Electrophysiological techniques and magnitudes of membrane potentials and resistances in *Nitella translucens*. *J. Exp. Bot.* **21**, 617–627.

Stein, W. D. (1967) *The Movement of Molecules across Cell Membranes*. Academic Press, N.Y. and London.

Thompson, T. E. and Henn, F. A. (1970) Experimental phospholipid model membranes. Chapter I in *Membranes of Mitochondria and Chloroplasts*. Ed. E. Racker, Van Nostrand/Reinhold, N.Y.

Träuble, H. (1971) The movement of molecules across lipid membranes: a molecular theory. *J. Membrane Biol.* **4**, 193–208.

Tsofina, L. M., Liberman, E. A. and Babkov, A. V. (1966) Production of bimolecular protein-lipid membranes in aqueous solution. *Nature* **212**, 681–683.

five

Links between cellular metabolism and ion transport

In this chapter we will consider in more detail the relationships between metabolism and ion pumps. Ion pumping can be heavily dependent on formation and hydrolysis of adenosine triphosphate (ATP) and on electron-flow; both processes occur during photosynthesis and respiration. The separation of positive and negative charges across the membranes of cytoplasmic organelles will be shown to be a fundamental property of electron transport systems of very great significance in ion transport between intracellular compartments and across the plasmalemma. An outline of the important events in photosynthesis and respiration is given at the beginning of this chapter. Standard texts on biochemistry should be consulted for full details of the reactions involved.

Energy-flow and ATP production in photosynthesis

The cells of green plants are equipped with a light-energy trapping system in which the most important pigments are chlorophyll *a* and (in higher plants) chlorophyll *b*. The absorption of light-energy results in an excited state in the pigment molecule, which if not coupled to further reactions, declines with time, the trapped energy being lost as heat and fluorescence. In photosynthesis, the energy of excited chlorophyll molecules is passed down a series of steps and conserved by the formation of molecules of ATP which possess a terminal phosphate bond, the hydrolysis of which is accompanied by a release of free energy which can be used for metabolic purposes and for active transport.

Excitation of chlorophyll by light. Every atom has, outside its nucleus, a number of electrons which have characteristic distributions in space. The carbon atom, for example, has six electrons arranged in three orbitals. An orbital represents the shape and size of the space around the nucleus where there is the highest probability of finding a given electron. The orbitals are characterized by gradually increasing energy levels, the nearest to the nucleus, 1_s, accommodating two electrons, the middle one, 2_s, also accommodates two electrons, while the outermost orbital, 2_p, contains the two electrons of highest energy. In the ground state the electrons exist in this array but in the 'excited state' one of the electrons from 2_s shell is energized by the absorption of a quantum of light, or photon, sufficiently for it to move into the 2_p orbital, making the atom very much more reactive and raising its free energy. Specific carbon atoms in a molecule can be excited only by light of certain wavelengths. This is because electrons moving from the 2_s to the 2_p orbital must gain exactly the right amount of energy from the absorbed photon to make the transition; the energy of the photon is therefore utilized either completely or not at all. The chemical environment surrounding an atom will strongly influence the amount of energy necessary for the transition to the excited state since it affects the energy distribution of the electrons in the orbitals (see discussion of the inductive effect in chapter 4). Photons of different wavelengths have different energies which can match these modifications in the energy requirements for transition to the excited state in a wide variety of compounds.

Excited atoms in a molecule are unstable and will quickly return to the ground state, i.e., to their lower energy orbitals, but in so doing the energy originally present in a photon will be released either as heat or light. Such emission of light is called fluorescence. It is the business of photosynthesis to ensure that the energy of the photons absorbed by the chlorophyll is not lost in this way but is conserved for future use by the cell. This is achieved by the transduction of light energy into chemical energy.

Energy transfer reactions in photosynthesis. If we irradiate a purified solution of chlorophyll *a* with red light, photons will excite the molecules as we have described above, but energy is lost again by fluorescence. In the chloroplast, however, chlorophyll behaves rather more like a semiconductor in that the electron which moves into the outer orbital actually parts company with the atom and is passed through an ordered series of electron carrier molecules. At each step the energy of the electron is tapped off and gradually reduced.

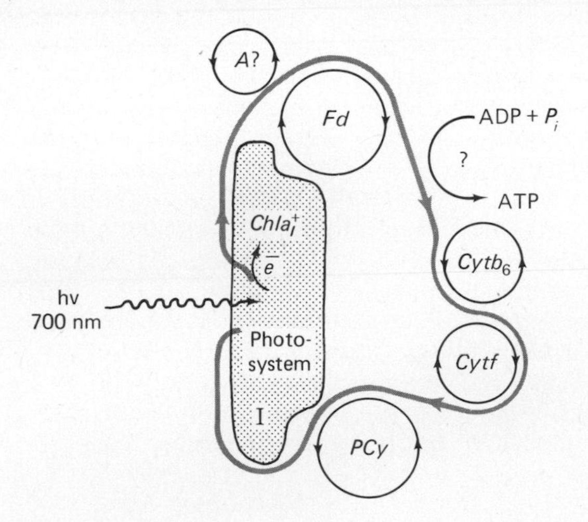

***FIG. 5.1** Cyclic electron-flow and photophosphorylation. The electron excited by the absorption of a quantum of light at 700 nm is passed from chlorophyll* a *around a circuit of electron carriers until it returns to its original orbital position. A = unidentified carrier, Fd = ferredoxin, Cyt b6 and Cyt f = cytochromes b6 and f, respectively, P Cy = Plastocyanin.*

Figure 5.1 illustrates a process known as cyclic electron flow. In chloroplasts, molecules of chlorophyll *a* are grouped together into photosynthetic units which absorb light maximally at 700 nm and have appreciable absorption in the far red end of the spectrum (above 710 nm). High energy electrons leave the excited chlorophyll *a* molecules (collectively known as photosystem I) and pass to an acceptor molecule A (the identity of which is uncertain) and then to an iron containing protein, ferredoxin, which becomes reduced since the acceptance of an electron changes its iron from the Fe^{3+} state to the Fe^{2+} state. Electrons pass from ferredoxin (which becomes reoxidized) to cytochrome b_6 and from thence to cytochrome f. The cytochromes are also iron-containing chromoproteins. Finally, the electrons return to chlorophyll *a* and having lost the energy they gained on excitation, they return to their original s orbitals. In this process, electrons of high potential are passed down a conducting system in which their potential is reduced stepwise until they return to their ground level. Each pair of electrons travelling round the circuit loses about 50 000 calories before returning to chlorophyll *a*. Some of this energy is conserved in coupled reactions which produce ATP; the process is known as photophosphorylation.

The complete process of photosynthesis is more complicated than the cyclic flow of electrons about chlorophyll *a*. In normal conditions electrons leaving photosystem I pass through ferredoxin to an acceptor molecule nicotinamide adenine dinucleotide phosphate (NADP) which becomes reduced to NADPH by the acquisition of a pair of electrons and a pair of protons (H^+) from the medium. NADPH is required to reduce CO_2 in the dark processes of photosynthesis which lead to the formation of polysaccharides (e.g., Calvin cycle). It is only when these dark reactions are inhibited that the cyclic process described in Fig. 5.1 represents a major pathway for electron movement. With the dark reactions operating, there is a net outflow of electrons from chlorophyll *a* to NADP. Clearly, this process cannot continue for very long unless there is some means of getting electrons back into chlorophyll *a*. Molecules of chlorophylls *a* and *b*, collected into groups in photosystem II, fill this role. Excited electrons from photosystem II flow down an electron transport chain *via* cytochrome f to photosystem I (chlorophyll *a*). Figure 5.2 illustrates the overall reactions in a simplified form. During their movement between the two photosystems at least one molecule of ATP is formed for each pair of electrons transported.

Electrons flowing from photosystem II to photosystem I are replaced by electrons which are derived from the photolysis of water. The actual mechanism of the reaction remains obscure but is summarized as follows:

$$H_2O \longrightarrow H^+ + OH^-$$
$$OH^- \longrightarrow \tfrac{1}{2}O_2 + H^+ + 2e^-$$

Water is therefore the primary source of reducing power. The temporary separation of protons and electrons from the hydrogen atoms in water is a device which provides a free-energy gradient sufficiently large to power the synthesis of ATP in the coupled reactions of photosynthesis. We shall discuss the matter in detail later in this chapter. The protons and electrons eventually recombine at the point where NADP is reduced. To complete the pathway from water to NADP two light quanta are required for each electron; since water supplies two electrons there will be four light quanta absorbed for each water molecule split up.

Dark reactions and the recycling of NADPH. The product of the two photosystems, NADPH, donates its hydrogen to 3-phosphoglycerate reducing it to 3-phosphoglyceraldehyde. The oxidized form of the carrier, NADP, is then available once more as an acceptor of electrons from photosystem I. The dark reactions of photosynthesis involving the enzymic fixation of CO_2 into the C–5-phosphorylated sugar, ribulose-1,5-diphosphate by carboxydismutase, and subsequent steps leading to the formation of glucose, do not seem to be directly involved in regulating ion transport (see page 141).

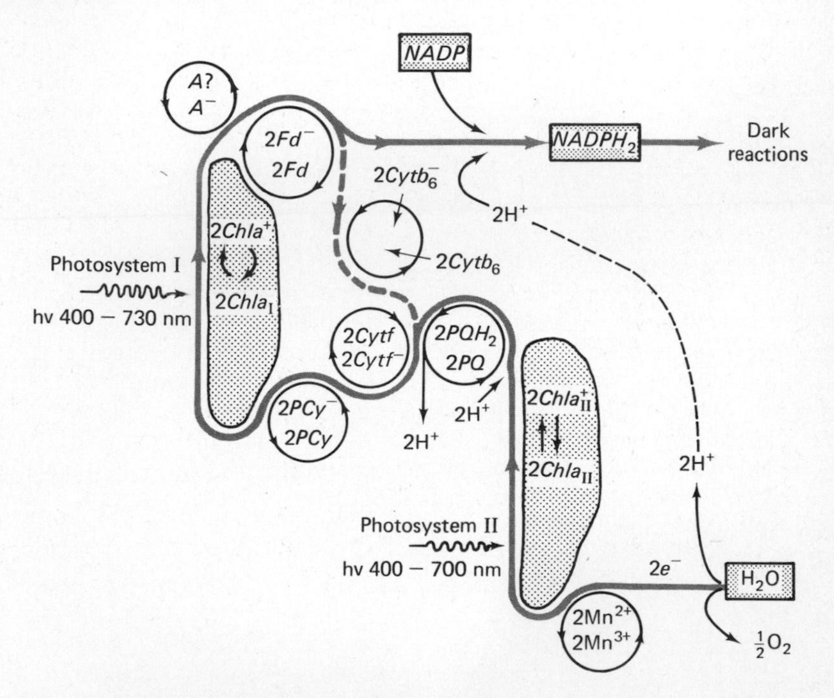

FIG. 5.2 Electron-flow in photosynthesis. Solid lines represent non-cyclic electron flow, broken line represents cyclic (as in Fig. 5.1). Bold line represents path of electrons; thin line represents recycling of oxidized form of redox carrier. Photosystem I contains chlorophyll a*; photosystem II contains chlorophyll* a *together with accessory pigments such as chlorophyll* b*. The exact number and sequence of intermediates between the two photosystems is not resolved, nor is the exact location of the coupled phosphorylation.*

PQ = plastoquinone, Mn = manganese ions, $Chla_{II}$ = chlorophyll of photoreaction two which brings about primary photochemical reaction; $Chla_I$ = chlorophyll of photoreaction two which brings about the primary photoreaction. Other abbreviations as in Fig. 5.1.

Respiration and its associated electron transport reactions are, however, directly dependent on substrates provided by photosynthesis. In non-photosynthetic cells, electron flow and ATP production by respiration are the only potential sources of energy for ion transport.

Energy-flow in respiration

Substrates for respiration are produced by the dark reactions of photosynthesis. In higher plants these are translocated through the phloem to the cells of the root; in the cells of algae they are used *in situ*. The free energy of these substrates is liberated during respiration in mitochondria, some of it being conserved in a more readily available form by the formation of ATP which is used for various types of cellular work. Taking a broad view, therefore, respiratory processes are a second stage in the conservation of the energy originally present in the photons trapped by chlorophyll. The conservation is achieved during electron transport in much the same way as in photosynthesis.

The substrates, carbohydrates, fats, etc. are broken down in a series of dehydrogenation reactions in the Krebs cycle (Fig. 5.3) in which the hydrogen atoms are removed enzymically and passed to the hydrogen-carrier NAD. In its reduced form, NADH passes electrons through to the electron transport chain and liberates protons into the medium. Electrons are passed down series of carriers which are alternatively reduced and oxidized. At each step the energy of the electrons is reduced. The identity of most of the carrier molecules in the chain is established but some uncertainty remains about their order in the sequence and to what extent the chain may be branched. At the end of the chain the electrons and the protons reassociate and reduce molecular oxygen to water in a reaction promoted by cytochrome oxidase.

During their passage down the electron transport chain each pair of electrons (separated from pairs of hydrogen atoms derived from the parent substrate) is involved in three coupled reactions which phosphorylate ADP to form ATP. The nature of coupling is likely to be similar in both respiration and photosynthesis. In the normal course of events this coupling is obligatory. If, for instance, a cell-free suspension of mitochondria is depleted of inorganic phosphate, P_i, electron transport will not continue and oxygen consumption by the preparation ceases. Similarly, there is no phosphorylation in the absence of respirable substrate except in exceptional circumstances (see page 159). The precise places at which the ATP is formed are still a matter for debate and Fig. 5.4, which outlines the process, is designed to be as generalized as possible.

All respiratory substrates are prepared for entry into the Krebs cycle by breakdown into 2-carbon pieces. The Krebs cycle accepts only an 'activated' derivative of acetic acid, acetyl co-enzyme A, as its starting material. This molecule now enters the Krebs cycle (Fig. 5.3) where it undergoes a series of dehydrogenation reactions which

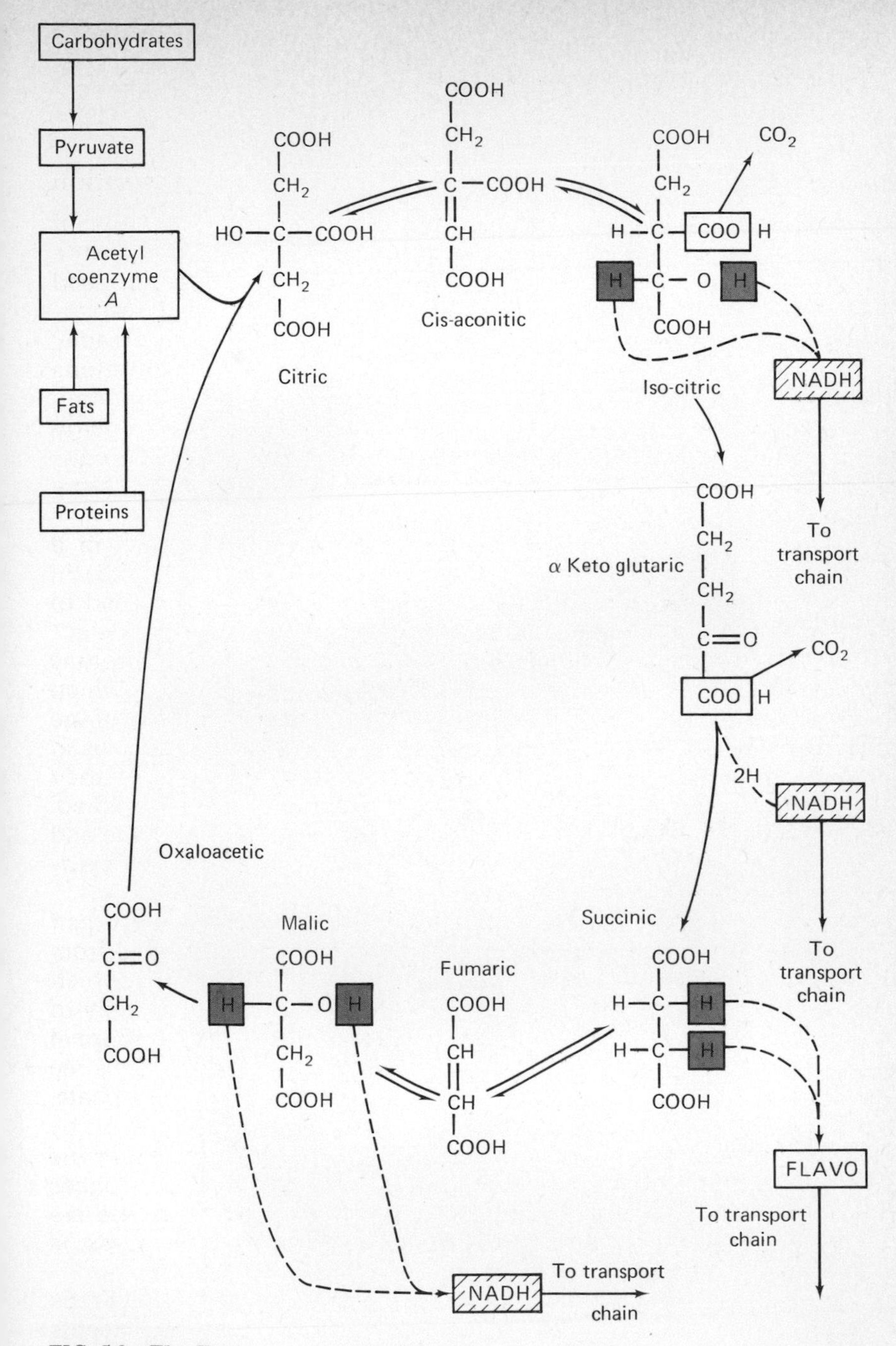

FIG. 5.3 The Krebs or tricarboxylic acid cycle. All substrates are prepared for entry by the formation of acetyl coenzyme A which reacts with oxaloacetic acid to form citric acid; subsequent dehydrogenation reactions provide protons and electrons for the electron transport system (see Fig. 5.4). Note that three pairs of hydrogen atoms are carried to the transport chain by NAD molecules, while those removed from succinate travel via reduced flavoprotein.

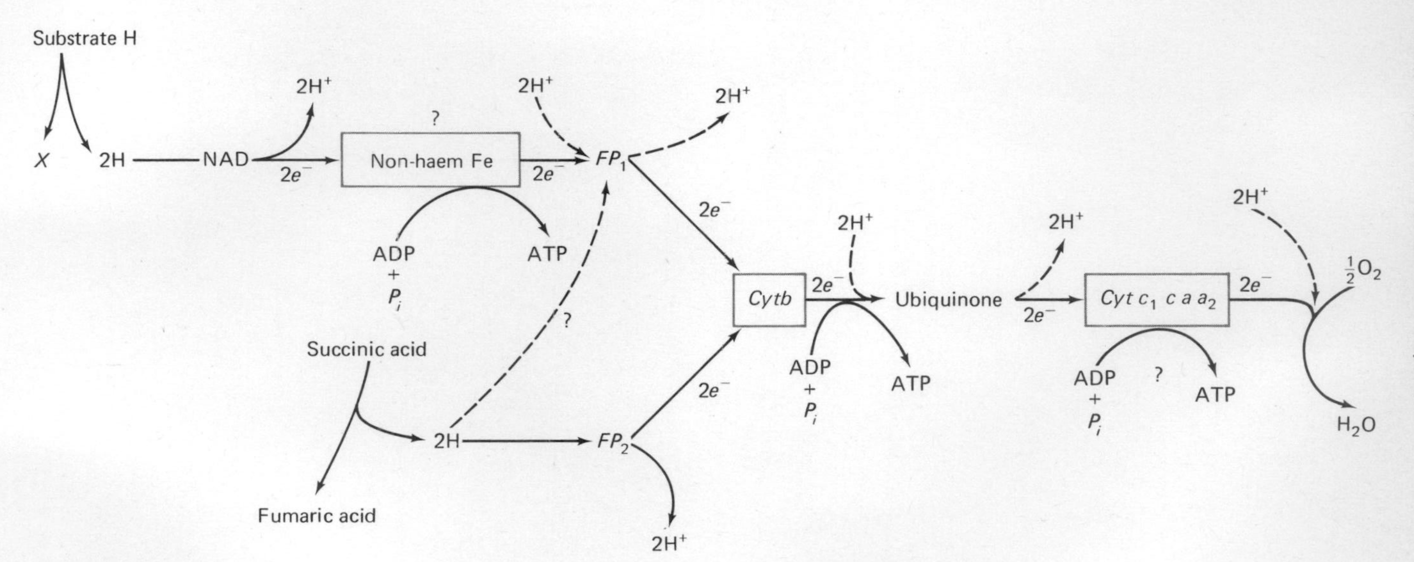

FIG. 5.4 ***Transport chain for electrons derived from the Krebs cycle. The question marks indicate places where the pathway is obscure. The exact location of the phosphorylation site is not certain. Components of the chain in heavy type are hydrogen carriers whilst those in lighter type are electron carriers. FP_1 and FP_2 are flavoproteins; other abbreviations as in Figs. 5.1 and 5.2. (The significance of the alternation of hydrogen and electron carriers is discussed on page 152 and illustrated in Fig. 5.10.)***

donate eight hydrogen atoms to NAD. Thus, four pairs of electrons enter the electron transport chain and each pair of electrons yields about three molecules of ATP, therefore 12 molecules of ATP are produced for each revolution of the Krebs cycle. Table 5.1 sums up the molecules of ATP produced from the oxidation of a mole of glucose. The first steps involve a number of reactions which are omitted in this text (students should consult standard works on biochemistry for the steps involved in the breakdown of the carbon skeleton).

Table 5.1

Reaction sequence	Pairs of electrons produced	Molecules ATP produced
(1) Glucose → 2 moles pyruvate		8
(2) 2 pyruvate → 2 acetyl-CoA	2	6
(3) 2 acetyl-CoA → CO_2 + H_2O	8	24
Glucose → CO_2 + H_2O		38

The formation of ATP in respiration has the same effect of conserving energy as we found in photosynthesis. In the case of respiration, what is happening is that the free energy stored in carbohydrates is being partly dissipated in entropy changes, partly liberated as heat and partly converted into a more readily usable form in the shape of the terminal phosphate bond of ATP. It is necessary to provide an input of approximately 7 kcal . mole^{-1} to accomplish the phosphorylation of ADP

$$\text{ADP} + P_i \rightleftharpoons \text{ATP}$$
$$\Delta G = \pm 7000 \text{ cal . mole}$$

Table 5.1 shows that 38 molecules of ATP are formed in this way from one mole of glucose requiring an input of 38 × 7000 = 266 000 calories. The standard free-energy change for the complete aerobic oxidation of glucose is −686 000 cal . mole^{-1}, thus a little over 40 per cent of the energy is conserved in the terminal bond of ATP. It is conserved because, upon hydrolysis, the terminal phosphate bond is broken and the free energy released; 7000 cal . mole^{-1} may be utilized for the working of the cell. We should remember that the energy conserved by the ATP produced in respiration is ultimately derived from the photons trapped by chlorophyll in the chloroplast, either in the same cell, or in a different cell from the same organism or from another organism.

Dependence of ion pumping on ATP and electron-flow

Electron transport and ATP synthesis occur in both photosynthesis and respiration. If either of these processes were limiting the rate of active ion transport in a photosynthetic cell kept in darkness, we might expect that the extra ATP or the additional electron transport occurring when the cell was illuminated would stimulate the operation of the transport process. Light-stimulated increases in the rate of ion movement across the plasmalemma of cells are, in fact, commonly found and are conveniently studied. In the non-photosynthetic cell the experimenter must find ways of disrupting or promoting respiratory production of ATP and electron transport when he wishes to examine their effects on ion transport. This is usually done by maltreating the cell, either physically or chemically, in some way and the obvious weakness in this approach is that the experimenter may disturb more systems than he intends by his treatment. In the photosynthetic cell, the additional machinery for ATP production and electron transport can be brought in and out of operation merely by switching the light on and off, thus avoiding drastic modification of the environment in the cell. In this account we shall be examining light-dependent ion fluxes and their wavelength dependence in our test case organism, *Hydrodictyon*. At the present time there seems every reason to believe that active ion transport in non-green cells depends on respiration for the supply of the same fuels as are supplied, by photosynthesis and respiration, in green cells.

Light-stimulated ion transport in *Hydrodictyon*. We established in chapter 3 that the following fluxes of the three major ions in the cell were against gradients of electrochemical potential and were therefore designated active transport processes:

$$K_{influx}, Cl_{influx}, Na_{efflux}$$

The respective fluxes in the opposite direction were probably by passive diffusion through imperfect 'leaky' membranes. The active fluxes should be more sensitive to what is happening in cellular metabolism than the passive fluxes. The graphical presentation of data taken from Raven (1967) makes it clear that this expectation is correct (Fig. 5.5). Ion fluxes in the light are represented by the whole circumference of the circle, the shaded sector being a corresponding flux in the dark. The magnitude of light stimulation of the fluxes is seen from the lighter coloured sector of the circles. It is clear that chloride efflux is little affected by light, whereas the chloride influx is nearly three times greater in the light than in the dark. Similar results are found for potassium and the reverse situation is found with

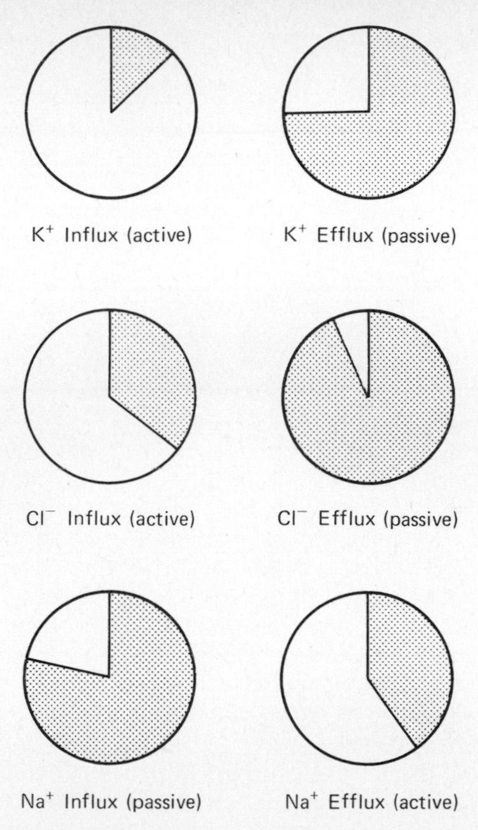

***FIG. 5.5** Relative magnitude of active and passive ion fluxes in* **Hydrodictyon africanum** *in light (whole circumference) or darkness (shaded sector). Size of the unshaded portion indicates the light-stimulation of the flux. (Data from Raven, 1967b.)*

sodium where the efflux is strongly stimulated by light, while influx is little affected.

The stimulation of active fluxes by light suggests that either photosynthetic electron transport and phosphorylation support the increased rates of pumping or that some product of CO_2-fixation is responsible. This latter possibility can be eliminated because light-stimulated fluxes are not altered when the cells are maintained in a solution aerated with CO_2-free air (Table 5.2).

This experiment indicates that the light stimulated ion fluxes depend on the two light reactions of photosynthesis (photosystems I and II) for which chlorophyll *a* and chlorophylls *a* and *b* are the photoreceptors. MacRobbie (1965) made use of the fact that the photosystems had different absorption in the far red end of the

***Table 5.2** Active ion fluxes in illuminated cells of* **Hydrodictyon africanum** *at 14 °C*

Flux	Air	CO_2-free air
	pmoles cm^{-2} s^{-1}	
Cl_{in}	2·04	2·09
K_{in}	1·24	1·04
Na_{out}	0·37	0·36

(Data from Raven, 1967b.)

spectrum to analyse further the dependence of active transport on the light reactions. The rationale in designing the experiment was as follows. If a process requires both photosystems for its efficient operation, then the progressive elimination of shorter wavelength red light will inhibit the process because photosystem II absorbs light at wavelengths greater than 703 nm very weakly. Photosystem I, however, utilizes longer wavelength red light and far red light with high efficiency, and can continue to do so even when photosystem II is not absorbing photons and providing a supply of electrons to photosystem I. It is in these circumstances that we get cyclic electron-flow about photosystem I (see Fig. 5.1) during which ATP synthesis can occur. If some physiological process could be supported by this ATP or some other unknown product of cyclic photophosphorylation then we would expect to find that its rate was not inhibited markedly when shorter wavelength red light was filtered out.

Raven (1967) used an appropriate set of filters to provide illumination for *Hydrodictyon africanum* with increasing depletion of red light with wavelengths shorter than 703 nm. The active influx of chloride into the cells illuminated by three light sources is shown in Fig. 5.6 and it is clear that filters 3 and 2 provide light which is utilized with low efficiency. To achieve the maximal rate of light-stimulated chloride influx approximately three times, and more than 10 times, the quantum flux would be required under filters 2 and 3 respectively than is required under filter 1. This is characteristic of a system which requires both photosystems working in tandem, or at least some product derived from photosystem II for its support.

The situation was quite different if either potassium influx or sodium efflux were observed under a similar set of conditions (Fig. 5.7). Here light transmitted from the longer wavelength transmission filter (filter 3) was marginally more efficient at supporting these fluxes than that from filter 1. We can conclude that these active

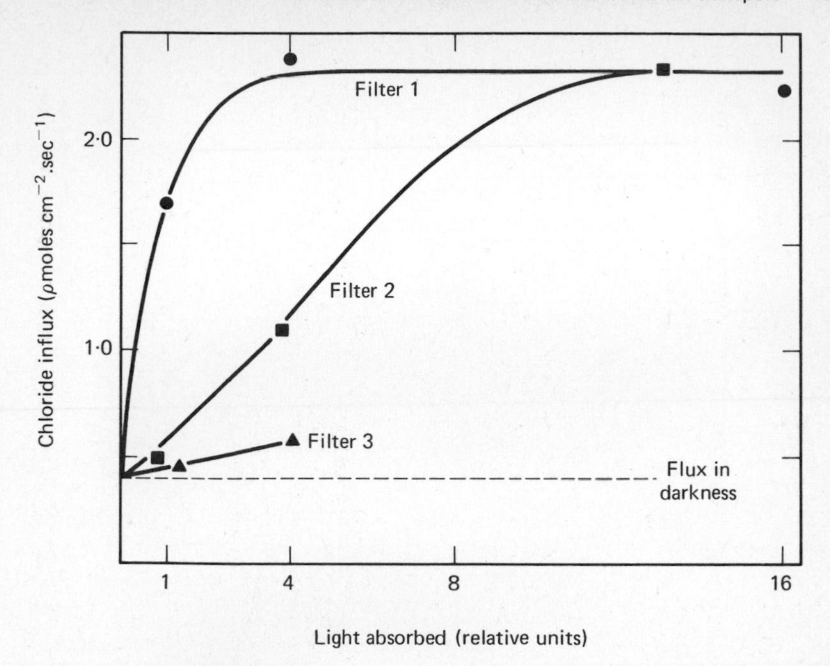

FIG. 5.6 The effect of light transmitted by filters on the light-stimulated influx of chloride in* Hydrodictyon africanum. *Filters 2 and 3 exclude progressively red light with wavelengths shorter than 703 nm. (Redrawn from Raven, 1967b with the kind permission of the author.)

fluxes of K^+ and Na^+ can be supported by photosystem I operating on its own.

The use of certain metabolic inhibitors provides some indirect evidence about those products from photosynthesis which are involved in supporting the active fluxes. CCCP (carbonyl cyanide chlorophenyl hydrazone) inhibits the synthesis of ATP during electron flow; its function is as an *uncoupler* which has its effect without inhibiting electron-flow (the mechanism of uncoupling will be described on page 159). A second inhibitor, DCMU (dichlorophenyldimethyl urea) has an important role in preventing the transfer of electrons from photosystem II to the electron transport chain: it therefore inhibits non-cyclic electron-flow between the two photosystems but does not necessarily inhibit cyclic electron-flow about photosystem I. The action of DCMU is analogous to filter 3.

Uncoupling ATP synthesis by CCCP in illuminated cells of *Hydrodictyon* severely inhibited the cation fluxes (Table 5.3) but had a much smaller effect on the chloride influx.

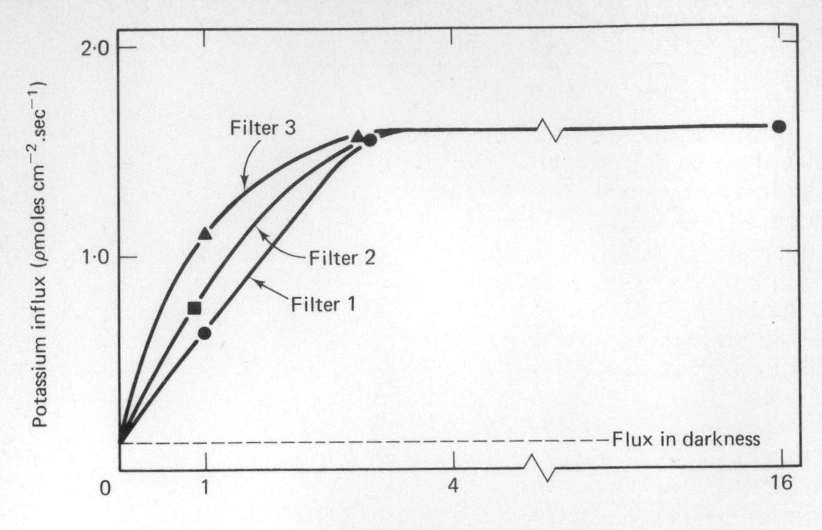

***FIG. 5.7** The effect of light transmitted by filters on the light-stimulated influx of potassium in* **Hydrodictyon africanum.** *Compare these results with those in Fig. 5.6 where the same filters were used. (Redrawn from Raven, 1967b, with the kind permission of the author.)*

***Table 5.3** Effect of the uncoupler, CCCP, on active ion fluxes in* **Hydrodictyon** *in the light*

Concentration of CCCP (μM)	Ion flux relative to that in the absence of uncoupler (%)		
	Cl^- influx	K^+ influx	Na^+ efflux
0	100	100	100
1	107	73	68
5	106	66	35
10	88	45	38

Note that the Cl^- influx is relatively insensitive in comparison with the cation fluxes. (Data from Raven, 1967b.)

In some organisms uncouplers actually increase the chloride influx since photophosphorylation normally imposes a rate limitation on electron-flow. The implication of this experiment is that the cation fluxes are dependent on the ATP synthesized during electron-flow, but that chloride is not. The dependence of the chloride flux on electron-flow between photosystems II and I is demonstrated by the inhibitory effect of DCMU; note that the cation fluxes are relatively insensitive to this inhibitor because they can be supported by cyclic electron-flow.

This elegant detective work by MacRobbie and by Raven indicates that there might be quite fundamental differences in the nature of cation and anion pumps. Both photosynthesis and respiration produce ATP needed to support cation pumps and, as we shall see, considerable progress has been made in understanding how the ATP is utilized. The transport of anions, and particularly of chloride, is more problematical and the mechanisms are still obscure. The results discussed above illustrate one of these puzzling features. Electron-flows between photosystem II and I and cyclic flow around photosystem I both involve common intermediates of the electron transport chain. The inhibitor experiments indicate that chloride influx depends on electron-flow, or at least some product other than ATP. Now, electron-flow occurs when photosystem I is activated by light on its own. The unanswered question is, therefore, 'What is so special about non-cyclic electron-flow?' Perhaps one explanation of the difference between cyclic and non-cyclic electron-flow might be forthcoming if it were found that although common intermediates were involved the transport chains were actually separated physically from one another in different parts of the chloroplast. Another approach to the nature of anion pumps comes from considerations of charge separation during electron-flow in photosynthesis which would provide an electrochemical gradient directed into the chloroplasts down which anions might move. As we shall see, in the discussion of this topic later in the chapter, one of the current difficulties in this hypothesis is understanding how events occurring across the membranes of organelles can be communicated to the primary barrier to anion entry into the cell, the plasmalemma.

The apparently separate energy requirements for pumping chloride and cations in *Nitella* and *Hydrodictyon* are not characteristic of all organisms. Using similar experimental approaches, studies of *Chlorella pyrenoidosa* (a unicellular green alga), showed that chloride influx was as sensitive to the action of uncouplers as cation fluxes (Barber, 1968). Further conflicting results have been obtained in the aquatic higher plant, *Elodea canadensis*, in which light-stimulated chloride transport is supported by cyclic photophosphorylation (Jeshke, 1967). A systematic study by Lüttge, Ball and Von Willert (1971) showed that active chloride transport was actually stimulated, when photosynthesis was uncoupled by FCCP, in strips of leaf tissue from *Atriplex spongiosus*, *Zea mays* and *Amaranthus caudatus*; in all of these species assimilated CO_2 follows the C4 pathway of photosynthesis (CO_2, first fixed in phosphoenolpyruvate to form oxaloacetic acid, Hatch and Slack, 1970). In leaf strips from

Atriplex hastata, Spinacia oleracea and *Oenothera* sp. both chloride transport and photosynthesis were inhibited to about the same extent by the uncoupler; this latter group of plants is characterized by the more common Calvin cycle of CO_2-fixation. Thus, in the former group as in *Nitella translucens* (MacRobbie, 1965) and *Hydrodictyon africanum*, active chloride transport does not depend directly on ATP or some other intermediate of photophosphorylation, while in the latter groups ATP is essential. The exact significance of this distinction between C_4-syndrome plants and the C_3 plants (those following the Calvin cycle) is not yet clear but it is interesting that C_4 plants are characterized by rapid translocation of products of photosynthesis between cells and organelles. This property might make it possible to move some other product of electron-transport rapidly to the plasmalemma which could be used in the active transport mechanism.

While the work of Lüttge and his co-workers was performed with great care, it should be stressed that in many experiments with metabolic inhibitors insufficient precautions are taken to ensure that the inhibitor really does the job it is said to do. Smith and West (1969) found that several uncouplers at low concentrations substantially inhibited ion uptake by *Chara australis* without having any perceptible effect either on ATP synthesis or electron-flow. In these experiments and possibly in other published work, the 'uncouplers' gave no indication of having 'uncoupled' anything and their effects on the plasmalemma may have been mediated more directly. Experiments, the interpretation of which depends entirely on the presumed effects of metabolic inhibitors, should always be approached with caution. In the work on *Nitella* and *Hydrodictyon*, the conclusions drawn from the wavelength dependence of active transport of chloride are not subject to this criticism.

The movement of other anions, e.g., NO_3^-, $H_2PO_4^-$, into plant cells has been less extensively studied although their transport is metabolically-dependent; active phosphate transport in *Nitella translucens* was studied by Smith (1966) but no mechanism was put forward. This represents a major gap in our knowledge since it is clear from our examination of membrane potentials and electrochemical activity gradients (page 62) that most of the energy expended on active transport is devoted to moving anions into cells. The omission is greatest with respect to plant roots which, in non-saline environments, encounter nitrate, sulphate and phosphate as the principal anions, chloride being present in trace amounts only.

Evidence for coupling of K and Na fluxes. The data in Table 5.4 show that any change in the rate of light-stimulated influx of K^+ was closely paralleled in magnitude by changes in the efflux of Na^+. The impression that the two fluxes may be linked in some way is strengthened by the results of experiments using the inhibitor *ouabain*. This inhibitor interferes with active K^+ influx, not by inhibiting ATP synthesis (cf. CCCP) but by mimicking part of the ATP molecule, becoming bound to the pumping mechanism and literally 'jamming' it up. Jamming the K^+ influx with ouabain produced a proportionately similar reduction in Na^+ efflux (Table 5.4).

***Table 5.4 Influence of 0·5 mM ouabain on active and passive fluxes of potassium and sodium in* Hydrodictyon africanum**

Flux	Control	+ Ouabain	% inhibition
	pmoles cm^{-2} s^{-1}		
K^+_{in} Light	1·53	0·56	63
Na^+_{out} Light	0·76	0·21	72
K^+_{out} Light	1·80	2·30	0 (+ve)

Data for Na^+ influx in the light in presence of 0·5 mM ouabain not available. (Data from Raven, 1967a.)

Neither K^+ efflux not Na^+ influx, both passive diffusion processes in response to electrochemical activity gradients, were influenced significantly by ouabain. This result on its own does not prove that the two active fluxes are coupled; they could be driven by separate mechanisms which have similar sensitivity to the inhibitor. However, the fact that the Na^+ efflux depends on the presence of K^+ in the external medium would be difficult to explain by such a suggestion; the evidence in Table 5.4 supports, therefore, the idea that the movements of these two ions are truly coupled.

In plant cells there are a growing number of examples of K/Na pumps of this kind, e.g., the Na^+ extrusion by *Porphyra* (a red alga) is dependent on external K^+ in the medium, and other coenocytic algae have similar light-stimulated coupling of Na^+ and K^+ fluxes (see Gutknecht and Dainty, 1968 for review). It is, however, from animal cells that we have details of the mechanism by which this coupling may be brought about and we shall now consider the best known example of such a pump in mammalian red blood cells.

Na/K transport by red blood cells. The plasma in which red blood cells are bathed resembles sea water in that its concentration of sodium greatly exceeds that of potassium. The red blood cells resemble organisms living in the sea since their internal sodium is low in relation to their internal potassium concentration. The sodium concentration inside the cell is lower than would be expected from its electrochemical potential and is therefore pumped out of the cell by an active transport process. It is possible to rupture red blood cells by osmotic shock so that most of their contents are lost, and then to reform the membrane by restoring the medium to its usual osmolarity. Whittam (1962) showed that these reformed membranes, which are known as 'ghosts', possess the same capacity for sodium extrusion as the intact cells and thus provide a very convenient experimental material. First of all one can look at the ghost itself, analyse what remains of its structure and biochemical composition. Secondly, the experimenter can reform the ghosts in a medium of defined and controllable composition thus being in a position to determine the environment both inside and outside the cell. (A similar, but much more complex situation can be achieved by perfusion of the vacuole of giant algal cells, see page 94.)

Investigations by Whittam and by many others reveal that the enzyme ATPase is bound to the membranes of the ghost; the activity of this enzyme determines the activity of a pump which moves sodium in an outward direction and potassium in an inward direction. The substrate for the enzyme is ATP which in the intact cell is supplied by respiration, but which must be supplied to the ghost by the experimenter. Omission of ATP from the medium in which the ghosts are reformed stops the pump and this omission cannot be made good by subsequently supplying ATP to the outside medium. The ATPase can accept its substrate only from the inside of the cell and must therefore be orientated with its active centre facing inwards. The ATPase has an obligatory requirement for Na^+ and Mg^{2+} on the inside and K^+ on the outside and will not work if these conditions are not provided.

At this point, we will recall the structure of the membrane revealed by freeze-etching in the electron microscope (Fig. 2.18). The particles which we saw embedded in the membrane lipid matrix are largely composed of protein and it is tempting to imagine that at least some of them contain the ATPase and hence are the sites of ion pumping. Digestion of these particles with a proteolytic enzyme eliminates the ion pumping capacity of the membrane.

The principal features of the operation of the pump are generally agreed to be as follows. ATP and Na ions are bound to the active site of ATPase in the presence of Mg ions; the terminal phosphate bond of ATP is hydrolysed to produce ADP and a phosphorylated form of the enzyme, a *phosphoenzyme*, which undergoes a conformational change exposing the site(s) to which sodium is bound to the external medium where the sodium is exchanged for potassium; a second conformational change results from the binding of potassium which exposes the bound potassium to the internal environment where conditions are such that it is released along with the phosphorus atom. The active centre of the enzyme is now dephosphorylated and is ready to accept another molecule of ATP and to repeat the cycle.

It is apparent from Fig. 5.8 that the actual details of what happens are somewhat more complex than the above description. As we have described it the reader may get the impression that there is an exchange of sodium for potassium in exact equivalents. In practice this was not found to be the case; extensive investigations by Whittam and Ager (1965) showed that the ratio $Na_{efflux}:K_{influx}$ was

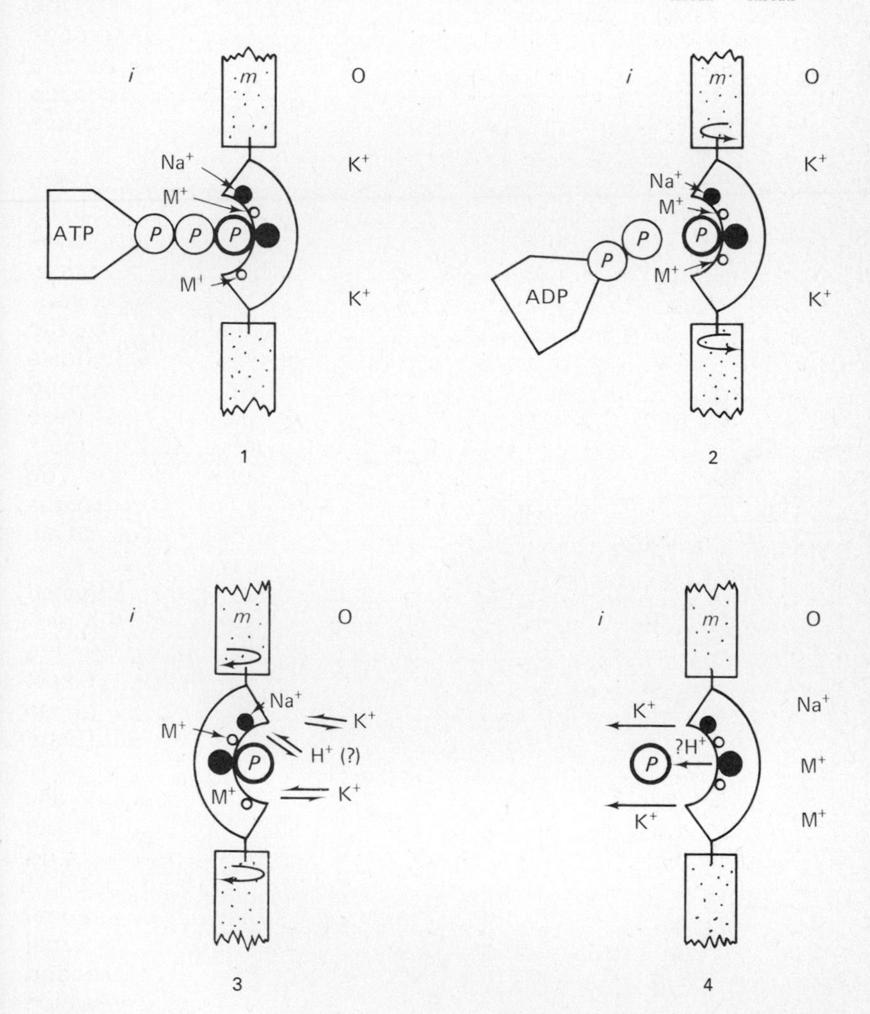

***FIG. 5.8** A representation of the mechanism of the NA/K pump in the red blood cell membrane. ... = active centre of ATPase; ● = site specific for Na^+ from inside the cell; ○ = site accepting monovalent cations; i, m and o represent outside, membrane and inside respectively. (Based on a hypothesis of Middleton, 1970.)*

about 1·6 over a wide range of Na/K concentrations. The scheme used in Fig. 5.8 is derived from kinetic analyses of ion binding of ATPase by Middleton (1970) which indicate that ATPase has three binding sites at the inside of the membrane, one of which is highly specific for sodium and two of which may be occupied by either sodium or potassium; these latter sites are distinguished by binding the unspecified ion M^+. The conformational change, which makes the active site accessible to K from the outside, also alters the selectivity of the binding site so that two of them become highly specific for potassium. The ratio of $Na_{efflux} : K_{influx}$ will thus vary according to whether the non-specific sites at the inside of the membrane are occupied by Na^+ or K^+ ions.

The proposed model indicates that the ions transported are those which (a) effect the binding of the substrate to the active centre of ATPase and (b) effect the release of P_i from the phosphoenzyme. Na^+ and K^+ are therefore essential cofactors in an enzyme reaction. A whole range of ion pumps can be envisaged which embody this principle, although few can be described in detail. Mitchell (1961) pointed out that all enzymes which are structurally orientated in a membrane will have the property of accepting reactants and cofactors from certain directions. Ion pumping is, therefore, merely a special instance of the vectorial component in metabolism. We can illustrate this by reference to the inner mitochondrial membrane where electron and hydrogen carriers are arranged in an orderly sequence; an electron actually moves in a fixed direction from the point where reduced NAD is oxidized by flavoprotein to the point where oxygen finally accepts the electron. Transport processes across membranes can be thought of in the same way except that the main vector of the transported entity is across the membrane rather than along its surface.

It is most important to stress that the nature of the conformational change, which alternately exposes the active centre of the phosphoenzyme to interior and exterior environments, is quite unknown. Figure 5.8 almost certainly does not represent what actually happens during ion pumping. It is unlikely that major movements of the protein occur within the lipid matrix of the membrane. Quite subtle changes in the orientation of the active site relative to the membrane matrix may be all that is required to alter the accessibility of the sites. It is worth recalling in this connection that the protein sub-units embedded in the membrane are wider than the hydrophobic interior so that the sub-unit might represent a hydrophilic core, or pathway, running across the membrane along which ions could diffuse until they reach a critical part of the pump which prevents their further movement.

Transport ATPases in other organisms. There are numerous well documented sodium extrusion pumps in animal cells and micro-organisms which are operated by membrane-bound ATPase. Not all of them are as tightly coupled with K^+ influx as in the red blood cell; indeed, in micro-organisms the choice of counter ion is quite catholic and ammonium, hydrogen and amino cations have all been shown, in different organisms, to partner sodium.

In plant cells evidence for transport ATPase systems, particularly those associated with the plasmalemma, is much less abundant and critical biochemical studies are generally lacking. Dodds and Ellis (1966) found that cell wall preparations contaminated by plasmalemmata from *Zea mays* possessed cation-stimulated ATPase activity but that, unlike most transport ATPase from animal tissue, it was insensitive to the action of ouabain. Fisher and Hodges (1969) estimated that there was sufficient cation-stimulated ATPase activity associated with membrane fractions from the oat root (*Avena sativa*) to account for the observed K^+ movement into the root. Even though it is uncertain how much of this ATPase is associated with the plasmalemma, this result indicates that there may be a considerable excess of ATPase over requirements for active K^+ transport; electrochemical considerations suggest that, in plant roots, little active K^+ transport is required to account for the maintenance of the internal K^+ concentration (see chapter 7, page 191).

More recently Hodges *et al.* (1972) have shown for the first time that purified membrane fractions from oat roots, consisting largely of plasmalemma fragments, contain appreciable amounts of ATPase, the activity of which is stimulated by monovalent cations. The membrane fragments had a high cholesterol:phospholipid ratio which, in animal cells (Emmelot *et al.* 1964), is characteristic of the plasmalemma.

Histochemical tests also indicate that the plasmalemma of cells may have considerable ATPase activity. In the root tip of *Zea mays*, Hall (1969) showed that the activity of this ATPase is affected by the general salt status of the cells and by the ionic composition of the external medium; it is, however, not certain whether the observed ATPase is actually concerned with ion transport. Rather similar kinds of evidence have been advanced by other authors, e.g., McClurkin and McClurkin (1967). Biochemical studies are complicated in plant material by the presence of the cell wall which makes the separation of membranes for examination difficult. A useful approach to this problem is suggested by the studies of Mayo and Cocking (1969) who prepared naked protoplasts from cells in tomato fruits by digesting the cell walls away with a fungal cellulase preparation. If the protoplasts could be lysed and then reformed, a system analogous to the red blood cell ghosts would be available.

Charge separation and ion transport

The Na/K–ATPase in red blood cells shows how one of the principal products of metabolism can be utilized in ion pumping. It is certainly a convenient way of harnessing the free energy conserved from the excitation of chlorophyll by photons and the degradation of respiratory substrates. The conserved energy will be available to mechanisms in the cell outside the immediate orbit of the organelles in which ATP is formed, since ATP can diffuse to the sites where it is required. In this section we shall consider a more general feature of electron-flow in mitochondria and chloroplasts which can provide gradients of potential down which ions can move across membranes.

Figure 5.9 makes a point about the movement of electrons in photosynthesis and respiration which was not emphasized in our preliminary discussion of these processes. The molecules which make up the electron transport chain in both processes are of two types both of which can undergo reduction-oxidation (redox) reactions. The first type, hydrogen carriers, are reduced by the acquisition of two hydrogen atoms and are oxidized when they lose them; plastoquinone, the first molecule of the photosynthetic electron transfer

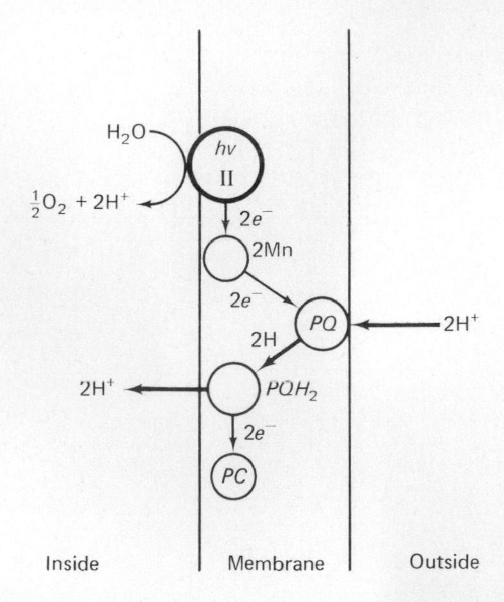

***FIG. 5.9** Possible mechanism transporting protons across a chloroplast membrane. The reduction of Plastoquinone (PQ) requires two electrons, received from Mn^{2+}, and two protons which are accepted only from the outside. Oxidation of PQ occurs at the inner face of the membrane; two protons released to the inside and two electrons passed to plastocyanin (PC). (Based on Mitchell, 1966.)*

chain which is actually bound to the lamellae of the chloroplasts, is of this type, as are flavoprotein, ubiquinone, etc. The second class of molecules have a metal atom of variable valency in their structure. The acquisition of an electron reduces, and the loss of an electron oxidizes, the carrier molecule; manganese, iron and copper atoms are involved in redox reactions of this type.

Let us now consider the photosynthetic electron transport chain in somewhat greater detail. Electrons liberated from photosystem II are passed *via* manganese ions to plastoquinone. The redox reaction with manganese involves a simple valency change but the next step in the process, the reduction of plastoquinone, requires, in addition to the two electrons it receives from manganous ions (Mn^{++}), two protons (H^+) to form two hydrogen atoms; the reaction is therefore

$$2Mn^{2+} + PQ + 2H^+ \longrightarrow 2Mn^{2+} + PQH_2$$

The next reaction in the chain involves the reduction of the copper containing molecule, plastocyanin, which requires only electrons to reduce cupric (Cu^{3+}) copper to cuprous (Cu^{2+}). The two protons from plastoquinone are released and resume their life as dissociated ions:

$$PQH_2 + Cu^{3+} \text{ (Plastocyanin)} \longrightarrow PQ + 2H^+ + 2Cu^{2+} \text{ (Plastocyanin)}$$

Having established the fact that in redox reactions protons are alternately bound to and released from hydrogen carriers we shall now make what may seem to be a digression.

It was suggested many years ago by Lundegårdh (1939) that the transport of electrons in respiration and the transport of ions into cells were intimately related since treatment of tissues with potassium cyanide inhibited both processes to a similar extent; it had also been found that respiration of tissue slices could be stimulated by supplying salts to the external medium. This process, known as *salt respiration*, was a major preoccupation of earlier workers on ion transport. Lundegårdh suggested that, when cytochrome oxidase was prevented by cyanide from passing electrons through to oxygen, the release of protons from the cell in exchange for other cations was also inhibited. Central to Lundegårdh's hypothesis was the idea that the separation of charge in respiratory electron-flow could provide the driving force for ion accumulation by the cell. More recent studies have shown that this view is substantially correct although Lundegårdh's suggestions about how the process might occur are no longer regarded as probable. The research of Robertson (1968) and Mitchell (1966) has given us a greater understanding of how the process of charge separation is related to ion movement, at least at the level of intracellular organelles. The most important advance has come with an appreciation of the role of membrane systems both as supporting matrices for electron-transport chains and as barriers across which protons and electrons may be effectively separated.

Figure 5.10 shows in a more general way how charge separation can occur during the redox reactions of electron-flow. For the system to work effectively there are two important provisos. The first is that the space marked in the figures as 'inside' should be a compartment completely bounded by membrane. The second is that hydrogen carriers like plastoquinone should accept protons from one side of the membrane and discharge them at the other. We may think of this process operating in a similar way to the conformational changes in the Na/K–ATPase brought about by the binding and exchange of ions to the active centre of the phosphoenzyme. In the case of plastoquinone, we can suppose that reduction of the molecule alters its relationship with its surrounding chemical environment in such a way as to change its orientation in the membrane. In its second orientation reduced plastoquinone loses electrons to plastocyanin and releases two protons into the internal space. The oxidized plastoquinone then reverts to its initial orientation and is ready for another cycle. Plastoquinone under these conditions is therefore operating as a pump for hydrogen ions which are derived from the ionization of H_2O on the outside of the membrane.

The operation of such a pump will have two results; first the hydrogen ion concentration of the interior space will increase, i.e., it will become more acid, and second, if the membrane were not permeable to cations moving in an outward direction or anions moving inwards, a large electric potential difference between the two compartments would quickly build up to a dangerous level and eventually break down the membrane. Since the latter situation clearly does not occur during either photosynthesis or respiration we should enquire why not.

Proton extrusion by mitochondria. The theory of proton transport which I have outlined was prompted by observations on mitochondria which are found to make the medium in which they were suspended more acid when supplied with respiratory substrates and oxygen. Using mitochondria from rat liver, Mitchell (1961) found that two substrates, β-hydroxybutyrate and succinate resulted in the formation of different amounts of acid in the external medium. The reason for this can be seen if we re-examine Fig. 5.3 which shows that succinate additions to the external medium will bypass part of the Krebs cycle and that hydrogen atoms from succinate pass directly to flavoprotein missing out NAD. By contrast the oxidation of β-hydroxybutyrate involves the whole of the Krebs cycle and hydrogen atoms are accepted by NAD. The scheme for electron-flow

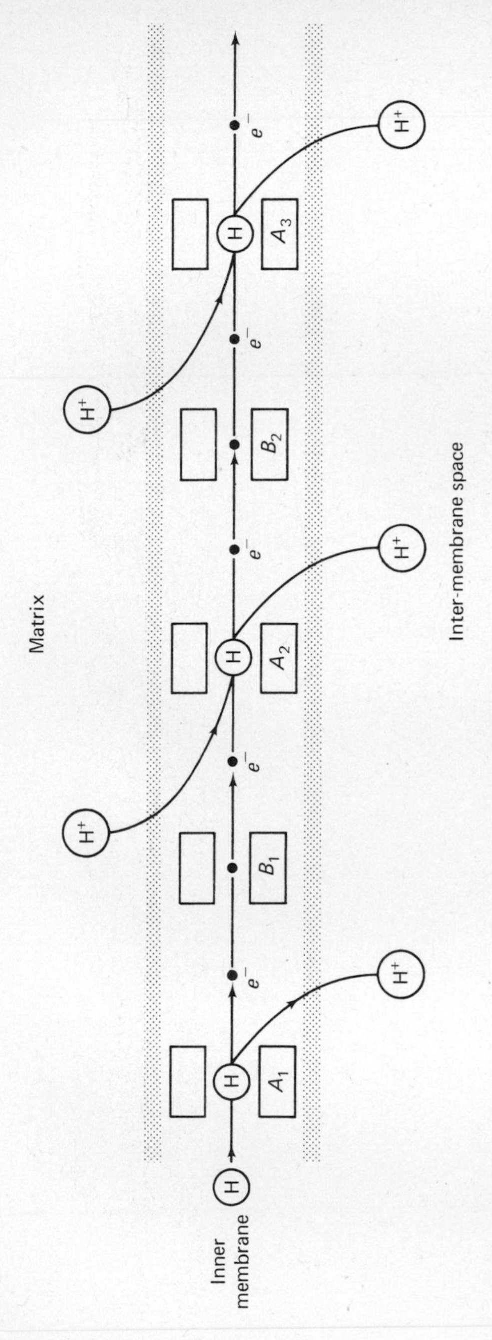

***FIG. 5.10** General scheme to show how alternating hydrogen and electron-carriers may 'pump' protons across a mitochondrial membrane.* H^+ *= proton;* e^- *= electron; H = molecular hydrogen;* A_{1-3} *= hydrogen carriers;* $B_{1,2}$ *= electron carriers.*

(Fig. 5.4) shows that, in passing hydrogen atoms directly to flavoprotein, the oxidation of succinate misses out one of the stages where a reduced hydrogen carrier passes on electrons to an oxidized electron carrier. These steps are presumed to be the sites at which protons are extruded, thus the oxidation of β-hydroxybutyrate, which involves three of these steps, results in the extrusion of six protons per oxygen atom reduced (i.e., two protons per site, as illustrated) while the oxidation of succinate results in the extrusion of only four protons per oxygen atom reduced.

Proton influx into chloroplasts. The hydrogen ion activity gradient (pH gradient) created during electron-flow in photosynthesis is in the opposite direction to that formed during respiratory electron-flow. Unbuffered medium in which chloroplasts are suspended rapidly becomes more alkaline on illumination but tends to reach a steady pH value after 30 seconds (Fig. 5.11). When the light is turned off the pH of the medium slowly returns to its initial value suggesting that protons accumulated inside the chloroplasts (or more usually chloroplast fragments) can diffuse back into the outer medium. If some back flow of protons did not occur then the interior would not only build up a dangerous excess of positive charge, but would also become so acid that all the proteins inside the chloroplast would begin to denature. To avoid the consequences of

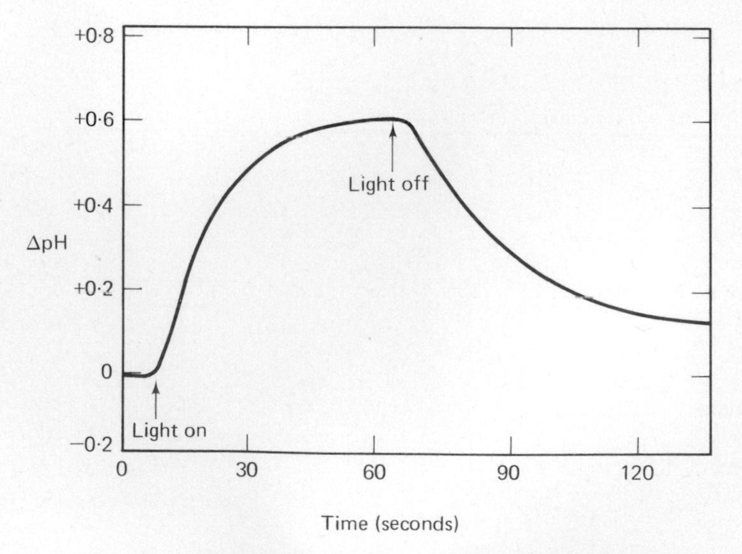

FIG. 5.11 The effect of light on the pH of an unbuffered suspension of chloroplasts. The upward deflection indicates increasing alkalinity in the medium. (Redrawn from Neumann and Jagendorf, 1965.)

excessive charge separation the entry of protons must either be accompanied by a coupled movement of anions in the same direction, or by a counter flow of cations into the outer medium. The latter process appears to operate and it was shown that the sum of the cation equivalents released from chloroplasts almost completely neutralized the charge on the protons taken up (Dilley and Vernon, 1965). The cations released were not exclusively of one ionic species; depending on pretreatment, either sodium or potassium or a mixture of them was released, both ions being accompanied invariably by magnesium. The equivalence of cation efflux and proton influx does, of course, give rise to speculations of the 'chicken and the egg' type. Could it not be that the observed proton influx was actually a response to cation pumping *out* of the chloroplast? This suggestion would be more plausible if only one ion species were involved in the exchange for protons but, as we have seen, the specificity of the extruded cations is not very high. Ion pumping mechanisms, on the other hand, are generally found to be highly specific for the ions which they handle. The results also show that there is more proton movement than is balanced by any single cation species. A further point against the alternative suggestion is that the pH rise in the external medium slightly precedes the first detectable arrival of exchanged cation.

The pH gradient will not, of course, be affected by the exchange of protons for monovalent cations and some interesting calculations by Jagendorf and Uribe (1967) indicate what the pH might be inside the chloroplast compartments. When illuminated chloroplasts were titrated with dilute acid to keep the suspension at its initial pH value it was found that 0·65 μmole of H^+ had been taken up per milligram of chlorophyll present. Using assumptions which we will not go into, they calculated that the volume of the membrane-bound compartments in their chloroplasts might amount to 0·2 ml per mg chlorophyll. The influx of 0·65 μmoles of H^+ into such a small volume would result in a hydrogen ion concentration of 3·3 mM giving a pH of 2·5. It seems unlikely that the pH can be permitted to fall to this low level without seriously affecting the activities of enzymes within the compartment and the permeability properties of the bounding membrane. Macromolecules bearing fixed negative charges may bind some of the protons, but even if 90 per cent of the protons were removed from free solution the pH of the compartment would still be 3·5. There is, however, a complementary series of reactions accompanying electron transport which may serve to reduce further the steep pH gradient between the outside medium and the interior of the organelles.

Charge separation and phosphorylation. In the earlier accounts of photosynthesis and respiration we learnt that ATP synthesis was coupled with electron-flow although we did not learn how the coupling was brought about. The synthesis of ATP is an energy-consuming process and in earlier attempts to explain the link between electron transport and phosphorylation an intermediate molecule was postulated which was part of, or was intimately associated with, the familiar chain of carrier molecules. Extensive searches by the most capable investigators have failed to identify or isolate this unknown molecule, which is usually referred to as ~X, although some workers still have faith in its existence. An alternative explanation of phosphorylation was put forward by Mitchell (1961, 1966), which has the great virtue of unifying several well documented phenomena in a theory which can be tested; this has become known as the 'chemi-osmotic' hypothesis.

Stated most simply, the theory predicts that the charge separation occurring during electron-transport sets up a pH gradient across the membrane, the potential energy of which is utilized in the phosphorylation of ADP. The reaction is activated by a membrane-bound ATPase which is demonstrably present in mitochondria and chloroplasts. (Our previous discussion of an ATPase showed it activating the hydrolysis of ATP, but under appropriate conditions the reaction, like that of all enzymes, can be reversed; the proton gradient provides these conditions.)

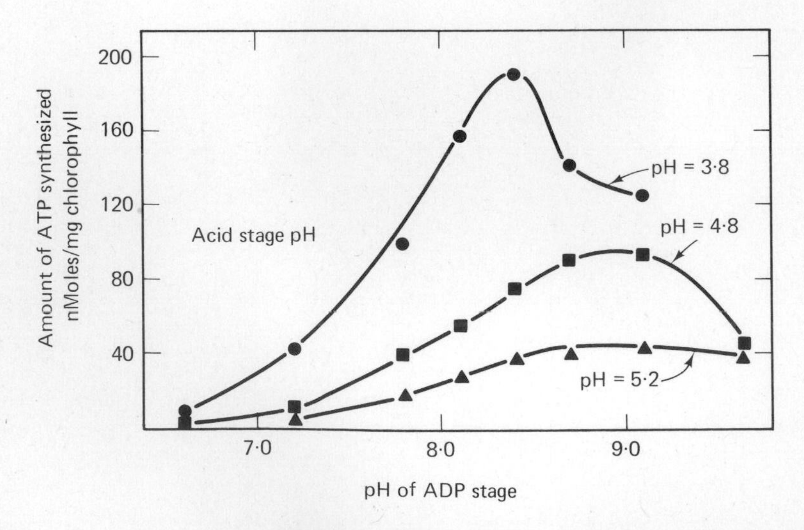

FIG. 5.12 Influence of the pH differential on ATP synthesized by chloroplasts in darkness. Chloroplasts equilibrated in acid media, pH 3·8–5·2, and then transferred to media of pH 6·5–10. Note that there seems to be an optimum pH differential between the acid and the phosphorylating stages. (Redrawn from Jagendorf and Uribe, 1966, with kind permission.)

The crucial test of this hypothesis is whether the presence of a pH gradient across a membrane can result in the synthesis of ATP. Experimentally it is necessary to prevent the electron-flow which normally creates the proton gradient and substitute for it a contrived set of conditions in which the organelles are pre-equilibrated in a solution at one pH and then quickly transferred to another with a markedly different pH value. It is much easier to do this in chloroplasts where normal electron-flow can be eliminated if they are kept in the dark. Hind and Jagendorf (1965) pretreated chloroplasts in the dark with an acid medium and then transferred them, still in darkness, into a medium buffered at pH 8·0 which contained ADP and inorganic phosphate. This acid–base transition resulted in the synthesis of a substantial amount of ATP. The only potential energy available for this synthesis came from the pH gradient. Jagendorf and Uribe (1966) found that the amount of ATP formed depended on the pH differential between the pretreatment and ADP containing media (Fig. 5.12). A spectacular reversal of the normal sequence of electron transport in isolated chloroplasts was achieved by Mayne and Clayton (1966) when an acid–base transition actually caused chlorophyll fluorescence in the dark.

The more difficult demonstration of ATP synthesis in response to a pH gradient was made by Reid, Moyle and Mitchell (1966) when mitochondria were suspended in a medium lacking respirable substrate, which was changed rapidly from pH 9 to pH 4·3.

Uncouplers and the chemi-osmotic hypothesis. Results from experiments with uncouplers also support the contention that the energy for phosphorylation is derived from charge separation and pH gradients. A number of molecules, some of them quite unrelated chemically, have the property of inhibiting the synthesis of ATP while permitting electron-flow to continue. All of these molecules have the common property of increasing the passive proton conductance both of biological membranes and of synthetic lipid bilayer membranes (see chapter 4, page 109). Proton pumping during electron-flow will continue in the presence of an uncoupler but the leakiness which is induced in the membrane does not permit the establishment of sufficiently large potential energy gradient (pH gradient) to drive the synthesis of ATP. Uncouplers speed up the flow of electrons through the transport chain; this can often be detected by a surge in oxygen uptake by tissues when uncouplers are introduced into the bathing medium. Without the formation of ATP, however, the free energy of the respiratory substrates runs to waste.

The more familiar types of uncoupler, e.g., 2,4-dinitrophenol (DNP) and CCCP increase the proton conductance of the membrane by interacting with the 'head' regions of positively charged phospholipids in the membrane matrix. Another way in which the membrane can be made leaky is to perforate it; this appears to be the way in which certain detergents, such as the widely used Triton-X, uncouple photophosphorylation in isolated chloroplasts (Neumann and Jagendorf, 1965). The addition of Triton-X to a suspension of chloroplasts which have established a pH gradient in the light and are then returned to darkness greatly accelerates the decline of the gradient and subsequently diminishes the steady state pH when the light is switched on again (Fig. 5.13). In the presence of Triton, very little ATP is formed during the second illumination.

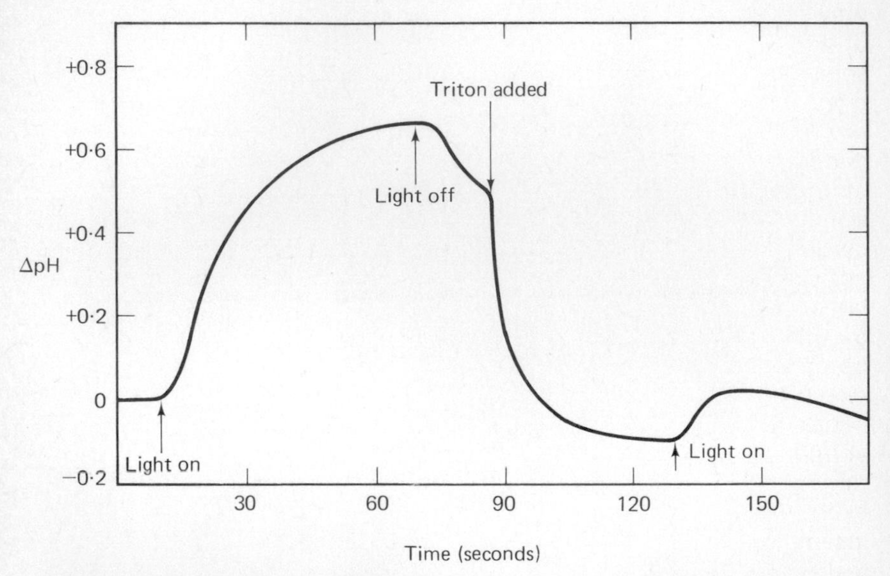

FIG. 5.13 Effect of Triton X-100 on pH changes in an unbuffered suspension of chloroplasts. The detergent speeds up the return of the pH to its original dark value and largely eliminates the pH rise during the subsequent illumination. (Redrawn from Neumann and Jagendorf, 1965, with kind permission.)

Charge separation and the transport of ions at the plasmalemma. Charge separation in organelles can bring about the counter movement of ions as we have seen, and it also provides the energy for the synthesis of ATP which is used in some of the ion pumps which operate at the plasmalemma. It is, therefore, a fundamental process in ion transport but certain formidable difficulties remain in explaining exactly how these processes in organelles control the events which are known to occur at the plasmalemma. In those instances where active transport does not seem to be dependent on direct utilization of ATP, e.g., light-stimulated chloride influx in *Hydrodictyon* and *Nitella* and the leaves of C_4-plants, the problems are particularly acute. It should be remembered that the plasmalemma does not have the components of the electron-transport chain associated with it. Charge separation of the kind envisaged by the chemi-osmotic model probably does not occur although there are reports of proton extrusion pumps at the plasmalemma. Chloroplasts and mitochondria are separated from the plasmalemma by an appreciable thickness of highly buffered cytoplasm. It is improbable, therefore, that the protons extruded by the mitochondria will exert much influence on the pH of the whole cytoplasm and it is therefore unlikely that a pH gradient across the plasmalemma can be used in quite the same direct way as in mitochondria. But there are undoubtedly ways, as yet undiscovered, in which charge separated in organelles is kept separate in the extra mitochondrial environment. Robertson (1968) cites the example of gastric mucosa in which positive charge (as protons) and negative charge (as bicarbonate) separated in mitochondria are actually pumped out from opposite sides of this epithelial tissue; the protons pass into the stomach and the bicarbonate into the blood stream. Robertson also points out that the cytoplasm often contains large numbers of small membrane-bound vesicles which can sometimes appear both to fuse with other, and larger membrane systems, especially the endoplasmic reticulum and the plasmalemma, and to be derived from them. These vesicles can be moved about the cell quite rapidly by cyclosis and protoplasmic streaming and it is by no means improbable that a small vesicle whose contents are enriched with protons recently extruded from a mitochondrion might be moved to the plasmalemma where its contents might be discharged in exchange for other cations and anions. This exchange of protons for nutritionally useful cations essentially resembles the early picture drawn out by Lundegårdh; what is now realized is that electron transport is restricted to organelles and is not a general feature of the cytoplasm itself.

It must be admitted that the evidence for vesicles running a shuttle service between various cellular compartments is all circumstantial but the process, as we have described it above, is essentially the reverse of pinocytosis (see page 49), the existence of which is well documented, and it has much in common with the movement of vesicles of wall building precursors which are cut off from the golgi apparatus, subsequently move to the developing cell wall and, of necessity, must cross the plasmalemma, or at least discharge their contents through it (see Heyn, 1971).

The concept of the cytoplasm and the membranes as a dynamic system is an important advance in our way of thinking about ion transport and there are experimental results which are difficult to explain in any other way than a kind of pinocytotic system. MacRobbie (1969) working with the related algal coenocytes *Nitella translucens* and *Tolypella intricata* found that the influx of chloride across the tonoplast into the vacuole could be resolved quite distinctly into a fast and a slow component. Chloride was transported into the vacuole by the fast component much more rapidly than could be the case if chloride entering the cell mixed with the chloride in the whole volume of the cytoplasm. The flux into the vacuole depended quantitatively on the flux at the plasmalemma in such a way that the results could be interpreted only by the entry of salt in pinocytotic vesicles at the plasmalemma. These vesicles are envisaged as either moving directly to the tonoplast, discharging their contents into the vacuole or moving to the endoplasmic reticulum. Transport across the cytoplasm through the cavity in the ER is probably rapid and tracer would enter the vacuole by subsequent vasicularization of the ER membrane and fusion with the tonoplast. In either system, there would be exchange of chloride between the vesicles and the bulk

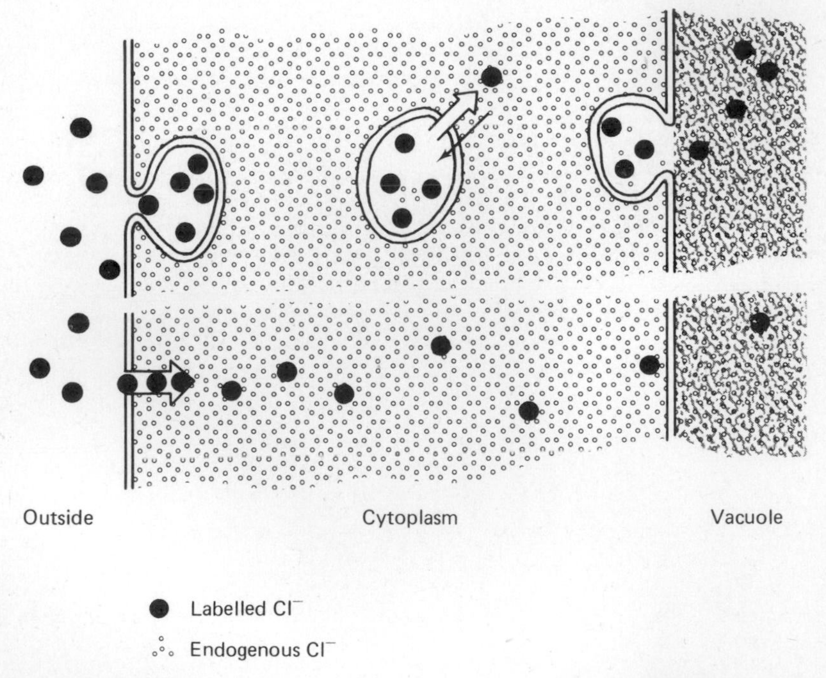

FIG. 5.14 Scheme comparing the Cl^- flux into the vacuole of Nitella *by way of vesicular transport and active pumping into the general cytoplasmic volume. Contents of the vesicles may exchange some labelled chloride with the bulk cytoplasm, but most of the labelled chloride is delivered to the vacuole. Both processes would operate at any time.*

cytoplasm during transit. Of the two schemes suggested, the direct movement of the vesicles to the vacuole is to be preferred since it makes fewer assumptions. I have attempted to illustrate this process in Fig. 5.14. It can be seen that labelled chloride ions entering by pinocytic vesicles will reach the vacuole in greater numbers and more rapidly than labelled chloride discharged directly into the bulk cytoplasm.

MacRobbie's data suggests that chloride ions will arrive in the vacuole in discrete batches; in other words the flux will be quantized. More recently Findlay, Hope and Walker (1971) have strongly disagreed with this suggestion finding no evidence of quantized fluxes in their experiments with *Chara australis*. Whether or not the flux is quantized seems to be much less important than the conceptual advance which MacRobbie's work has made. It is in line with much of the recent thinking about the dynamic processes in membrane synthesis and the transfer of membrane building blocks from one membrane system to another (see Franke *et al.*, 1971).

Bibliography

a Further reading

Robertson, R. N. (1967) The separation of protons and electrons as a fundamental biological process. *Endeavour* **26**, 234–239.

This simple account of charge separation and the chemi-osmotic hypothesis should be easily understood by any student and is thoroughly recommended.

Robertson, R. N. (1968) *Protons, Electrons, Phosphorylation and Active Transport.* Cambridge University Press.

This short book expands and thoroughly documents the basic hypothesis put forward in the article above.

MacRobbie, E. A. C. (1970) The active transport of ions in plant cells. *Quart. Rev. Biophys.* **3**, 251–294.

A very thorough and readable review, including an excellent account of the author's own opinions on dynamic features of transport processes such as pinocytosis.

Jagendorf, A. T. and Uribe, E. (1967) Photophosphorylation and the chemiosmotic hypothesis. In *Energy Conversion by the Photosynthetic Apparatus*, 215–241. *Brookhaven Symp. Biol.* **19**.

A very lucid account of photophosphorylation and its relationship with proton pumping.

b References in the text

Barber, J. (1968) Light induced uptake of potassium and chloride by *Chlorella pyrenoidosa*. *Nature* **217**, 876–878.

Dilley, R. A. and Vernon, L. P. (1965) Ion and water transport processes related to light dependent shrinkage of spinach chloroplasts. *Arch. Biochem. Biophys.* **111**, 365–375.

Dodds, J. J. A. and Ellis, R. J. (1966) Cation stimulated adenosine triphosphatose activity in plant cell walls. *Biochem. J.* **101**, 31–32.

Emmelot, P., Bos, C. J., Benedetti, E. L. and Rumke, P. L. (1964) Studies on plasma membranes. I. Chemical composition and enzyme content of plasma membranes isolated from rat liver. *Biochim. Biophys. Acta* **90**, 126–145.

Findlay, G. P., Hope, A. B. and Walker, N. A. (1971) Quantization of a flux ratio in chlorophytes? *Biochim. Biophys. Acta* **223**, 155–162.

Fisher, J. and Hodges, T. K. (1969) Monovalent ion-stimulated ATPase from oat roots. *Plant Physiol.* **44**, 385–390.

Franke, W. W., Morre, D. J., Deumling, B., Cheetham, R. D., Kartenbeck, J., Jarasch, E. D. and Zentgraf, H. W. (1971) Synthesis and turnover of membrane proteins in rat liver: an examination of the membrane flow hypothesis. *Z. Naturforsch.* **26**, 1031–1039.

Gutknecht, J. and Dainty, J. (1968) Ionic relations of marine algae. *Oceanogr. Mar. Biol. Rev.* **6**, 163–200.

Hall, J. L. (1969) A histochemical study of adenosine triphosphatase and other nucleotide phosphatases in young root tips. *Planta* (Berl.) **89**, 254–265.

Hatch, M. D. and Slack, C. R. (1970) Photosynthetic CO_2-fixation pathways. *Ann. Rev. Pl. Physiol.* **21**, 141–162.

Heyn, A. N. J. (1971) Observations on the exocytosis of secretory vesicles and their products in coleoptiles of *Avena*. *J. Ultrastructure Res.* **37**, 69–81.

Hodges, T. K., Leonard, R. T., Bracker, C. E. and Keenan, T. W. (1972) Purification of an ion-stimulated adenosine triphosphatase from plant roots: association with plasma membranes. *Proc. Nat. Acad. Sci.* (U.S.) **69**, 3307–3311.

Hind, G. and Jagendorf, A. T. (1965) Light scattering changes associated with the production of a possible intermediate in photophosphorylation. *J. Biol. Chem.* **240**, 3195–3209.

Jagendorf, A. T. and Uribe, E. (1967) Photophosphorylation and the chemiosmotic hypothesis. In *Energy Conversion by the Photosynthetic Apparatus*. Brookhaven Symposia in Biology, **19**, 215–241.

Jeschke, W. E. (1967) Die Cyclische und die Nichtcyclische Photophosphorytierung als Energiequellen der Lichtabhängigen Chloridion Aufnahme bi *Elodea*. *Planta* (Berl.) **73**, 161–174.

Lundegårdh, H. (1939) An electro-chemical theory of salt absorption and respiration. *Nature* **143**, 203.

Lüttge, U., Ball, E. and Von Willert, K. (1971) A comparative study of the coupling of ion uptake to light reactions in leaves of higher plants-species having the C_3- and C_4-pathway of photosynthesis. *Z. Pflanzenphysiol.* **65**, 336–350.

MacRobbie, E. A. C. (1965) The nature of the coupling between light energy and active ion transport in *Nitella translucens*. *Biochim. Biophys. Acta* **94**, 64–73.

MacRobbie, E. A. C. (1969) Ion fluxes to the vacuole of *Nitella translucens*. *J. Exp. Bot.* **20**, 236–256.

McClurkin, I. T. and McClurkin, D. C. (1967) Cytochemical demonstration of a sodium-activated and a potassium-activated adenosine triphosphatase in loblolly pine seedling root tips. *Plant Physiol.* **41**, 1103–1110.

Mayne, B. C. and Clayton, R. K. (1966) Luminescence of chlorophyll in spinach chloroplasts induced by acid-base transition. *Proc. Nat. Acad. Sci.* (U.S.) **55**, 494–497.

Mayo, M. A. and Cocking, E. C. (1969) Pinocytic uptake of polystyrene latex particles by isolated tomato fruit protoplasts. *Protoplasma* **68**, 223–230.

Middleton, H. W. (1970) Kinetics of monovalent ion activation of the (Na^+–K^+)-dependent adenosine triphosphatase and a model for ion translocation and its inhibition by the cardiac glycocides. *Arch. Biochem. Biophys.* **136**, 280–286.

Mitchell, P. (1961) Coupling of phosphorylation to electron and hydrogen transfer by a chemi-osmotic type of mechanism. *Nature* **191**, 144–148.

Mitchell, Peter (1966) Chemi-osmotic coupling in oxidative and photosynthetic phosphorylation. *Biol. Rev.* **44**, 445–502.

Neumann, J. and Jagendorf, A. T. (1965) Uncoupling photophosphorylation by detergents. *Biochim. Biophys. Acta* **109**, 382–389.

Raven, J. A. (1967a) Ion transport in *Hydrodictyon africanum*. *J. Gen. Physiol.* **50**, 1607–1625.

Raven, J. A. (1967b) Light stimulation of active transport in *Hydrodictyon africanum*. *J. Gen. Physiol.* **50**, 1627–1640.

Reid, R. A., Moyle, J. and Mitchell, P. (1966) Synthesis of adenosine triphosphate by a proton motive force in rat liver mitochondria. *Nature* **212**, 257–258.

Robertson, R. N. (1968) *Protons, Electrons, Phosphorylation and Active Transport*. Cambridge University Press.

Smith, F. A. (1966) Active phosphate uptake by *Nitella translucens*. *Biochim. Biophys. Acta* **126**, 94–99.

Smith, F. A. and West, K. R. (1969) A comparison of the effects of metabolic inhibitors on chloride uptake and photosynthesis in *Chara corallina*. *Aust. J. Biol. Sci.* **22**, 351–363.

Whittam, R. (1962) The asymmetrical stimulation of a membrane adenosine triphosphatase in relation to active cation transport. *Biochem. J.* **84**, 110–118.

Whittam, R. and Ager, M. E. (1965) The connexion between active transport and metabolism in erythrocytes. *Biochem. J.* **97**, 214–227.

six

Transport processes and membrane structure : a summary

In concluding this first part of the book we should now take stock of what we have learnt and see how near we are to understanding the ways in which plant cells control their internal ionic concentrations. In general terms, we see that this control is determined by the molecular structure of the cell membrane and by the linkage of certain parts of it to the metabolic machinery of the plant cell. There remain, however, some questions which cannot be answered and these will be dealt with at the end of the chapter.

Membrane structure

It is probable that the matrix structure of the membranes surrounding all types of cell and cellular organelle is similar. The structure of the phospholipid bilayer is such that the hydrophobic ends of two ranks of molecules mingle in the semi-liquid interior leaving the polar head regions of the molecule to face the aqueous phases on either side of the membrane. As we have seen this structure will form spontaneously and is relatively stable. This stable arrangement of molecules should not be confused with structural rigidity since the membrane can be bent and invaginated to form vesicles which become pinched off from the parent membrane. The surface area of a protoplast can be rapidly adjusted without loss of permeability control indicating that phospholipid molecules can be added to or withdrawn from the existing plasmalemma with great rapidity. There do seem to be certain situations where the bilayer can be interrupted by the formation of globular lipid micelles which have a hydrophobic core and a polar periphery; indeed vesicle formation and membrane fusion probably involve micelle transitions in the predominantly bilayered structure of the membrane.

Polar membrane surfaces

The outer and inner polar surfaces of membranes have proteins and glycoproteins associated with them; some of these molecules are electrostatically bound to the phospholipid heads, while other proteins in the membrane, especially those of the mitochondrial membrane, are hydrophobically bound suggesting deeper penetration of the protein into the bilayer matrix. The adsorbed protein at the membrane surface greatly reduces the interfacial tension between the membrane and the aqueous phases and may modify the charge distribution at the membrane surface in such a way as to confer some rudimentary ion selectivity by the membrane (page 106).

The phospholipid heads themselves can also influence the permeability-selectivity characteristics of membrane through their surface charge. The net electrical charge of the bilayer surface has a strong influence on the rate at which cations move across synthetic phospholipid micelles (page 105). In nature, a variety of phospholipid molecules is found which differ in the charge carried by the polar heads under normal physiological conditions; there are both positively and negatively charged as well as neutral phospholipid molecules and this diversity may be reflected in the differences, which are found between different types of cell, in the rates of passive ion transfer across the membrane. The rigid packing of the phospholipid heads is influenced by the presence of polyvalent ions which cross link adjacent molecules. Calcium ions are believed to exert their influence on membrane characteristics in this way.

Lipid interior of membrane

The hydrocarbon interior remains, at normal temperatures, in a relatively fluid condition although certain neutral lipid molecules, the most prevalent of which is cholesterol, appear to stabilize the structure of the lipid interior. We have encountered two situations in which the degree of structure, or order, of the lipid interior is of significance in determining the movement of water and solutes across the membrane. In synthetic bilayers made from pure phospholipids, the ionic conductance is so small ($<10^{-9}\ \Omega^{-1}\ cm^{-2}$) that the existence of any pores or gaps in the membrane filled with electrolyte solution is very improbable. In these situations, however, the diffusive movement of water remains anomalously high. In the theory of Träuble, we can see how small packets of water may move across the lipid interior in what are described as travelling structural defects. Once the water has filled one of these packets and commenced its journey across the membrane, it is no longer in direct communication with the water on either side of the membrane. It is, in a sense, analogous to pinocytosis at a molecular level. The existence of a hydraulic or osmotic pressure gradient across the membrane can have no influence on the size of the packets nor the rate at which they move across the lipid. If the coefficients for diffusive and osmotic movement of water across a pure bilayer are compared they are found to be the same (page 116). In biological membranes, however, it is invariably found that the osmotic permeability coefficient is appreciably larger than the diffusive coefficient and the simplest explanation of this is that the membrane has holes in it. A number of considerations sets the diameter of 0·4 nm as the upper limit of these pores. Electron microscopy has never revealed these pores and is unlikely to do so, firstly on technical grounds and secondly because their existence at a given location is transitory. They may be, in reality, transient gaps in the bilayer formed by molecular replacements or by transition from the bilayer to the micellar conformation of the phospholipids (page 33). It seems probable that any treatment which might stabilize the hydrocarbon interior will increase the lifetime of the pores and increase the hydraulic conductivity of the membrane and the experimental evidence on the effects of temperature and cholesterol is consistent with this view.

Ion-carriers

It is unlikely that the above factors alone can account for the diversity of membrane characteristics in nature. Neither the surface charge on the membrane nor the existence of pores satisfactorily explains the ion selectivity displayed by biological membranes. The addition of carrier molecules to synthetic lipid bilayers provides us with a simple model which resembles biological membranes sufficiently closely to make it seem certain that carrier molecules are an essential feature of the membrane infrastructure. The essential feature of carrier molecules is that they contain the ion in their hydrophilic interior while presenting a hydrophobic exterior of relatively high lipid solubility.

The larger elements of the membrane infrastructure which are revealed in freeze-etched fractures of specimens (page 44) may well be associated with active transport of ions. They may be the places at which ions are transported as cofactors in enzyme reactions, e.g., ATP hydrolysis (page 150). The frequency of these particulate structures is greatest in membrane systems which actively transport ions rapidly and in large quantities, and least in membranes which are highly resistant to ion movement, e.g., nerve myelin.

In chapter 5 we established that the transport of all ions across membranes depends ultimately on the metabolism of the cell or organelle. Active transport is essential in establishing the membrane potential and in its maintenance. It is likely, then, that the infrastructure of membranes associated with active ion pumping is the most important variable in determining their functional diversity.

Maintenance of steady state conditions

The coenocytes of *Hydrodictyon africanum* which we have considered are in a state of dynamic equilibrium in which the net fluxes of ions across the membrane are zero. For each of the three major ions crossing the membrane, viz., Na^+, K^+ and Cl^-, active transport is necessary to maintain one of the observed fluxes. In common with nearly all other organisms, the sodium concentration in the cell is lower than would be expected at a passive equilibrium, indicating that this ion is actively pumped out of the cell. With K^+ and Cl^- the influxes involve active ion pumping. In *Hydrodictyon* and in other organisms the operation of the Na-efflux pump is coupled with part of the active K^+ influx and is dependent on ATP synthesis in the cell. This is strong circumstantial evidence that plant cells contain a Na^+–K^+ activated, membrane-bound ATPase analogous to that studied and characterized in animal cells.

Chloride and other anions are transported into the cell up a steep gradient of electrochemical potential but the mechanism of this pumping remains obscure. In some instances the pump can be supported by electron-flow which has been uncoupled from ATP production; in such circumstances the Na/K pump is severely inhibited (page 144). In other algae and in the photosynthetic cells of higher plants, the active transport of chloride shows the same dependence on conditions which lead to the synthesis of ATP as Na/K transport, e.g., *Chlorella pyrenoidosa* and *Elodea canadensis*.

The vacuole, bounded by the tonoplast, is the largest compartment in the coenocyte of *Hydrodictyon* and in most other mature cells. The electric potential difference between the cytoplasm is small compared with that between the cytoplasm and the external solution and the distribution of ions between the cytoplasm and the vacuole is predicted fairly accurately by the Nernst equation (page 62). Thus, there is little, if any, active transport between the compartments. The electrical resistance of the tonoplast in *Nitella translucens* is much lower than, and the ionic permeability coefficients are usually higher than, corresponding values for the plasmalemma. In *Hydrodictyon* and in most other organisms, the plasmalemma is the principal barrier to the free diffusion of ions.

The steady state is controlled by events occurring at the plasmalemma and represents a fine balance between active ion transport and passive leakage under the influence of gradients of electrochemical potential.

Relationships between the plasmalemma and activity of cellular organelles

In general, the ionic relations of mitochondria and chloroplasts have been subjected to more detailed observations than the plasma membrane. While it is clear that the plasmalemma is the principal permeability barrier of the cell there is abundant evidence that active transport into the cell is strongly dependent on the activity of organelles, both for supplies of ATP for ion pumps and in other ways.

The membranes of both mitochondria and chloroplasts are equipped with electron-transport systems which can bring about charge separation. Proton pumping by chloroplasts is an immediate reaction to illumination and the electrogenic nature of this pumping is reflected by the rapid build-up of an electrical potential difference across the thylacoid membranes within the chloroplast (Junge, Rumberg and Schröder, 1970). These events are remote from the plasmalemma and yet they are very quickly paralleled by changes in the membrane potential between the bulk cytoplasm and the external medium (Vredenberg, 1969) and by changes in the active transport of chloride, e.g., *Acetabularia* (page 77). There is great uncertainty about the way in which communication is effected between chloroplasts and the ion pumps at the plasmalemma. It is, however, quite certain that our early assumption (that the cytoplasm is a homogeneous compartment) just will not do if we hope to understand what is going on. The concentration of certain ions, e.g., Ca^{2+} and $H_2PO_4^-$ in mitochondria and Mn^{3+}, Mg^{2+} and Cl^- in chloroplasts, is very much higher than in the bulk of the cytoplasm. The extent to which these local accumulations invalidate the idea of an average cytoplasmic concentration of ions depends, of course, on the proportion of the cytoplasmic volume which they occupy. In the leaves of pea plants (*Pisum sativum*) chloroplasts occupy about 40 per cent of the cytoplasmic volume (Lin and Nobel, 1971); and in the coenocyte of

Nitella axillaris the data of Diamond and Solomon (1959) suggest a value of 40–60 per cent.

There are other compartments in the cytoplasm in which the ionic composition is completely unknown, the most conspicuous being the appreciable volume enclosed in the cavity of the endoplasmic reticulum.

Although it does not show exactly how the communication between organelles and the plasmalemma is achieved it is helpful to have in mind the dynamic nature of both membranes and the cytoplasm itself. We have seen that membranes can be assembled and disassembled at great speed and that they can be convoluted to the extent that vesicles are cut off as in pinocytosis. We also learnt that membranes can fuse with one another so that a vesicle cut off from the plasmalemma might discharge its contents directly into the vacuole or the endoplasmic reticulum without its contents being mixed with the bulk cytoplasm. If the cytoplasm were a static gel such quantum transport would be very slow but it is readily seen to be in constant motion. In the streaming process it can be envisaged that small membrane-bound vesicles whose ionic composition differs markedly from that of the bulk cytoplasm can be carried around and impinge on areas of membrane remote from their point of formation. It seems possible that 'packets' of hydrogen ions and bicarbonate formed in organelles may reach the plasma membrane in this way and be exchanged for other ions from the external medium.

Bibliography

b References in the text

Diamond, J. M. and Solomon, A. K. (1959) Intracellular potassium compartmentation in *Nitella axillaris*. *J. Gen. Physiol.* **42**, 1105–1121.

Junge, W., Rumberg, B. and Schröder, H. (1970) The necessity of an electric potential difference and its use for photophosphorylation in short flash groups. *Eur. J. Biochem.* **14**, 575–581.

Lin, D. C. and Nobel, P. S. (1971) Control of photosynthesis by Mg^{2+}. *Arch. Biochem. Biophys.* **145**, 622–632.

Vredenberg, W. J. (1969) Light-induced changes in membrane potential of algal cells associated with photosynthetic electron transport. *Biochem. Biophys. Res. Commun.* **37**, 785–792.

Part II

seven

Ion transport in higher plants: the electrochemical approach

The second part of this book is concerned with higher plants and will deal with both ionic relationships between roots and the external environment and between the various tissues in the plant. Many of the problems we shall discuss differ from those dealt with earlier in the book. Large, multicellular organisms may have disadvantages when used for detailed study of ion-transport mechanisms, but they present interesting opportunities for studying the relationships between the various parts and tissues of the plant body. The greater complexity of such a system in comparison with micro-organisms requires a certain broadening of the scale in which we look at things, and also a loss of precision. In Part I, we spent much of our time thinking at a molecular and ultrastructural level. We shall find that our present knowledge of higher plants does not, generally, permit enquiry on such a precise scale. When we consider structure in relation to function in higher plants it is usually on a fairly gross level.

I shall begin by reviewing some evidence from higher plant cells which suggests that the basic features of membranes and of ion-transport processes in them differ only in detail and not fundamentally from those described earlier. For this reason I shall be giving what is, perhaps, less attention than is due to a large body of admirable work on higher plants which attempts to describe mechanisms of ion uptake and their relationships with metabolism. Much of this work has made use of tissue slices from storage roots; students wishing to pursue this matter can do so in excellent reviews by Robertson (1960, 1968).

Compartmentation and other complications in higher plants

In most higher plants, the organism is differentiated into an above-ground portion in which photosynthesis occurs, and a below-ground portion which anchors the plant and absorbs water and ions from the surrounding medium. Communication between these two portions is made possible by vascular tissue and either portion may, to some extent, regulate the activities of the other. Expressed, even in the simplest terms, higher plants have, therefore, three major compartments but within each of these are several tissues which may be specialized with respect to both structure and function. This clearly contrasts with the unicellular organism in which all metabolic activities are carried out within a very short distance of the plasmalemma and hence the outside environment. The second complicating factor is that in most situations which have any bearing on reality, various parts of the plant body increase in size with time; even during the course of an experiment. Clearly, this is a non-equilibrium situation where there will be a net influx of ions into the plant (i.e., $\phi_{in} > \phi_{out}$); this complicates a precise analysis of the driving forces in ion transport by the Nernst or Ussing–Teorell criteria which depend on the observation of a steady state. Because it is growing a given organ may contain cells at all stages of development from mitosis to senescence; it is probable that these stages differ in their capabilities for ion transport.

When plants lose water from their leaves by evapo-transpiration, water moves across the root and rapidly through the conducting tissue; in certain circumstances this net flux of water may have major interactions with the fluxes of ions into and across the root to the vascular tissue.

Cells of higher plants are generally very much smaller and less accessible than those of the algae, and although this has not prevented the application of electrophysiological procedures to tissues of higher plants, it has meant that crucial information on the concentration of ions in the cytoplasm has been estimated only by indirect methods subject to some uncertainties.

Ionic composition of intracellular compartments

It is not possible to sample directly the ionic composition of the cytoplasm of single cells from higher plants, though it has been possible to insert ion-selective electrodes into the vacuoles of cells in the cortex of *Zea mays* to measure the activity of potassium (Dunlop and Bowling, 1971). In some cases the authors have ignored any differences which may exist between the cytoplasm and the vacuole and have extracted tissues with hot water (e.g., Higinbotham, Etherton and Foster, 1967) working on the assumption that the error in calculating the vacuolar concentration is very small since more than 90 per cent of the cell volume is within the tonoplast. This procedure has very grave weaknesses if concentrations within the cytoplasm are inferred from the results. The direct detection of elements in either freeze-dried sections cut from frozen tissues (Läuchli, 1967) or freeze-substituted and plastic embedded sections (Läuchli *et al.*, 1970) using the electron probe micro-analyser provides a promising technique for the direct chemical assay of tissues, but suffers from the disadvantage that the probe can make no distinction between free ions and those which are bound in some way. Thus, a fragment of a calcium oxalate crystal in a section might give rise to the misleading impression that the calcium activity of a cellular compartment was very high.

A composite table of the ionic composition of higher plant cells could hardly serve any useful purpose since circumstances and species can cause enormous variations. It is generally true, however, that potassium is the most abundant cation in plant tissue but that dissolved sugars and organic anions can make up the major portion of the osmotic potential of the cell.

In cells of the oat (*Avena*) coleoptile it was found that parenchyma cells had an osmotic concentration of 410 mM of which only 22 mM was potassium and 1 mM was sodium, sugar being the major osmolyte (Goldsmith, Fernandez and Goldsmith, 1971). Kirkby (1968) has shown that the leaves of crop plants receiving nitrogen from the soil as nitrate may have as much as 45 per cent of their total osmolarity made up from organic cations, principally malate and citrate.

Generally speaking, the rather low concentration of potassium found by Goldsmith *et al.* (1971) is not typical of tissues and the familiar plants used in laboratory investigations usually have potassium concentrations in roots in the range of 50–100 mM and up to 200 mM in shoots, depending on species and nutritional treatment. Thus, the range of potassium concentrations is not unlike that seen in algae from both fresh and brackish waters. There is a marked discrimination against sodium absorption in most higher plants, the exceptions being those which either grow or have their origin in saline or maritime environments (e.g., sugar beet).

Compartment analysis

Estimates of the quantities and concentrations of ions in cellular compartments can be made by observing the kinetics of the absorption or elution of radioactive tracers by relatively homogeneous samples of tissue in which fluxes inwards and outwards are somewhere near a steady state. The basic theory of these methods presented in chapter 3, page 88, should be consulted before continuing further with this section.

Experimental procedure. We will consider first an elution experiment and see how much information we can get from it. A relatively homogeneous piece of tissue is excised from an organ and stabilized in a constant environment so that it approaches equilibrium with the bathing solution. Thin slices of storage roots such as carrot or beet, or narrow strips of leaves are favourite objects for experiments of this kind, but excised roots and root tips have also been used. Storage roots are popular because they contain appreciable amounts of carbohydrate which can sustain normal cellular activity for many days. After equilibration with the bathing medium, the tissue is placed in a solution of similar concentration in which one or more of the ion species is labelled with a radioactive tracer; this is sometimes known as the 'loading' period and it may last from several hours to several days. At the end of the 'loading' period the tissue is removed from the bathing medium, the superficial solution removed by blotting and then transferred successively to a number of vessels containing unlabelled bathing medium. The isotope eluted from the tissue into these solutions is determined and, at the end of the experiment, the radioactivity remaining in the tissue is determined. The sum of the radioactivity of all of the elution solutions plus the residue in the tissue gives the total radioactivity at the beginning of the elution, t_0. The amounts removed at various sampling periods are then deducted, stepwise, from this initial quantity, giving the amount remaining in the tissue at any given time.

Experimental precautions

1 Calcium ions. The experimental solutions for both equilibration and efflux phases of the experiment should contain calcium ions (usually $>10^{-4}$ M) since the maintenance of normal membrane function seems to depend strongly on their presence (see chapter 3, page 72). In certain algal tissues and in the roots of *Zea* and

Hordeum, the efflux of ions is much more rapid in the absence of calcium (e.g., Maas and Leggett, 1968; Weigl, 1969; Elzam and Epstein, 1965). This does not necessarily invalidate conclusions about the size and composition of cellular compartments drawn from experiments where calcium was omitted. These conclusions, as we shall see, are based on *differences* in the rate at which the compartments lose their labelled ions; the actual rates may not be of great significance. Having said this, it should be stressed that it is inadvisable to conduct any experiment on biological membranes in media lacking calcium.

It has been claimed that, where calcium concentration is adequate, ion movement across the plasmalemmata of higher plant cells is a unidirectional process; in other words there is no efflux (Epstein, 1972). Without doubt this is an extreme claim, for there is a considerable amount of evidence that passive permeation of the plasmalemma both by anions and cations does occur in the presence of calcium even though fluxes are slower than in its absence (e.g., Weigl, 1969).

2 Xylem exudate. Efflux experiments conducted with excised whole roots and with short root segments are subject to errors (serious ones in the latter material) because of the exudation of radioactive tracers from the cut ends of the xylem elements. The xylem exudate is very highly concentrated relative to the outside solution (see chapter 8, page 214 and chapter 9, page 245), thus, in spite of its small volume, it may add significant quantities of tracer to the outside solution which have not 'effluxed' but which have been actively 'secreted'. This source of error can be avoided if the cut ends of the xylem are isolated from the bulk of the solution into which efflux is taking place; a simple apparatus whereby this isolation can be effected is shown in Fig. 7.1.

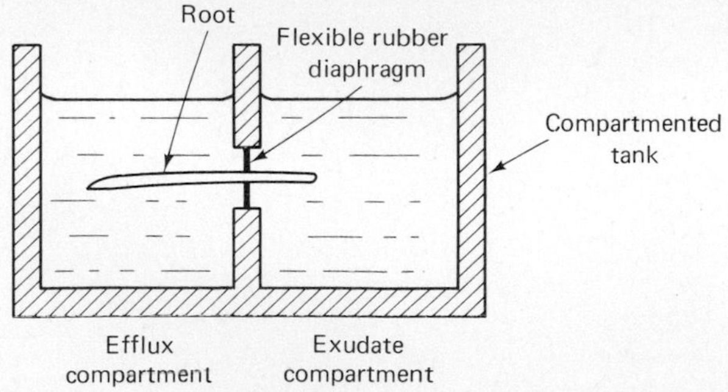

FIG. 7.1 Compartmental tank in which the efflux and exudate from excised root tips can be separately measured. The rubber diaphragm is perforated with holes smaller than the diameter of the root; many roots can be inserted into a single tank. (After Weigl, 1969.)

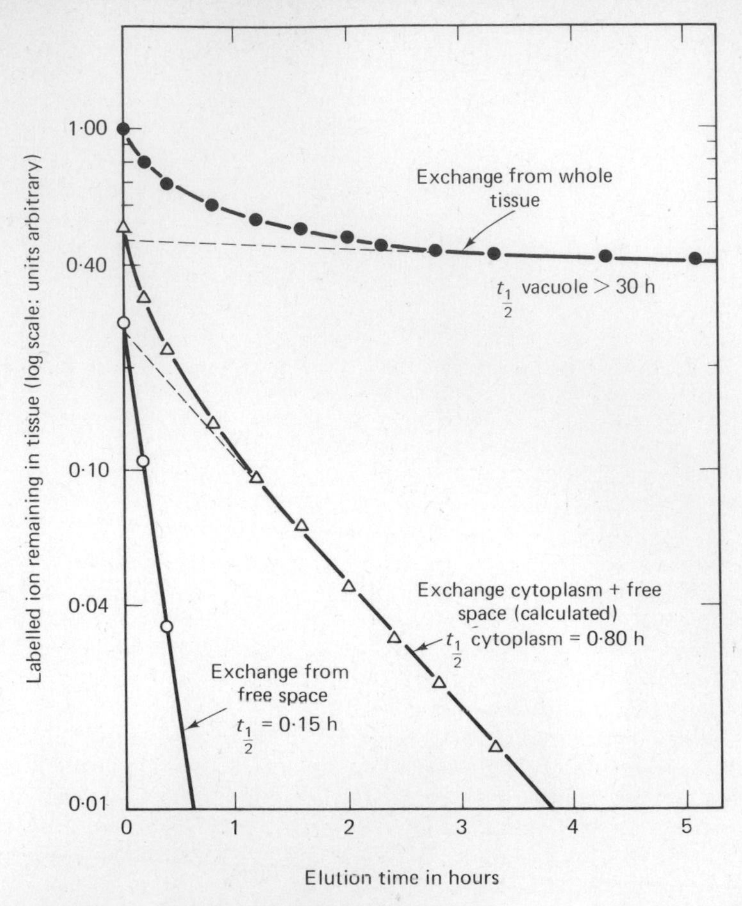

FIG. 7.2 Tracer efflux from labelled tissue. The data is contrived but rates of exchange for free space and cytoplasm are characteristic of root tissues. The half times for exchange are calculated by extrapolating linear parts of curves to the ordinate.

Interpretation of data. The elution is the sum of several exponential processes (see chapter 3, page 90) and if the logarithm of the radioactivity remaining in the tissue is plotted against time a relationship similar to the contrived curve in Fig. 7.2 is usually obtained.

Closer examination of the kinetics reveals that the uppermost curve can be resolved into three distinct components which have logarithmic rates of elution. When the radioactivity reaching the bathing medium is derived from more than one of these components the relationship is curved. By inspection we note that at later times the uppermost curve is linear, thus indicating that the rate of elution is determined by a single compartment which is usually assumed to be the vacuole. We now assume that radioactivity has been eluted at the same rate from this compartment throughout the experiment from t_0. If, therefore, the linear portion is extrapolated back to the ordinate we get a series of values which can be deducted from those in the curve above. These we now plot (the middle curve on Fig. 7.2) and discover that the relationship is still curved during the early stages of the elution and is due to the loss of radioactivity from two compartments, assumed to be the cytoplasm and the cell wall. After a few minutes, however, the elution rate becomes linear. If, again, this linear portion is extrapolated to the ordinate we can deduct from the middle curve the tracer eluted from the cytoplasm. This gives a third curve which is linear from t_0 with a very steep slope; this represents the rate of exchange from the cell wall.

Quantities of ions in the compartments. The points where the extrapolated linear portions of the curves meet the ordinate give measures of the quantity of tracer present in the compartments at the end of the labelling period (t_0). We can see how long it takes for one-half of the quantity in a given compartment to be exchanged from the external medium.

For the fastest compartment, the cell wall, the half-time for exchange is 5–10 minutes for beet disks, but is appreciably shorter for roots of cereal plants and for leaf strips. The rate of exchange from the walls is not influenced strongly by temperature but Pitman (1963) showed that in beetroot slices the half-time for potassium exchange in the cytoplasm was 40 minutes at 25 °C but more than two hours at 2 °C.

If the labelling period is long in comparison with the half-time for the cytoplasmic exchange (in Pitman's 1963 experiments it was 18 hours and so more than five times the length of the cytoplasmic half-time), tracer in the cytoplasm will have approximately the same specific activity as the external medium (this is in fact a slight oversimplification: see Pitman, 1963). Thus, when the intercept of the linear portion of the cytoplasmic elution curve with the ordinate gives a fair approximation to the total quantity of the ion in the cytoplasm, the half-time for the exchange of vacuoles in any tissue is much longer than the exchange of the cytoplasm since its volume is much greater. This is true even though ion fluxes across the tonoplast may be more rapid. Pitman (1963) estimated for beet slices that 70 per cent of the total volume of the slice was accounted for by vacuoles and that only 5 per cent was cytoplasm; the remaining volume was enclosed by the free space and damaged cells.

Factors controlling the rates of elution. At this point we should consider the factors which determine the rates of tracer elution from the various compartments. The insensitivity of elution from the cell wall to changes in temperature suggests that ions are liberated from it by exchange; in such circumstances the length of the path which the exchanged tracer has to diffuse before it enters the external solution can become the limiting factor in the observed rate of elution. Unless thin slices of tissue are used it is not possible to resolve accurately the half-time for the exchange of the cell wall compartment since the slope of the curve will be determined by diffusion through the 'free space' (see chapter 8, page 205).

We learnt in chapter 3 that a passive flux in a given direction depends on both the permeability of the membrane and the physical driving forces acting on the ion under consideration. Electrical resistance measurements (see page 183) show that the tonoplast has a much greater ionic conductance than the plasmalemma and hence permeability coefficients for most ions will be appreciably greater than for the plasmalemma. The physical driving forces on major ions are usually, however, much more strongly directed across the plasmalemma than those operating across the tonoplast. These opposing features tend to make the fluxes of ions across the plasmalemma and tonoplast more nearly the same than one might predict from permeability considerations.

Let us consider the situation in which a labelled potassium ion is present in the cytoplasm at a higher electrochemical potential than in the outside solution but is in approximate equilibrium with the potassium ions in the vacuole. During the elution experiment, this labelled ion can move either to the external solution down a steep thermodynamic gradient or into the vacuole by self-diffusion in the absence of a driving force. As the ion moves from its original compartment its place will be filled either by an unlabelled ion from the outside solution, or by a labelled or an unlabelled ion from the vacuole. (The replacing ion from the vacuole usually comes from a solution of much lower specific activity than the cytoplasm, hence the probability of it being an unlabelled ion, which was present in the cell before the experiment began, is quite high.) Both types of replacement (or exchange) will tend to reduce the abundance of tracer in the cytoplasm quite rapidly; the specific activity will fall to the point where it becomes similar to that of the solution in the vacuole, which becomes the only source of labelled ions. Thus, the vacuole keeps the cytoplasm 'topped-up' with labelled potassium at a rate determined by the flux across the tonoplast. Thus, we observe two phases on the elution curve characteristic of the rates of change of specific activity in the cytoplasm and in the vacuole.

There may seem to be a paradox in these events because, as we have said, ion fluxes across the tonoplast are faster than across the plasmalemma. How then can we account for the more rapid exchange of ions in the cytoplasm? The paradox is resolved if we bear in mind the volumes of the two compartments and realize that a slower flux will empty a small compartment *relatively* more quickly than a more rapid flux can empty a much larger one.

Validity of compartment analysis. Compartment analyses of several kinds can be conducted. For instance, data on the ionic concentration of, and fluxes into, compartments can be made from short term influx data (see Cram, 1968, 1969). Analysis of this kind frequently involves quite sophisticated reasoning and, although we cannot discuss such matters here, it is usually based on the assumption that the observed influx or elution is from three compartments arranged in series. Thus, an ion leaving the vacuole enters first the cytoplasm and then the cell wall during its journey to the external medium. While there is no argument about the latter stage, we might recall that MacRobbie (1969) showed that some of the chloride reaching the vacuole apparently bypassed the cytoplasm, presumably due to transport within vesicles. There seems to be no reason why this process should not occur in the opposite direction and in a penetrating review of compartment analysis MacRobbie (1971) has cited a number of experiments where the results are distinctly at odds with the 'three compartments in series' model. If vesicular transport accounts for only a minor portion of the ions reaching, or leaving, the vacuole then the analysis we have applied will be not far out, but in circumstances where it is large, the analysis will cease to be valid. At the present time, too little is known of vesicular transport to predict when either of these conditions will apply.

Transmembrane potentials in cells of higher plants

Electric potential differences between the interior of cells and the external medium can be measured using fine tipped glass electrodes as described in chapter 3, but it is seldom possible to determine the potential difference across the plasmalemma alone. The layer of cytoplasm in most cells is so thin that it is virtually impossible to puncture the plasmalemma without piercing the tonoplast as well. Most measurements of transmembrane potentials from higher plant cells are for the vacuole relative to the outside medium, i.e., with the tonoplast and plasmalemma in series. A selection of these measurements has been collected in Table 7.1. The values are in the same range as those obtained from the brackish water and marine algae we considered earlier.

In the coenocyte of *Nitella translucens* and *Chara australis*, it was found that the tonoplast had a much lower electrical resistance than the plasmalemma so that most of the potential difference measured

Table 7.1 ***Some selected values of membrane potentials measured between the vacuole and the external medium***

Plant material/conditions	Potential difference (mV)	Concentrations of K^+ and Ca^{2+} in medium (mM)		Reference
		K^+	Ca^{2+}	
Pisum: Cells of root cortex	−110	1·0	1·0	(1)
Epicotyl	−119	1·0	1·0	(1)
Avena: Coleoptiles	−102 to −109	1·0	1·0	(2)
Roots	−71 to −84	1·0	1·0	(2)
Zea: Sections cut from intact root	−90 to −104	1·0	0	(3)
	−76 to −84	1·0	0·1	
	−96 to −99	0·1	0·1	
Beta vulgaris (red beet): Aged disks of storage root	−153	0·6	0	(4)
Vigna sequipedalis (bean): Parenchyma cells Hypocotyls of seedlings	−55 to −60	0·1	0	(5)
Vicia faba (broad bean): Seedling roots	−130	1·0	0·5	(6)

(1) Higinbotham, N., Etherton, B. and Foster, R. J. (1967).
(2) Etherton, B. (1963).
(3) Dunlop, J. and Bowling, D. J. F. (1971).
(4) Poole, R. J. (1969).
(5) Katou, K. and Ikamoto, H. (1970).
(6) Scott, B. I. H., Gulline, H. and Pallaghy, C. K. (1968).

between the vacuole and the external solution was across the plasmalemma. This also seems to be the case in the cells of higher plants, although the data are rather meagre. It may seem a surprising subject for microelectrode work, but root hairs are perhaps the only accessible cells in roots which have a layer of cytoplasm thick enough for the investigators to be certain that the electrode has not penetrated the tonoplast. Some of the earliest potential measurements were made on the root hairs of *Avena* by Etherton and Higinbotham (1960); by inserting the electrode obliquely through the terminal cap of cytoplasm in the root hair they were able to measure the potential first across the plasmalemma and then across the plasmalemma and tonoplast in series as the electrode was inserted more deeply. Their results showed quite clearly that the main potential drop was across the plasmalemma. By the insertion of a current electrode into the vacuole (see chapter 3, page 91) Greenham (1966) measured the electrical resistance of the two membranes in the root hair cells of cucumber, maize and oats (Table 7.2) and showed that

Table 7.2 Membrane surface resistance of root hair cells
(R_p = plasmalemma resistance, R_{p+t} = plasmalemma and tonoplast resistance, Ω cm^2)

Species	R_p	R_{p+t}	R_{p+t}/R_p
Cucumber	2930 ± 420	3680 ± 500	1·29 ± 0·10
Oats (cv. Victory)	2610 ± 600	3190 ± 300	1·12 ± 0·16
Maize (cv. Fitzroy)	3410 ± 550	3740 ± 500	1·20 ± 0·17

Growth/experimental medium contained 1·0 mM KCl, 0·5 mM $CaCl_2$ and included 0·1 M mannitol in the medium for oats and maize roots to minimize bursting of the root hairs when the micro-electrode was inserted into the cell. In experiments with root hairs from cucumber, mannitol has no effect on the measured resistance. (*Data from Greenham, 1966.*)

the tonoplast resistance was small compared with that of the plasmalemma. Fortunately, we can generalize from these rather specific examples because other investigations have established that the vacuolar potential in all of the cells of a given root is similar (Fig. 7.3) and it seems unlikely that the properties of the tonoplast will differ greatly in cells from different tissues. Scott, Gulline and Pallaghy (1968) measured the vacuole potential in the intact roots of broad bean seedlings, inserting their electrodes into cells at all stages of their development from the meristem to the point where they became fully vacuolated; they found no significant deviations from a mean value of −130 mV. While it is not strictly true to describe cells

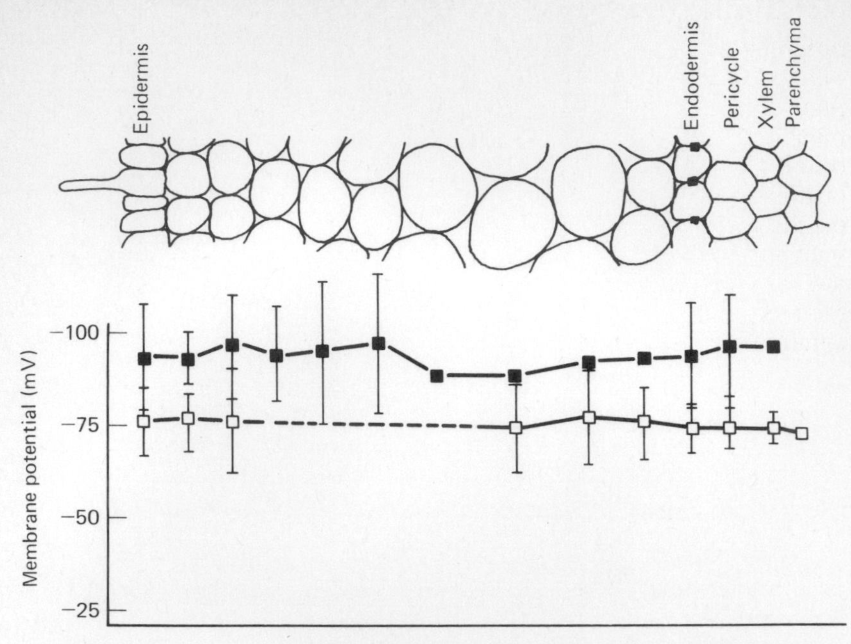

FIG. 7.3 The membrane potentials of various maize root cells in 1·0 mM KCl. Open symbols, tissue bathed in $CaCl_2$ solution; closed symbols, tissue bathed in nutrient medium. Vertical bars indicate 95 per cent confidence limits where five or more replicate measurements were made. (From Dunlop and Bowling, 1971, with kind permission.)

recently cut off from the meristem as non-vacuolated it seems improbable that the electrode tip would have penetrated one of the numerous small vacuoles of the cells; the electrode would, therefore, be measuring the potential difference across the plasmalemma. In these instances also the potential did not differ from those where the electrode was undoubtedly inserted in the vacuole and indicates further that there is little, if any, electric potential difference between the vacuole and the cytoplasmic compartments of the cells. More detailed observations on the vacuole potential in different tissues of excised maize roots have been made recently by Dunlop and Bowling (1971) and Davis (1972). Some of these observations are shown in Fig. 7.3; their significance in connection with ion movements in root tissues will be discussed in a later section (see page 232).

Origin of membrane potential in cells of higher plants

We have seen that in some circumstances the Goldman equation (chapter 3, page 71) can be used to predict the transmembrane potential of cells; the equation shows that the potential is determined by diffusion coefficients of the passively diffusing ions and their concentration ratios across the membrane. Generally, the permeability coefficient for potassium is much larger than that for the other ions (the exception being hydrogen ions in some cases) and it is, therefore, to be expected that this ion will have a marked influence on the membrane potential in those circumstances where the Goldman equation is applicable. We saw in chapter 3, however, that there are major differences between species, often between quite closely related ones, in the response of the membrane potential to the external concentration of potassium which left us in some reasonable

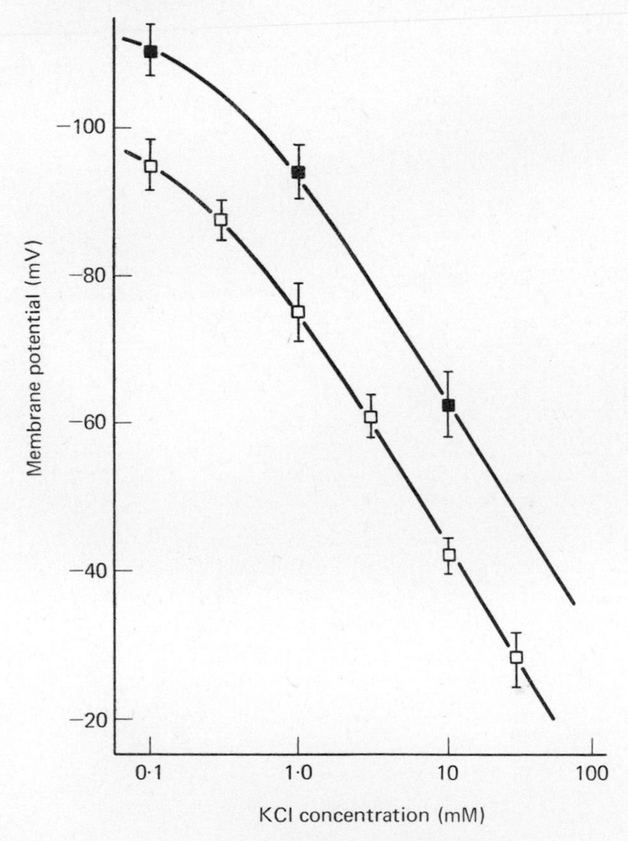

FIG. 7.4 Dependence of the electrical potential difference between the vacuole and the external solution of maize root epidermal cells on the KCl concentration of the bathing solution. Open symbols, tissue bathed in $CaCl_2$ solution: closed symbols, tissue bathed in nutrient medium. Vertical bars indicate 95 per cent confidence limits. (From Dunlop and Bowling, 1971, with kind permission.)

doubt as to the general applicability of the Goldman equation in describing the membrane potential. The same shadow of doubt is cast by the results from the tissues of higher plants.

The potential difference between the vacuole and the medium in the epidermal cells of excised maize roots is strongly dependent on the external potassium concentration; even in the presence of 0·1 mM calcium chloride (Fig. 7.4) the cells were depolarized by 33 mV by a tenfold increase of potassium concentration from 1–10 mM. By contrast, the presence of a similar concentration of calcium prevented completely the depolarization of cells of pumpkin seedlings over a similar range of potassium concentrations (Fig. 7.5); in the absence of calcium in medium, however, the cells became depolarized by

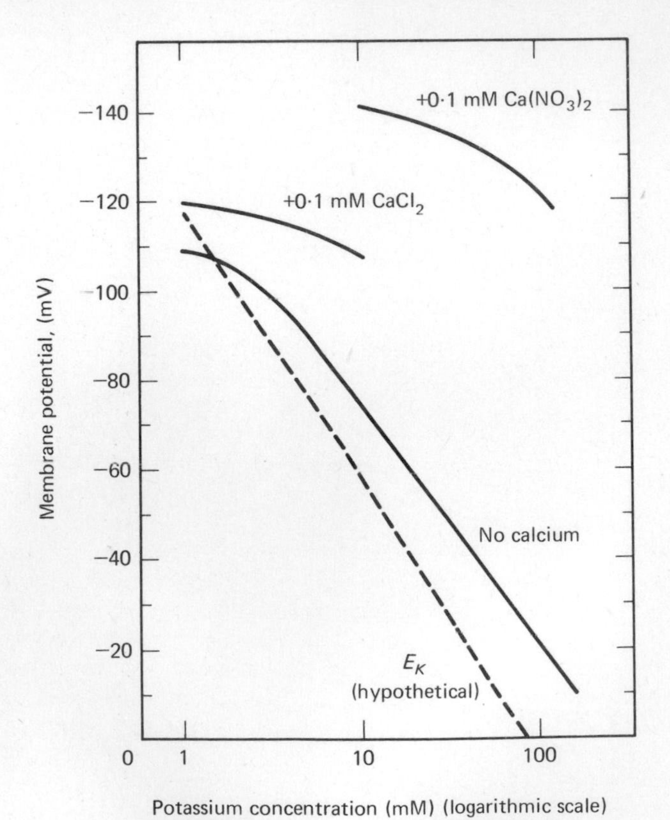

FIG. 7.5 The membrane potential between the vacuole and the outer solution in the cortex of intact roots of Cucurbita *seedlings at different potassium concentrations in the presence or absence of calcium salts. The hypothetical curve, E_K, is the depolarization of a perfectly K-selective membrane calculated from the Nernst equation. (After Sinyukhin and Vyskrebentseva, 1967.)*

52 mV thus behaving almost like a perfect K-selective electrode (Sinyukhin and Vyskrebentseva, 1968). Although it is known that calcium ions are essential for membrane stability it is difficult to advance any more precise explanation for this interaction and impossible to explain the differences between species. It may be that there is some kind of species-dependent critical ratio of calcium and potassium activities below which depolarization occurs. There is, however, no evidence to support this view from the electrochemical standpoint.

Electrogenic ion pumping

If the potential difference is not dependent on the concentration ratio of the most rapidly diffusing ion then we must look for other ways of explaining its origin. We learnt in chapter 3 that in some organisms, e.g., the alga *Acetabularia* and the fungal hyphae of *Neurospora*, the membrane potential was heavily dependent on electrogenic ion pumps which effectively separate charge across the membrane. One of the criteria used in detecting an electrogenic pump is that the inhibition of the pump by metabolic inhibitors causes an immediate depolarization of the cell. The cell only returns to its resting potential when the inhibitor is washed out and the pump begins to work again. Let us now examine the evidence for such electrogenesis in higher plants.

Higinbotham and his co-workers (1970) have made a special study of this problem. They found that in the coleoptiles and roots of oat seedlings there was very little relationship between the cell electropotential difference and the external potassium concentration. Some of their observations are collected in Table 7.3.

It can be seen that where the potassium concentration was increased from 0·1 mM to 10 mM there was no significant change in the transmembrane potential. The third column of this table shows values for the membrane potential calculated from the Goldman equation and the measured concentration ratios of the major cations across the membrane. The fourth column shows the difference between the measured and the calculated values which diverge progressively as the external potassium concentration rises so that when potassium in the outside solution is at 10 mM the measured potential is 58 mV more negative than that predicted by the Goldman equation. Similar results were recorded from the roots of the oat seedling, although, in this case, moving from a solution of 1 mM potassium to one of 10 mM produced a small depolarization of the cell. The difference between the measured and the calculated values is far too great to be accounted for by changes in the values of the permeability coefficients of the ions although the authors concede that small changes are likely to occur over wide ranges of concentration.

Table 7.3 Relationship in oat seedling tissues of measured and calculated potential differences between the vacuole and the external medium

Tissue	Potassium concentration in external medium[a]	Measured p.d.	Calculated p.d.[b]	Difference
Coleoptiles	0·1	−109	−116	+7
	1·0	−105	−89	−16
	3·0	−109	−67	−42
	10·0	−102	−44	−58
Root	1·0	−84	−79	−5
	3·0	−89	−66	−23
	10·0	−71	−45	−26

(a) All of the other nutrients in the medium were increased in proportion to K. Thus the solution containing 10·0 mM K had 100 times greater concentrations of calcium, magnesium, sodium, phosphate, chloride and nitrate than in the 0·1 mM K solution.
(b) Calculated from the Goldman equation using measured values of the internal concentrations of K, Na and Cl^- and the permeability coefficients relative to K, i.e., P_K 1·0, P_{Na} 0·68, P_{Cl} 0·037.

(Data from Higinbotham, Graves and Davis, 1970.)

In both *Avena* coleoptile tissue and epicotyl segments from *Pisum sativum* (pea), treatment of the cells with 0·1 mM potassium cyanide resulted in a dramatic and reversible depolarization (Fig. 7.6). The half-time for the depolarization was less than 1 minute; a result which suggests that an electrogenic pump is contributing directly to the potential difference across the membrane. Cyanide blocks the transfer of electrons from cytochrome oxidase in the respiratory electron transport to molecular oxygen, thus inhibiting respiration. The severity of this inhibition can be monitored by the rate of oxygen uptake by the tissue. Figure 7.7 shows that increasing cyanide concentration progressively reduced oxygen uptake but that the depolarization of the cell was correlated with this reduction only at the lower concentrations of cyanide. At 0·3 mM KCN the potential appeared to stabilize at a value near to −70 mV. This result is of special significance because this is the predicted value of the membrane potential calculated from the Goldman equation using the measured ion concentration ratios and estimates of the relative ionic permeability coefficients (see Table 7.3). The stable nature of the diffusion potential at a higher cyanide concentration shows that the initial depolarization was not caused by a general increase in membrane permeability allowing anions to leak from the cell. Additional confirmation of this point comes from chemical analysis of cyanide

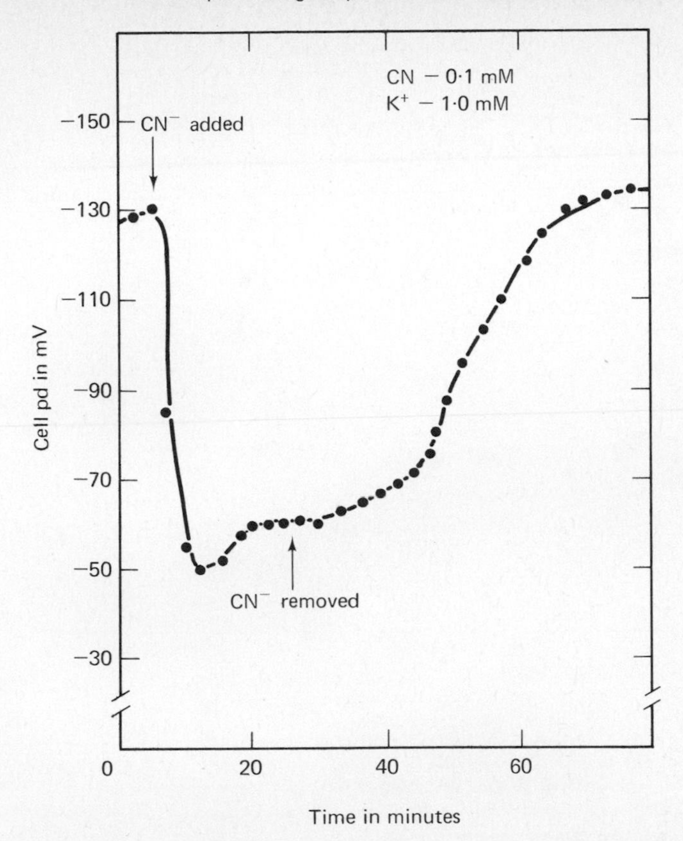

***FIG. 7.6** Time course of depolarization of membrane potential in* Pisum *epicotyl tissue by cyanide, and subsequent recovery. (From Higinbotham* et al.*, 1970, by kind permission of the publisher.)*

treated and controlled tissue. After four hours in 0·1 mM KCN the potassium content of the tissue was 43·5 μeq. g^{-1} (fresh weight) compared with 48·4 μeq. g^{-1} in the control; the values for chloride were 2·93 and 3·09 respectively. These small losses are nowhere near large enough to account for the depolarization.

The experiment shows that, in normal cells bathed in concentrations of potassium chloride greater than 1 mM the membrane potential is made up from two components, a cyanide sensitive portion which can account for as much as 60 per cent of the total potential and a cyanide insensitive portion, the value of which can be predicted from the Goldman equation and is, therefore, a diffusion potential. The data also indicate that the cyanide sensitive portion is controlled by external potassium concentration. Table 7.3 showed

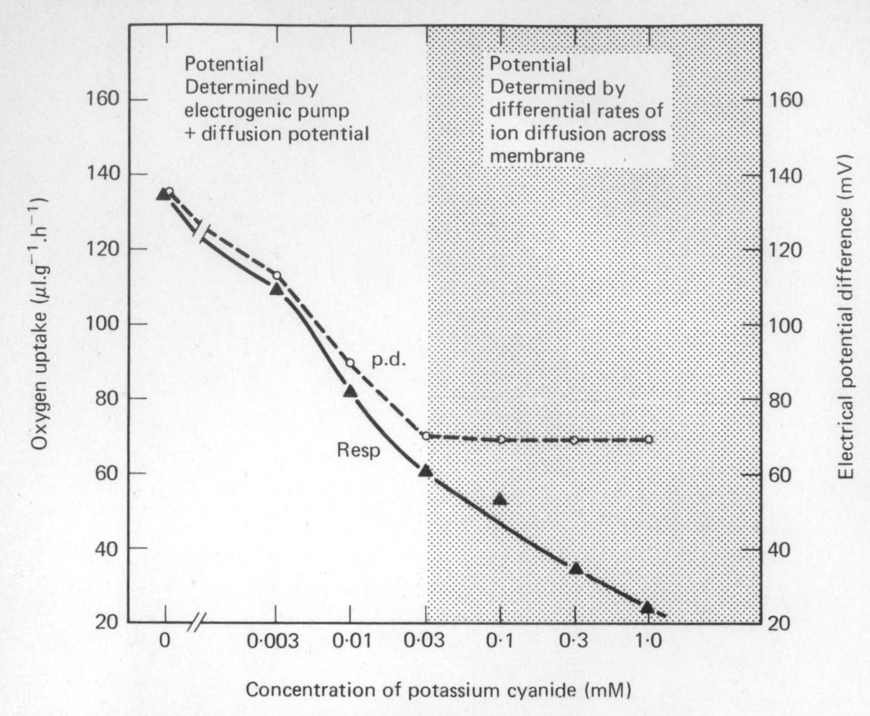

FIG. 7.7 The effect of cyanide on membrane potential in* Pisum *epicotyl tissue. The apparent insensitivity of the potential to respiration at higher concentrations of cyanide demonstrates the two major components of the membrane potential. The potassium concentration in the medium was 1·0 mM. (After Higinbotham* et al.*, 1970, by kind permission of the authors.)

that when potassium concentration in the external medium was low the diffusion potential calculated from the Goldman equation was equivalent to the measured potential. This suggests that the electrogenic pump (cyanide sensitive) need not operate where the external potassium concentration is less than 1 mM. The depolarizing effect of increasing potassium in the medium is therefore offset by increased electrogenic pumping activity. There is no indication how such a self-regulating system would operate; since the potassium moves out of the cell at high external concentrations against the physical driving forces, the control of the potential is possibly through an electrogenic K-efflux pump.

Direction of passive driving forces

The values for the electrical potential between the vacuole and the outside medium given in Table 7.1 indicate that any anion more concentrated in the cell interior than in the surroundings will move into the cell up a steep gradient of electrochemical potential and thus must be transported actively. Some data from Higinbotham, Etherton and Foster (1967) on excised oat roots showed that the directions of passive ion fluxes in this material were substantially similar to those in *Hydrodictyon* (chapter 3). They took excised tissues and held them in media of constant composition and, by measuring the amounts of ions removed by hot water treatment, showed that the tissue was at equilibrium with the surroundings and that there were no net fluxes in either direction. This flux equilibrium satisfies one of the conditions necessary for the proper application of the Nernst equation. The electrical potential differences between the vacuole and the outside were measured in the usual way. Some of their data are given in Table 7.4; the predicted concentrations of ions are those which would be expected if, in each case, the measured electric potential was the Nernst potential of the ion in question.

Table 7.4 ***Comparison of observed and predicted[c] steady-state concentrations of ions in excised roots of* Avena *(oat)***

	×1 strength solution[a]		×10 strength solution[b]	
	Observed	Predicted[c]	Observed	Predicted[c]
	(mmoles/l tissue water)		(mmoles/l tissue water)	
K^+	66	27	73	159
Na^+	3	27	3	159
Mg^{2+}	8·5	175	11	638
Ca^{2+}	1·5	700	1·5	2550
NO_3^-	56	0·0756	38	1·250
Cl^-	3	0·0378	4	0·625
$H_2PO_4^-$	17	0·0378	14	0·625
SO_4^{2-}	2	0·0004	5·5	0·010
Membrane potential	$E_{vo} = -84$ mV		$E_{vo} = -71$ mV	

Notes

(a) Ionic composition mM: K^+, 1·0; Na^+, 1·0; Mg^{2+}, 0·25; Ca^{2+}, 1·0; NO_3^-, 2·0; Cl^-, 1·0; $H_2PO_4^-$, 1·0; SO_4^{2-}, 0·25.

(b) As above but all concentrations increased tenfold.

(c) Predicted values obtained by substitution of E_{vo} for the Nernst potential in the following equation

$$C_j^i = C_j^o \exp \pm \left(\frac{Z_j FE}{RT}\right)$$

E = the electric potential difference, E_{vo}. C_j^i and C_j^o = concentrations of the ion j inside the cell and in the outside solution. Z_j = valency (algebraic) of ion j. F = the Faraday constant. T = absolute temperature and R = gas constant.

(Data from Higinbotham, Etherton and Foster, 1967.)

The tissue had much less sodium in it than predicted on the above assumption and this indicates that sodium inside the cell is at a lower electrochemical potential than outside. Thus, its transport into the cell is passive and the equilibrium must depend on an active sodium extrusion mechanism. Both calcium and magnesium are also at very much lower electrochemical potentials inside the cell; in these cases, the failure to reach equilibrium appears to be due to very effective exclusion of calcium and magnesium by the plasmalemma. There is no evidence for a very rapid efflux of calcium from the protoplasts which would support the idea of active divalent cation pumping. By contrast, it is clear that the maintenance of high internal concentrations of anions depends on active influxes combined with low passive membrane permeability coefficients. In the more dilute solution, the potassium concentration in the roots is higher than predicted by the Nernst equation, hence at least part of it may be transported actively into the cell. In the more concentrated solution, however, potassium is at a lower electrochemical potential inside the cell than outside; thus potassium seems to be pumped out of the cell. From these data, we might conclude that the electrogenic pumping at high external potassium concentrations, as described by Higinbotham *et al.* (1967) (see section above) might be due to a cation extrusion pump which was switched on only at high external salt concentrations.

The data also make the point very elegantly that the internal ionic composition of the root tissue was strongly buffered against changes in the external ionic environment. It is worthwhile bearing this in mind because, as we shall see in chapter 10, in this range of concentration it is frequently shown that the absorption of tracer by excised roots increases. Clearly, it is most important to distinguish influx and net flux in these circumstances although this is not always done. The data in Table 7.4 suggests that this increased influx is countered by increased efflux and that energy has then to be expended both in cation extrusion and anion absorption.

This brief account should have suggested that the relationships between the cells of higher plants and the solutions which bathe them are similar to those in algae. In the next chapter we shall consider the situation in cells which are remote from the simple external bathing medium but depend on other cells in the organism for the ions they receive.

Bibliography

b References in the text

Cram, W. J. (1968) Compartmentation and exchange of chloride in carrot root tissue. *Biochim. Biophys. Acta* **163**, 339–353.

Cram, W. J. (1969) Short term influx as a measure of influx across the plasmalemma. *Plant Physiol.* **44**, 1013–1015.

Davis, R. F. (1972) Membrane electrical potentials in the cortex and stele of corn roots. *Plant Physiol.* **49**, 451–452.

Dunlop, J. and Bowling, D. J. F. (1971) The movement of ions to the xylem exudate of maize roots. 1. Profiles of membrane potential and vacuolar potassium across the root. *J. Exp. Bot.* **22**, 434–444.

Elzam, O. E. and Epstein, E. (1965) Absorption of chloride by barley roots: kinetics and selectivity. *Plant Physiol.* **40**, 620–624.

Epstein, E. (1972) *Mineral Nutrition of Plants: Principles and Perspectives.* Wiley (N.Y.).

Etherton, B. (1963) Relationship of cell transmembrane electropotential to potassium and sodium accumulation ratios in oat and pea seedlings. *Plant Physiol.* **38**, 581–585.

Etherton, B. and Higinbotham, N. (1960) Transmembrane potential measurements of cells of higher plants as related to salt uptake. *Science* **131**, 409–410.

Greenham, G. C. (1966) The relative electrical resistances of the plasmalemma and tonoplast of higher plants. *Planta* (Berlin) **69**, 150–157.

Goldsmith, M. H. M., Fernandez, H. R. and Goldsmith, T. H. (1972) Electrical properties of parenchymal cell membranes in the oat coleoptile. *Planta* (Berlin) **102**, 302–323.

Higinbotham, N., Etherton, B. and Foster, R. J. (1967) Mineral ion contents and cell transmembrane electropotentials of pea and oat seedlings tissue. *Plant Physiol.* **42**, 37–46.

Higinbotham, N., Graves, J. S. and Davis, R. F. (1970) Evidence for an electrogenic ion transport pump in cells of higher plants. *J. Membrane Biol.* **3**, 210–222.

Katou, K. and Okamoto, H. (1970) Distribution of electric potential and ion transport in the hypocotyl of *Vigna sesquipedalis*. I. Distribution of overall ion concentration and role of hydrogen ion in generation of potential difference. *Pl. and Cell Physiol.* **11**, 385–402.

Kirkby, E. A. (1968) Influence of ammonium and nitrate nutrition on the cation–anion balance and nitrogen and carbohydrate metabolism of white mustard plants grown in dilute nutrient solutions. *Soil Science* **105**, 133–141.

Läuchli, A. (1967) Untersuchungen über Verteilung und Transport von Ionen in Pflazengeweben mit der Röntgen-Microsonde. I. Versuche an vegetativen Organen von *Zea Mays. Planta* (Berlin) **75**, 185–206.

Läuchli, A., Spurr, A. R. and Wittkopp, R. W. (1970) Electron probe analysis of freeze-substituted, epoxy-resin embedded tissue for ion transport studies in plants. *Planta* (Berlin) **95**, 341–350.

Maas, E. V. and Leggett, J. E. (1968) Uptake of ^{86}Rb and K by excised maize roots. *Plant Physiol.* **43**, 2054–2056.

MacRobbie, E. A. C. (1969) Ion fluxes to the vacuole of *Nitella translucens. J. Exp. Bot.* **20**, 236–256.

MacRobbie, E. A. C. (1971) Fluxes and compartmentation in plant cells. *Ann. Rev. Pl. Physiol.* **22**, 75–96.

Pitman, M. G. (1963) The determination of salt relations of the cytoplasmic phase in cells of beetroot tissue. *Aust. J. Biol. Sci.* **16**, 647–658.

Poole, R. J. (1969) Carrier-mediated K^+ efflux across the cell membrane of red beet. *Plant Physiol.* **44**, 485–490.

Robertson, R. N. (1960) Ion transport and respiration. *Biol. Rev.* **35**, 231–264.

Robertson, R. N. (1968) *Protons, Electrons, Phosphorylation and Active Transport*. Cambridge University Press.

Scott, B. I. H., Gulline, H. and Pallaghy, C. K. (1968) The electrochemical state of cells of broad bean roots. I. Investigations of elongating roots of young seedlings. *Aust. J. Biol. Sci.* **21**, 185–200.

Sinyukhin, A. M. and Vyskrebentseva, E. I. (1967) Influence of potassium ions on the resting potential of the root cells of *Cucurbita pepo*. *Sov. Plant Physiol.* **14**, 553–557.

Weigl, J. (1969) Efflux and Transport von Cl^- und Rb^+ in Maiswurzeln Wirkung von Aussenkonzentration, Ca^{++}, EDTA und IES. *Planta* (Berlin) **84**, 311–323.

eight

Ion movement within the plant: radial ion migration in roots

With a few exceptions higher plants acquire ions principally from the environment surrounding their roots. It is logical, then, to consider first how ions destined for the shoot move through the root tissues and enter the xylem sap.

Ions moving across the root to the vascular tissue can follow two paths, one in which migration occurs from protoplast to protoplast and another in which some protoplasts are bypassed by the movement of ions through channels in the cell walls, or extracellular *free space*. The former pathway has small links between adjacent protoplasts known as plasmodesmata, the structure and distribution of which will be discussed in relation to intercellular communication. The component protoplasts in this pathway are known collectively as the *symplast* or *symplasm*. We shall review the evidence for ion movement both in the symplast and in the free space, before considering the factors which determine the preferences of given ions for the two pathways.

The ultrastructure of both protoplasts and walls changes as the cells mature and it will be shown that these changes may bring about a marked differentiation in the capacity of different parts of the root axis to transport certain ions into the xylem.

In the final part of the chapter we will consider where the major permeability barriers to ion transport reside in the root and also discuss further evidence for active ion pumping in roots.

Some aspects of root anatomy and cell structure

The detailed anatomy of plant roots is subject to variation between different species so that in this brief treatment I shall restrict myself to some general statements which apply to nearly all roots. Thus, the simplified diagrams of the monocot and dicot roots in Figs. 8.1a and 8.1b are generalized and do not represent named species. There is usually a difference in the arrangement of the vascular tissue at the centre of the root in each case; this becomes more pronounced during the secondary development of the dicot root. Let us, however, consider their more obvious similarities. In both types we have a central core of vascular tissue separated from the peripheral cortex by a uniseriate layer of cells known as the endodermis. The vascular tissue effects communication between various parts of the plant, particularly between the growing zones of the root and the source of carbohydrate synthesis in the shoot, and conducts water and solutes to the shoot from the root system. The root cortex and the outer layer of

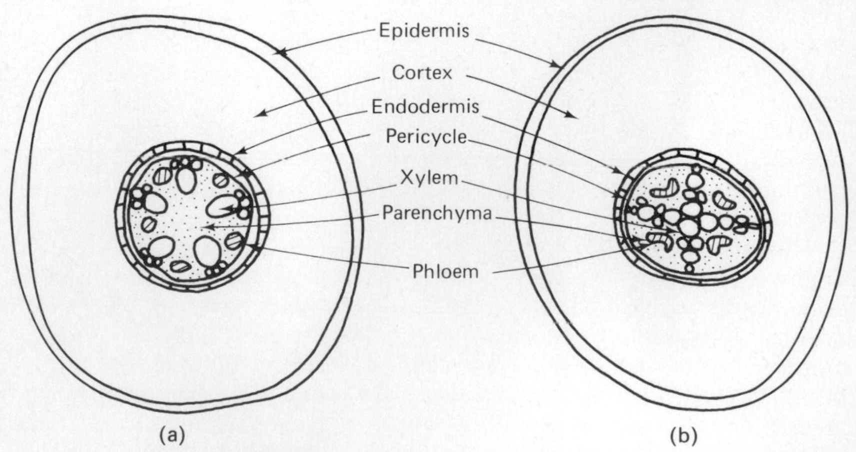

***FIG. 8.1a** Typical monocotyledon root seen in transverse section. Apart from thickening of the cell walls in the vascular tissue and the endodermis there are few changes as the root matures and ages.*

***FIG. 8.1b** Typical young dicotyledon root seen in transverse section. As the root ages secondary development occurs by the division of cambial cells between the xylem and the phloem. Thus, the root increases in diameter and the cross-section of the vascular tissue increases relative to that of the cortex.*

epidermis make up the major part of the root volume in roots where there is no secondary cambial activity. It is well equipped with intercellular spaces which may be either gas or water filled depending on conditions; in the former state they may play an important role in the aeration of the root tissues.

Before we consider the various tissues in greater detail it is necessary to appreciate the general features of cell walls and their pattern of development.

Cell wall structure

When cut off from the apical meristem the daughter cells become separated by a plate of carbohydrate material which forms the matrix in which the primary cell wall is subsequently laid down. The matrix of the primary cell wall consists mainly of pectic substances (polymers of galacturonic acid and its methylated derivatives) whose structure is stabilized by cross-linkage of adjacent carboxylic acid groups by divalent ions (mainly by calcium). In this matrix, fibrils of cellulose are built up to form a loosely woven mesh (Fig. 8.2) and frequently the fibrils are seen to be arranged transversely to the long axis of the cell. In this primary condition, the wall is not a rigid structure but can accept additional wall building materials and can accommodate the stretching which occurs during cell enlargement. The deposition of the cellulose fibrils is not uniform and, in particular, they are much less tightly woven in areas of the wall in which groups of plasmodesmata occur. The plasmodesmata connect adjacent cells (see page 219) and where they are grouped they define pits, or thin areas, in the primary wall. During secondary wall development these pit areas never thicken to the same extent as the rest of the wall, and frequently plasmodesmatal connections are maintained even in quite old cells (see page 229).

The differentiation of the secondary wall involves major changes in chemical composition and considerable thickening which is readily apparent at the level of the light microscope. More detailed examination in the electron microscope reveals that the cellulose fibrils are much more tightly packed in the matrix, sometimes in parallel bundles. The structure is, however, made up from many layers, each layer being inclined at a rather precise angle to those above and below it. There are changes in the matrix as well. Pectic substances are not generally found in the secondary matrix, the bulk of which in the early stages of development is made up from carbohydrate material known as hemicellulose (e.g., mannans and xylans). At this stage of secondary development, the cell wall remains a highly permeable structure. Indeed, the spaces between the microfibrils are often referred to as microcapillaries, the dimensions of which are large in comparison with those of water molecules and hydrated ions. Direct estimates of the permeability of cellulose walls to water are rare in the literature but Briggs (1967) considers that values are unlikely to be less than 10×10^{-6} cm s^{-1} bar^{-1}; the hydraulic con-

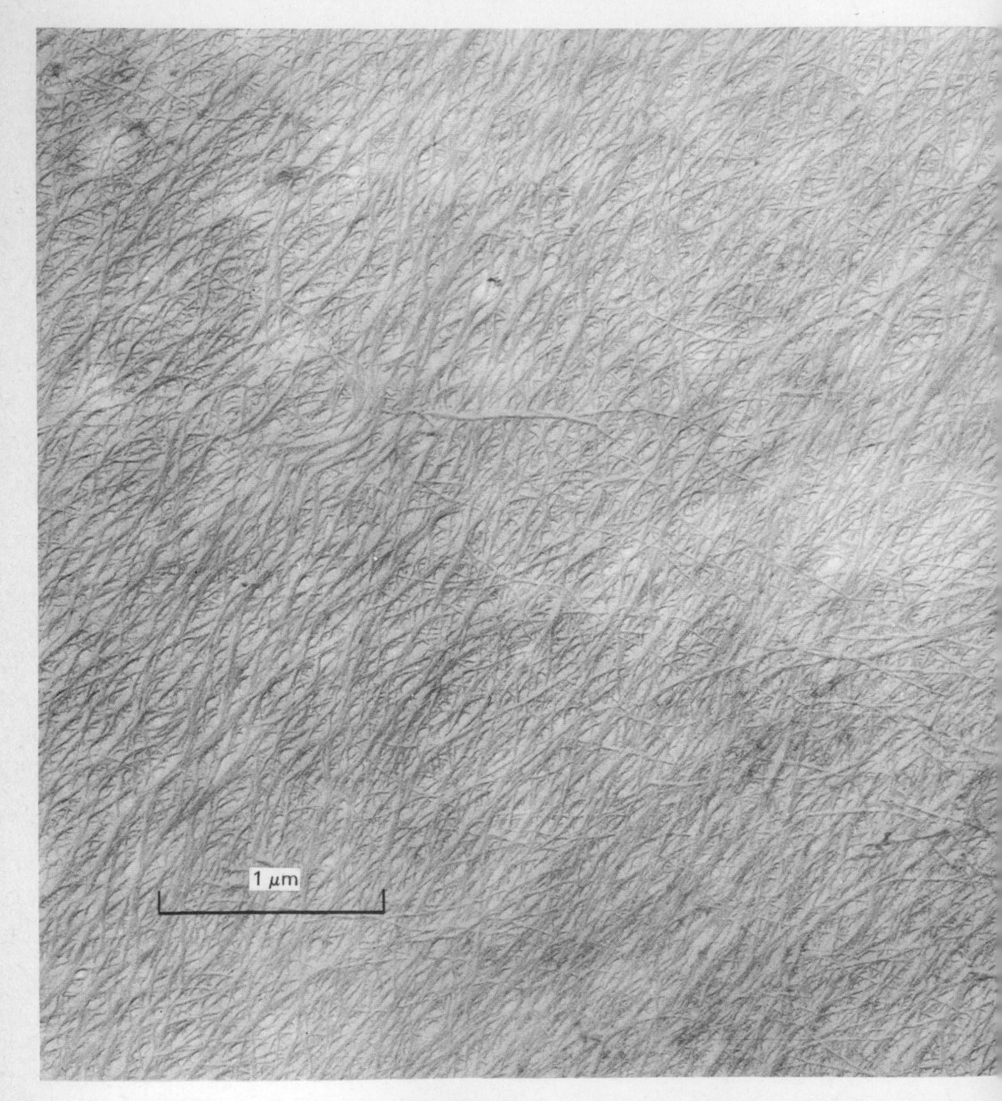

FIG. 8.2 An electronmicrograph of the surface of a primary cell wall in parenchyma from the bulb of **Allium cepa** *(onion). The cellulose appears as fibrils with substantial areas of amorphous matrix between them. (By courtesy of Dr A. W. Robards.)*

ductivity of cell membranes is usually in the range of 0·1–0·6 × 10^{-6} cm s^{-1} bar^{-1}.

Lignification. Further stages of secondary wall development may, however, reduce the general permeability of the cell wall, either by blocking the microcapillaries with lignin or by impregnating the matrix with hydrophobic substances such as suberin. Lignin is formed by the hydrolysis of phenolic glucosides for which β-glucosidase is the enzyme. It is present in woody tissue but from our point of view its early appearance in the endodermis is of particular significance (see page 201). Lignified cell walls are not completely impermeable to water and small solutes since there is evidence that at least some microcapillaries remain open. The exact dimensions of these remaining pores seem to vary widely. Estimates in a review by Preston (1965) range from 5 to 100 nm. The general resistance to water permeation is, however, much increased in the narrow pores. The flow of a liquid through a pore under a given pressure is dependent on the fourth power of the pore radius as shown in eq. (8.1).

$$v = \frac{\pi r^4}{8l} . \eta P \tag{8.1}$$

where v = flow in cm^3 s^{-1}; r = radius (cm); η = viscosity (Poise); l = length of pore (cm) and P = pressure difference at the ends of the pore (dyn cm^{-2}).

Briggs (1967) quotes a value of L_p for lignified cells in *Pinus* of 0·6 × 10^{-6} cm s^{-1} bar^{-1} which is only one-seventeenth of the lower limit for cellulosic walls but is about equivalent to the plasmalemma of living cells.

Suberization. Suberin is a term for a number of different long chain hydroxy fatty acids, e.g., trihydroxyoctadeconoic acid, molecular weight 339. These acids are rather inert, strongly hydrophobic substances, whose presence in the cell wall enormously increases the resistance to water flow. Suberin tends to be deposited in thin lamellae at the junction of the secondary and tertiary cellulosic walls (see page 227 and chapter 9, page 248).

For the purposes of our general treatment, we should regard the cell wall as a system of freely permeable capillaries which offer relatively little resistance to water or solute movement until phenolic or fatty acid substances are laid down in them. We should also recall that the free carboxyl groups in the matrix offer exchange sites for both monovalent and divalent cations, although the latter will be preferred. By the same token, the matrix will tend to exclude anions by charge repulsion.

Epidermal cells

The epidermis occurs at the root periphery and consists of closely packed elongated cells typically in one layer. In monocot roots, which do not increase their diameter greatly once the cells are mature, the epidermis is long lived and the walls may become modified by the deposition of cutin, a water repellent material allied to suberin.

Root hairs. Epidermal cells are capable of great expansion of the outer tangential wall to form tubular protrusions known as root hairs which extend the root surface into the surrounding medium. The function of root hairs has been assumed with little direct evidence. The most extreme oversimplification is that they are responsible for much of the nutrient supply to the plant, the implication being that they are especially endowed as absorbing structures. Manifestly this is not the case since roots of plants grown in aerated water cultures frequently lack root hairs but absorb nutrients from solution at much the same rate as roots possessing root hairs. The life of root hairs is subject to much variation; some authors report death within a few days of formation (McElgunn and Harrison, 1969) while others, e.g., Weaver (1925) show that hairs may remain alive for up to 10 weeks on wheat roots in soil. In view of reports that they can become cuticularized it could be that they cease to be readily permeable shortly after formation (Scott, 1963). The greatest density of root hairs is usually found slightly back from the tip extending roots; this distribution, plus a belief in their special properties, has been the basis, in part, for the suggestion that roots have an 'absorbing zone'. As we shall see later, physiological observations fail to support this suggestion. In the soil environment, however, root hairs may, by virtue of their position, be better placed than other cells in the root to intercept incoming ions whose supply is severely limited (see chapter 11). It is clear that root hairs may have an important mechanical function in anchoring the root and permitting the growing tip to force its way into the soil.

Cortical cells

These cells account for the major portion of the cross-sectional area in monocot and unthickened dicot roots. In barley seminal axes and young primary roots of marrow, the cortex plus epidermis account for approximately 86 and 90 per cent of the cross-sectional area of the root respectively. They are large cells elongated parallel with the main axis with a very thin layer of cytoplasm (1–5 μm) surrounding a central vacuole which occupies more than 90 per cent of the volume of the protoplast.

The walls of cortical cells undergo secondary cellulose deposition but only in a few roots do they become lignified. They remain in a permeable condition and ion exchange studies indicate that the 'free space' volume of barley roots does not alter appreciably as the root ages. In certain species, the walls of the outer cortex may become lignified to form an exodermis, or may have cutin deposited over the walls of the outermost cells.

Endodermis

This uniseriate rank of cells seems to be especially equipped to form a barrier between the vascular and cortical tissues. After cell expansion has ended it is possible to detect the presence of a band of material in the radial and transverse walls of the endodermis which gives a positive staining reaction for lignin; this is known as the Casparian band. Viewed in the electron microscope this region does not show any particular structure, but one would not expect this since lignin is thought to impregnate the amorphous pectic substance of the matrix. The band of lignification decreases the porosity of the wall in this zone, greatly increasing the resistance to water flow through the microcapillaries. The electron micrograph (Fig. 8.3) shows that the plasmalemma is particularly prominent in the vicinity of the Casparian band. It seems to stain more darkly and to be pressed to the wall which is quite smooth in outline by contrast with the irregular wall on either side of it.

Plasmolysis of endodermal cells. It was noted by Bryant (1934) that plasmolysis of the endodermal cells was not accompanied by the usual pattern of shrinkage, but that the plasmalemma remained attached to the Casparian band, thus forming a septum across the cell. The early notion that this attachment was promoted by numerous plasmodesmata in the Casparian band is now known to be incorrect; in fact, plasmodesmata are seen only rarely in the radial wall. It seems more likely that the close association between the structures depends on some linkage which is strong enough to withstand the considerable forces involved in plasmolysis. Possibly, there is an association of the protein of the cell wall and the outer surface of the plasmalemma. An alternative suggestion comes from plasmolysis experiments in *Spirogyra* (an alga) in the presence of aluminium salts. This study, by Böhm–Tüchy (1960), shows that a polyvalent cation can cause cross-linkages powerful enough to prevent the plasmalemma separating from the wall, so that on plasmolysis the protoplast fragments into a number of discrete pieces.

Tertiary development. In the roots of monocots further changes occur in the wall structure of the endodermis. At the point on the axis where lateral roots emerge from the cortex, a tertiary phase of wall building commences. This is most pronounced on the inner tangential and radial walls. The matrix becomes heavily lignified and the suberin lamellae are laid down. This process can continue until the endodermis has a most formidable appearance and the whole cell is surrounded by a highly impermeable layer. It can be seen, however,

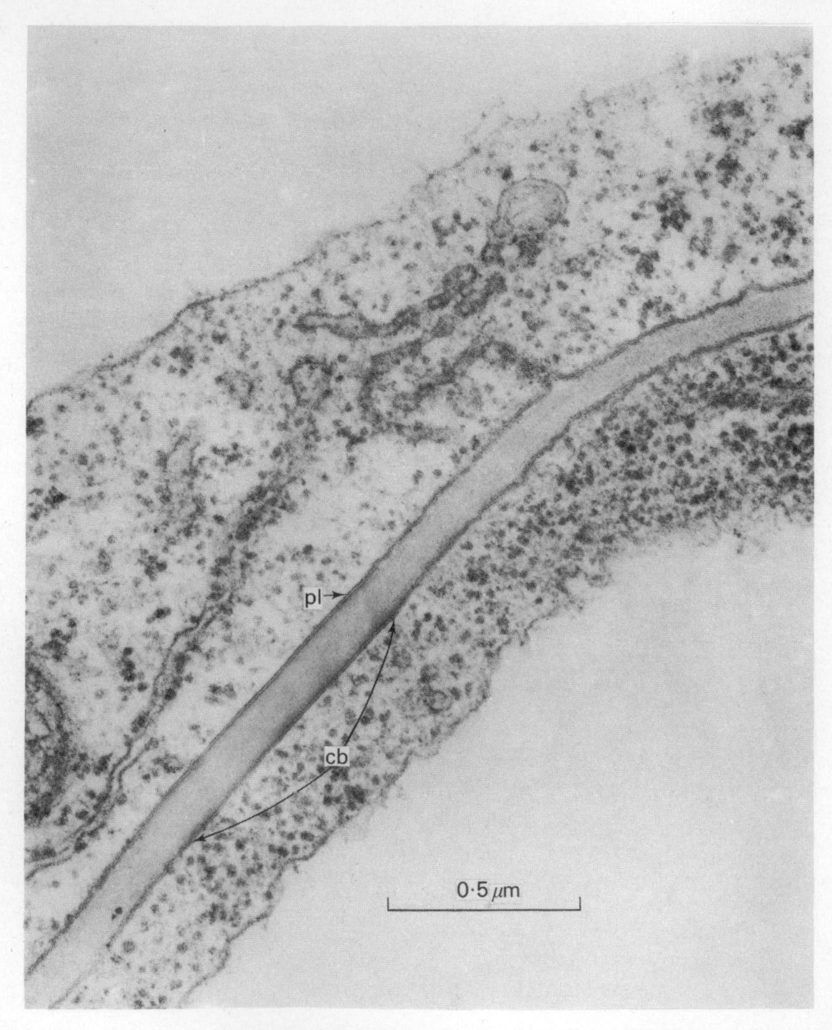

FIG. 8.3 An electronmicrograph of the radial wall of the endodermis in a barley seminal root axis. The Casparian band (cb) is seen in transverse section in the centre of the picture. Note the prominent plasmalemmata (pl) in the vicinity of the casparian band. (By courtesy of Dr A. W. Robards.)

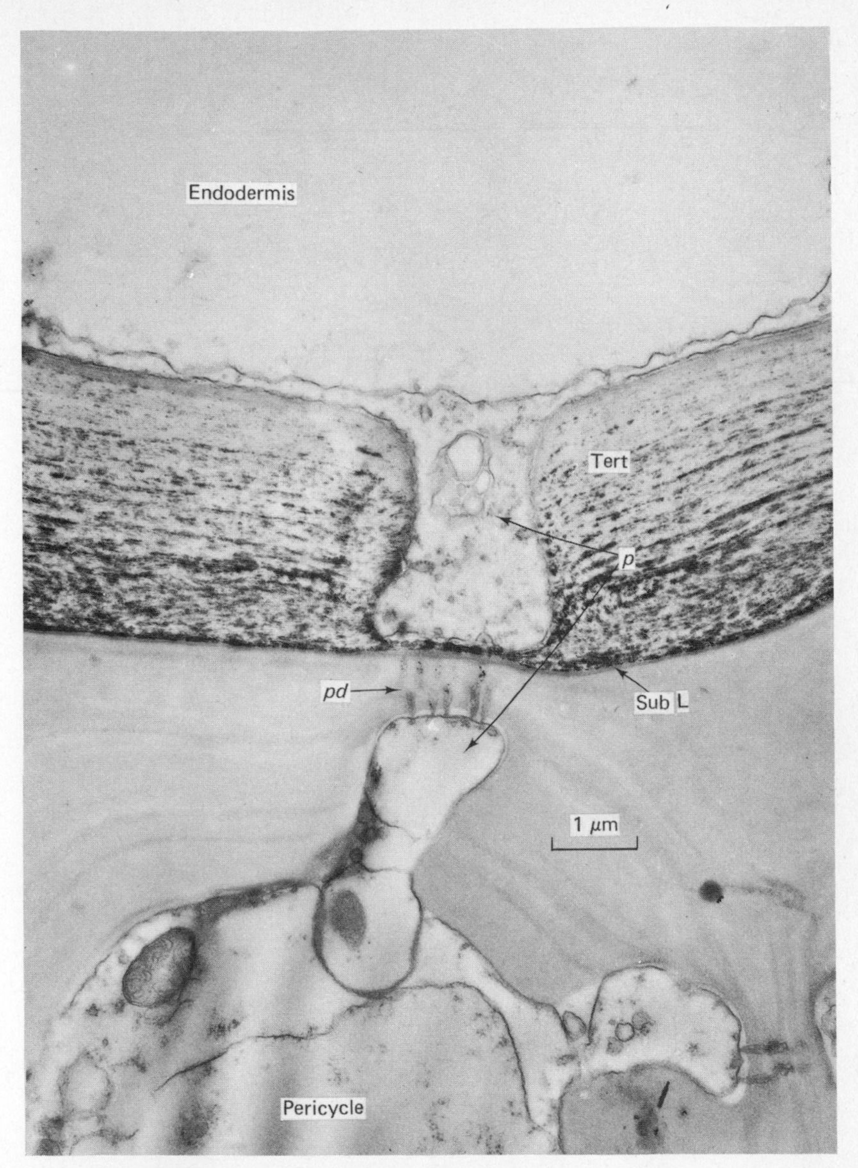

FIG. 8.4 An electronmicrograph of the junction between the endodermis and pericycle in a barley seminal root axis. The section was taken about 40 cm from the root tip and shows the heavily thickened tertiary wall of the endodermis (tert), a pit pair (p), the base of which on the endodermal side is underlain by a thin layer of tertiary wall and suberinized lamellae (sub. 1). Plasmadesmata (p.d) link the pits. (By courtesy of Dr A. W. Robards.)

that there are deep pits in the inner tangential wall which appear to link with others in the pericycle (see Fig. 8.4); communication between these pits is effected by plasmodesmata (see page 229).

Passage cells. Tertiary wall building does not commence at the same moment in all of the cells of the endodermis. In the seminal roots of barley, unthickened cells are frequently seen alongside thickened ones at distances of 12–15 cm from the apex. These cells have been referred to as 'passage' cells. Whether or not they do act as 'passages' is a matter for debate but it is quite certain that they do not remain in this unthickened condition indefinitely, because more than 20 cm from the apex they are rarely, if ever, encountered. In adventitious roots from barley, passage cells may persist for longer periods of time and may be found up to 30–40 cm from the root tip (Robards *et al.*, 1973).

Tissues of the stele

Immediately interior to the endodermis is the pericycle from which lateral root primordia develop. This uniseriate tissue becomes lignified during secondary wall development. In the developed stele in

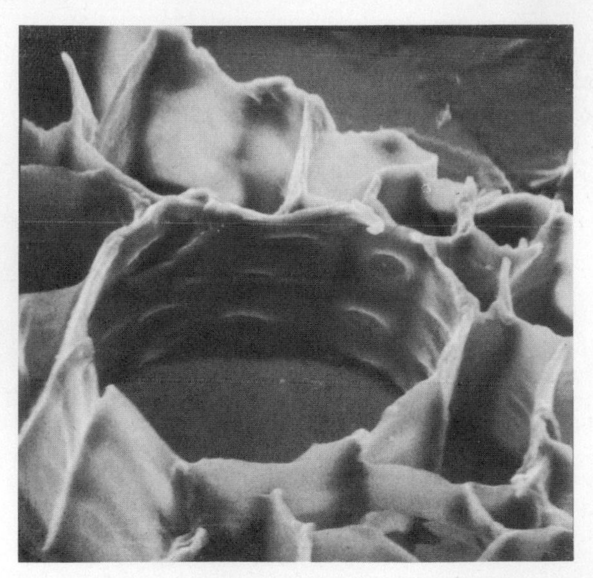

FIG. 8.5 A scanning electronmicrograph of a metaxylem vessel in the root of maize, (Zea). The section was cut from a frozen root and freeze dried on a carbon disc and observed at an angle of 45 degrees. Notice the regular series of elliptical patches on the walls of the vessel; these are the pits through which most of the fluid delivered to the vessel passes. Magnification about ×4500.

monocots special cells in the pericycle, which are smaller and contain more dense cytoplasm than the rest, abut on the groups of smaller xylem elements and are known as protoxylem poles.

Another group of living cells of the stele is the xylem parenchyma which can be seen to surround the larger xylem vessels almost like an epithelial tissue. Communication between the xylem elements and the parenchyma is maintained through regular files of large pits (Fig. 8.5). The cells of the phloem are of three kinds, the sieve tubes, companion cells and phloem parenchyma. The sieve tubes of the phloem transport carbohydrate and other substances from the shoot to the root. The mechanism of this transport is still the subject of much disagreement and it will be impossible even to review the major theories adequately in this text. There has been a recent monograph on this matter by Crafts and Crisp (1972) and there is also a recent very scholarly and thought provoking review by MacRobbie (1972).

The xylem vessels cease to have living cell contents at relatively short distances from the root tip so that for the greater part of their length they are empty capillaries with occasional cross walls with large perforations which allow continuity in the water column to be maintained. The walls of the xylem vessel are lignified and have ion exchange sites in them which selectively bind multivalent cations.

Movement towards the stele *via* the free space

We have seen that in secondary cellulose cell walls which have not been lignified there is a system of microcapillaries which offers little resistance to water and solute movement. Thus, a proportion of the root volume can be freely penetrated and is directly accessible to the outside solution. This proportion is known as the *apparent free space* which is defined as that volume into which a solute can move without the restraint imposed by membrane permeability barriers. The word 'apparent' is added to emphasize that the volume differs for various solutes and for varying conditions.

One way of measuring the free space is to observe the apparent uptake of a substance which is known not to cross biological membranes to any great extent. If it is assumed that the concentration of this substance in the cell walls and other compartments of free space is the same as in the external solution, then the volume occupied by the substance can be simply inferred by how much of it is recovered in the tissue. This, of course, is most easily done using a radioactively labelled substance, i.e., ^{14}C-labelled mannitol, when it is assumed that the specific activity in the free space is the same as in the outside solution. With positively charged ions, however, there will be a certain quantity of ion which will become associated, by electrostatic binding, with the fixed carboxyl groups of the wall matrix in addition to that quantity which is in free solution in the micro-capillaries of the wall. Ions associated with the wall are said to be held in the Donnan free space. We have seen, in chapter 7, how the amount of material associated with the Donnan free space can be calculated from the kinetics of washout experiments.

By most conventional methods of apparent free space measurement it is found that 10–15 per cent of the volume of the root is freely accessible to impermeable solutes; the fraction of the total volume occupied by cell walls is the same or somewhat greater than this.

The inner limit of the free space. The composition of the xylem sap can be quite different from the external solution and constant electric potential differences can be measured between electrodes inserted in the sap and in the external solution (see page 231). There must, therefore, be some barrier to free diffusion into the xylem; if there were not, potentials of the magnitude normally found would not occur.

The endodermis as the limiting structure. The electron micrograph in Fig. 8.6 shows small particles of lanthanum hydroxide which have been drawn into a young barley root by mass flow, along with the water which was moving into the transpiration stream. Since the particles are electron dense they can be visualized directly in the electron microscope. The particles have moved across the root as far as the midpoint of the radial wall of the endodermis, i.e., the Casparian band. With time, there is an accumulation of lanthanum hydroxide particles in the endodermal walls exterior to the Casparian band. This distribution is consistent with the Casparian band being the inner limit of the free space.

A second, less precise, method also leads to this conclusion. When a plant is treated with radioactive calcium (^{45}Ca) a large part of the total uptake (approximately 65 per cent) remains rapidly exchangeable, suggesting that it is outside the protoplasts and in the free space. If the roots are treated with solutions which have a high enough specific activity of ^{45}Ca it is possible to visualize the distribution of tracer in the root by making micro-autoradiographs from sections of frozen tissue. This technique is described in detail by Appleton (1972) and the special problems encountered in working with plant material are reviewed by Sanderson (1972). The density of developed silver grains in the autoradiographic image at any point is proportional to the radioactivity of the underlying tissue.

The histograms in Fig. 8.7 show the grain density over selected tissues after labelling with ^{45}Ca and again after exchange with unlabelled calcium for 10 minutes. In all of the tissues there was some decline in radioactivity during the exchange period, but proportionally this was very much greater in the epidermis and cortex. The falls over the pericycle and xylem parenchyma were small and were probably accounted for by loss to the xylem sap rather than by exchange with the outside solution (Clarkson and Sanderson, in press b). The

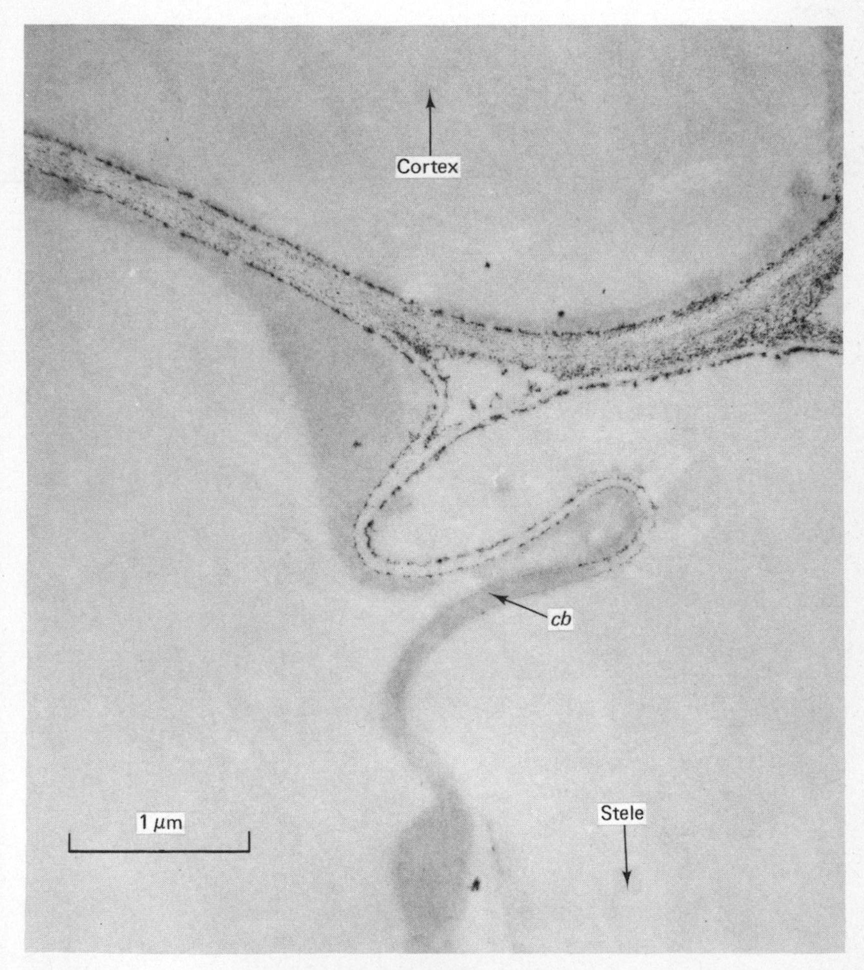

FIG. 8.6 An electronmicrograph showing the distribution of small particles of colloidal lanthanum hydroxide in the vicinity of the radial wall of an endodermal cell in the seminal root axis of barley. This section has been fixed in glutaraldehyde but not stained in any other way. Fine particles can be seen in the radial wall as far as its mid point where the Casparian band (cb) is developed. (By courtesy of Dr A. W. Robards.)

radioactivity in the cortex and epidermis declined by about 60 per cent and again this is consistent with the cortical cell walls and intercellular spaces providing the exchange sites for the Donnan free space with a barrier at the endodermis.

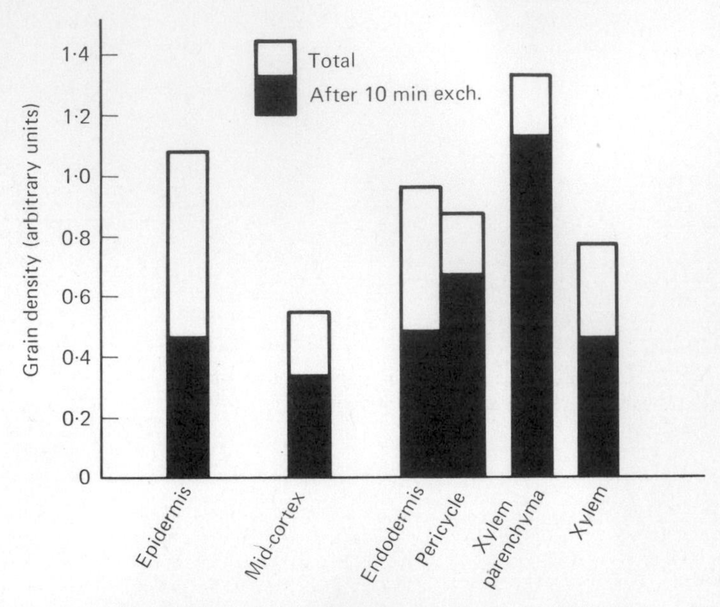

FIG. 8.7 The effect of a brief period of ion-exchange on the distribution of labelled calcium in the seminal root axis of barley. The results were obtained by micro-autoradiography of unfixed, frozen sections cut about 1 cm from the root tip. Notice that proportionally the decline in ^{45}Ca was much greater over the epidermis, mid-cortex and endodermis than over the pericycle and xylem parenchyma. (From Clarkson and Sanderson, 1973.)

Consequences of damage to the endodermis. If the endodermis does limit the free space then damage to it should extend the free space and reduce the control which can be exerted over the ionic composition of the stele and the xylem sap. In monocot roots the stele can be separated from the cortex quite readily either by using a pair of modified wire strippers or by manual stripping, which can expose lengths of stele (up to 10 cm in barley) which are left protruding from the intact axis. It can be shown that in portions of stripped stele which remain attached to a transpiring plant (see Fig. 8.8) water movement through the segment is greatly accelerated as is the translocation of labelled phosphate to the shoot (Table 8.1). The accumulation of labelled phosphate in the stripped stele is very low, in marked contrast with the situation in intact roots.

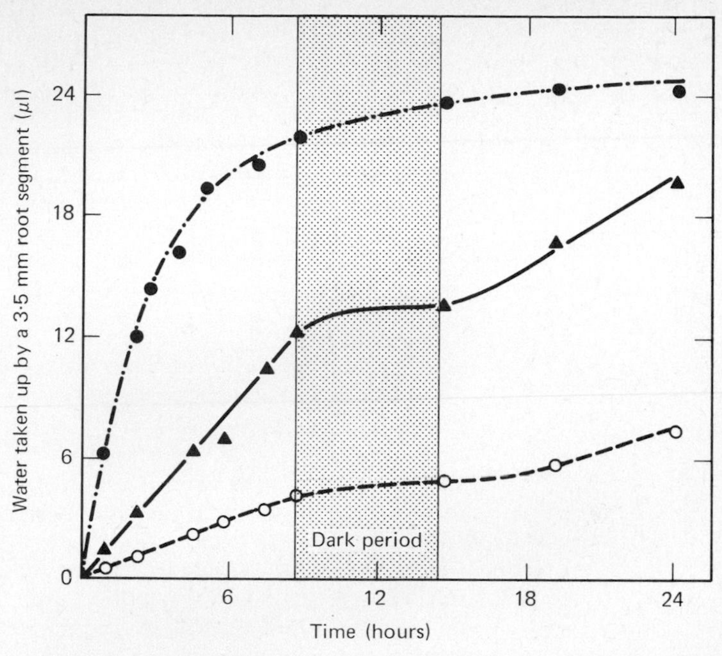

FIG. 8.8 Water uptake through short segments of intact seminal axis of barley. ● = stripped stele (see text) 1 to 2 cm from the root tip. ○ = an undamaged segment > 40 cm from the root tip. ▲ = an undamaged segment 5 cm from the root tip. The night period is indicated by shading. (From Clarkson and Sanderson, in press.)

Table 8.1 *Effect of stele stripping on the time course of the uptake and translocation of phosphate and water by segments of barley seminal axis 1·5–2·0 cm from the root tip*

Uptake period (h)	State of segment	Phosphate uptake T. Seg.	Transloc. (pmoles seg^{-1})	Total	Water uptake seg^{-1}	Phosphate equivalent of water uptake pmoles
0–0·25	Intact	3·4	0·3	3·7	0·16	0·5
	Stripped	0·3	3·5	4·0	1·19	3·6
0–4	Intact	21	2·5	24	2·60	7·6
	Stripped	0·3	54	54	13·5	41

Examination of stripped steles under the microscope shows that the structural integrity of the endodermis is destroyed during the stripping, probably by shearing along the planes of the Casparian band in the radial and transverse walls, so that the pericycle becomes the outermost band of intact cells. If the cortical cells are carefully dissected from a barley root, leaving the endodermis intact, then control over phosphate accumulation in the stele is retained (Clarkson and Sanderson, in press a) and water uptake is hardly affected.

Thus, it seems that removal of the endodermis, but not of the cortex, disrupts the control over radial movement of water and ions into the stele.

The structure of the endodermis in relation to its function. We have already considered the general features of endodermal structure, but here we ought to think again about its design in relation to water and solute movement into the stele. Water and solutes probably encounter less resistance to movement in the walls of the cortical cells than across their plasmalemmata until they reach the Casparian band in the endodermis. At this point in the young endodermal cell, further movement through the cell wall is slow and the possibility of squeezing past the Casparian band, by pushing aside the plasmalemma, is firmly resisted by the attachment of these two structures. Water and any solute which has not crossed the plasmalemma and entered a protoplast in its journey to the stele will be required to do so at the Casparian band. This is borne out by the fact that the hydraulic conductivity, L_p, of roots and of the plasmalemma of single cells, e.g., *Nitella*, is of the same order, both being in the range $0{\cdot}1\text{–}0{\cdot}6 \times 10^{-6}$ cm s^{-1} bar^{-1}.

Once within the protoplast of the endodermal cell the solute can either travel subsequently from cell to cell *via* plasmodesmata, or move back across the plasmalemma at a point on the inner side of the Casparian band and continue its journey through the cell walls to the xylem.

During the early tertiary stage of wall building in the endodermis, the deposition of an encircling band of suberin over the entire wall surface interposes a hydrophobic barrier between the cell wall and the plasmalemma over the entire surface. When this happens direct access to the endodermal protoplast from the wall is made very difficult. In both marrow (*Cucurbita pepo*) and barley roots it has been noted that the entry of calcium into the stele ceases at the point where these suberin lamellae are deposited (see page 227).

Movement towards the stele in the symplast

There has been a growing interest in ion movement through the symplast in recent years and, although largely indirect, different kinds of evidence point to this as a most important pathway. To summarize this evidence before it is presented may help to draw together what may seem to be rather fragmentary information. Briefly, then, the evidence for symplast transport is as follows:

(a) Ion movement can occur rapidly over considerable distances in the absence of any water conducting tissue and mass flow.
(b) Ions can be released from the protoplasts of cells into the xylem in circumstances which make it extremely unlikely that they could have been derived from the external medium, or could have moved *via* the cell wall pathway.
(c) There is evidence that cells in leaves and roots are coupled electrically through junctions of very low electrical resistance which will have, correspondingly, a high ionic conductance.
(d) Related to item (c), it is found that changes in the membrane potential at the root periphery are propagated very rapidly throughout the root.
(e) There is visible evidence of intercellular coupling through plasmodesmata and some suggestion that these structures link the endoplasmic reticulum of adjacent cells.

We shall begin by considering some classic work on symplast transport by Professor Arisz which, although it is related to leaves, provides very convincing evidence of intercellular ion movement.

Symplast transport in the leaves of Vallisneria

Vallisneria is a large aquatic plant with long narrow leaves in which specialized water conducting cells, i.e., xylem, and cuticle are absent; features which are clearly related to its submerged habitat. The leaf has rudimentary bundles of phloem which are separated by broad areas of parenchyma. Arisz (1964) devised a simple system in which lengths of leaf could be sealed over several watertight compartments. In experiments on chloride transport, the proximal 2·5 cm of the leaf strips were submerged in a solution containing 3·3 mM KCl labelled with ^{36}Cl. The distal parts of the leaf strip were kept in distilled water. At the end of 24 hours the compartments were emptied, the leaf strips freeze dried and subsequently divided into the labelled portion (section 1) and two further 2·5 cm lengths (sections 2 and 3). The radioactively labelled and total chloride were determined for each section and compared with the initial condition in which the chloride concentration throughout the leaf had been uniform. The results are summarized in Fig. 8.9.

The supply of chloride to section 1 increased the total chloride concentration in all three sections. In section 1 all of this increase was accounted for by labelled chloride. Indeed, it is clear that some of the unlabelled chloride originally present in the tissue was displaced, hence D1 has a negative value. The increment in total chloride in sections 2 and 3 was made up from both labelled and unlabelled chloride. The increments in the labelled chloride, D2 + D3, are equal to the decrease, D1, in section 1. This result indicates that labelled chloride mixes with endogenous chloride in section 1 and that a mixture of endogenous and labelled chloride is transported to the distal parts of the leaf. Careful monitoring of the water surrounding sections 2 and 3 indicated that no ^{36}Cl leakage from the tissue could be detected.

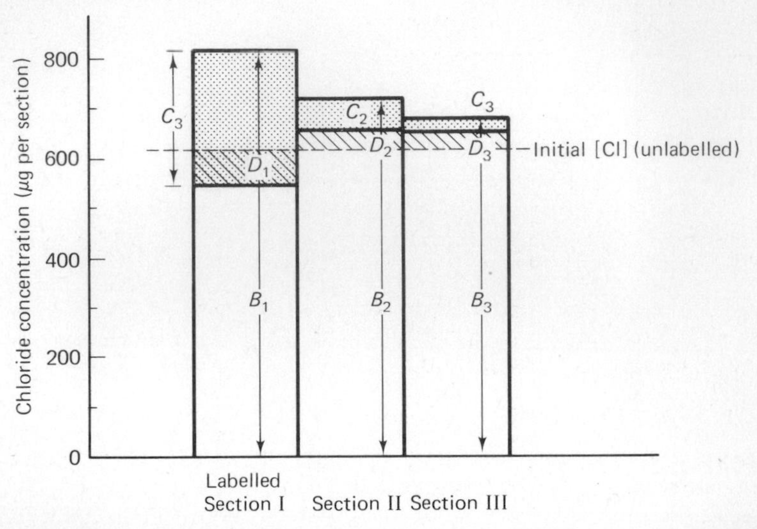

***FIG. 8.9** Distribution of labelled and unlabelled chloride in leaf sections of* **Vallisneria.** *Section I was treated with labelled chloride for 24 hours, while sections II and III were kept in distilled water. For further explanation of B, C, and D see text. (Data from Arisz, 1964.)*

If the chloride had been moving by diffusion through the free space of the leaf (in the cell walls) substantial movement of ^{36}Cl into the water would have occurred. Arisz found also that whereas the uptake of ^{36}Cl from the external solution was strongly inhibited by respiratory poisons, movement from cell to cell through the symplasm was far less sensitive. This was established by comparing the effect of the poison, either in section 1 or in the water surrounding the distal part of the leaf. It was this result which suggested that once chloride had been accumulated in the protoplasts of the labelled portion, its subsequent transport in the symplast was by passive diffusion, which showed only a very weak dependence on metabolism.

Rates of movement in the symplast. The rate of movement of solutes in leaves of *Vallisneria* can be quite impressive and gives some indication of the capabilities of the symplast. In suitable conditions chloride can move 100 cm in 24 hours with only a small chloride concentration gradient to drive it. It is instructive to see how far chloride would move by simple diffusion in an aqueous medium in this period of time. The diffusion coefficient, D, of chloride in a dilute solution with a viscosity similar to that of pure water (0·01 poise) at 20 °C is $2{\cdot}08 \times 10^{-5}$ cm^2 s^{-1}. We can estimate the mean linear distance which chloride would move in a given time from $\sqrt{(Dt)}$ where t = time in seconds. In 24 hours we have $\sqrt{2{\cdot}08 \times 10^{-5} \times 57{\cdot}6 \times 10^3} = \sqrt{1{\cdot}16} = 1{\cdot}08$ cm. Thus, the observed movement of an appreciable amount of chloride *via* the symplast, in the absence of mass flow, is a hundred times further than the mean distance expected from chloride diffusion. It is difficult to reconcile the observed rates of movement with simple diffusion. On the other hand, the reported insensitivity of movement to respiratory poisons, such as sodium arsenate and potassium cyanide (Arisz, 1954, 1964), is difficult to reconcile with transport dependent on rapid cyclosis or active chloride pumping.

Symplast transport in the roots of Zea mays

An interesting demonstration of ionic distribution within cells of the root of *Zea* was provided in experiments by Jarvis and House (1970). They took excised roots which had been grown previously in solutions of potassium and calcium chloride and maintained them in an atmosphere saturated by water vapour. For a period of 24 hours there was a steady accumulation of exudate at the cut end of the root though no salt was supplied externally (Table 8.2).

The chemical composition of this exudate was measured; potassium and calcium appearing in it must have been derived from the cells of the root itself. They found that the volume of exudate declined by about two-thirds but that the concentration of potassium and calcium in the exudate remained the same as in the excised roots which were bathed throughout the experiment in KCl and $CaCl_2$ solutions. Thus, the flux into the xylem vessels was reduced to one-third. The quantity of potassium and calcium moving into the xylem in roots suspended in water vapour was such that much of it must have originated from within the protoplasts of the cells. The authors argue that in the absence of externally supplied ions, movement from the cells into the free space and subsequent migration to the xylem in the cell walls was most unlikely. Their evidence is, therefore, consistent with the delivery of ions to the stele through the symplast.

This suggestion is supported by their observation that when cells

Table 8.2 ***Exudation rates and ionic composition of exudates from excised maize roots in water vapour and in solution with or without pretreatment with 0·8 M mannitol***

Pretreatment	External medium	Exudate flux (µl cm^{-2} h^{-1})	Ionic concentration (mM)		Net ionic influx (nM cm^{-2} h^{-1})	
			K	Ca	K	Ca
Minus mannitol	Water vapour	0·48	22·6	2·09	10·9	1·00
	Nutrient solution	1·71	20·9	2·56	35·8	4·37
Plus 0·8 M mannitol	Water vapour	0·11	20·3	1·78	2·29	0·20
	Nutrient solution	0·26	14·7	2·97	3·85	0·78

(Data from Tables 1 and 2 of Jarvis and House, 1970.)

were plasmolysed in 0·8 M mannitol before the roots were suspended in the saturated atmosphere, the fluxes of ions into the xylem and hence the volume of exudate were much reduced (Table 8.1). Plasmolysis of the cells was believed to break the plasmodesmatal connections between them; the reduced delivery of ions to the xylem would, therefore, have been due to the disruption of the symplast. The authors did not, however, examine their material directly and in view of the report of Burgess (1971) that plasmodesmata can stretch during plasmolysis and form bridges between quite severely shrunken protoplasts, it is possible that the disruption of the symplast may not have been as severe as they supposed.

Further indirect evidence of symplast transport in the roots of barley will be dealt with in a later part of this chapter (see page 227).

Electrical coupling between cells

If the current and voltage electrodes are inserted into one cell of a tissue and a small current applied to the cell, the electric potential difference between the cell interior and the bathing solution, E_{vo}, will change. The magnitude of the change is related to the electrical resistance of the tonoplast and plasmalemma in series (see chapter 3, page 91, chapter 7, page 183). If, in addition, we insert voltage electrodes into the surrounding cells, as in Fig. 8.10, we find that the current pulse in the vacuole of cell A produces hyperpolarizations in the adjoining cell vacuole, V_B, and in the vacuole of a more remote cell, V_C. This indicates a certain degree of electrical coupling between the cells. Spanswick (1972) has made measurements of this kind on the cells of the leaves *Elodea canadensis*, coleoptiles of *Avena sativa* and root cortical cells of *Zea mays*. In each case, he found that there was electrical coupling and that as much as 30 per cent of the current applied to cell A was conducted to B, C and surrounding cells rather than directly across the plasmalemma.

It should be pointed out at this stage that, in experiments reported by Goldsmith, Fernandez and Goldsmith (1972), no electrical coupling could be demonstrated between adjacent parenchyma cells in *Avena* coleoptiles. Since in both cases the work seems to have been very carefully done, it is difficult to decide how this discrepancy has arisen.

From his data Spanswick was able to calculate a value for the electrical resistance of the junction between the cells which contained plasmodesmata; the junctional specific resistance was 0·051 kΩ cm^2 whereas the specific resistances of the plasmalemma and tonoplast were 3·2 and 0·58 kΩ cm^2 respectively. Thus, the junction had only one-sixtieth of the resistance of the plasmalemma. Even so, the resistance is appreciably higher than would be expected from the frequency and dimensions of the plasmodesmata present if it is assumed that they were merely fine threads of cytoplasm passing through the wall. A similar discrepancy has been observed in the electrical resistance of plasmodesmata which link adjacent cells in *Nitella translucens* (Spanswick and Costerton, 1967). We might anticipate, therefore, that plasmodesmata are more than just open ended pipes with a core of cytoplasm running through them.

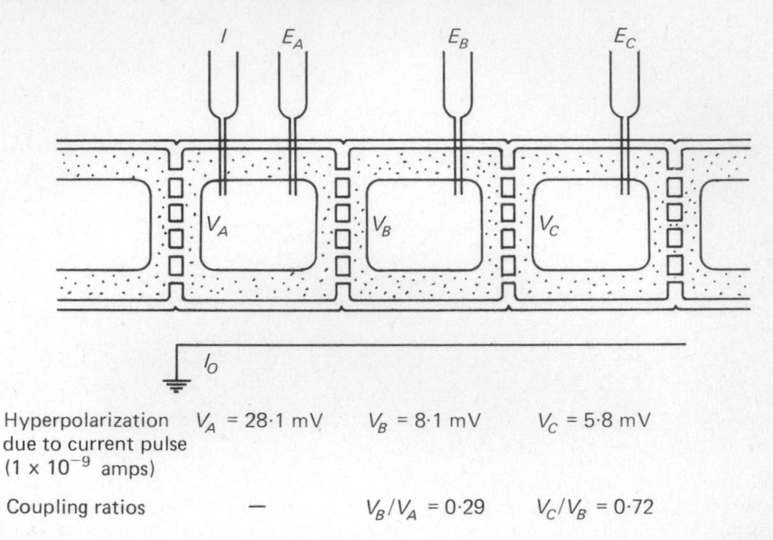

Hyperpolarization due to current pulse (1×10^{-9} amps)	V_A = 28·1 mV	V_B = 8·1 mV	V_C = 5·8 mV
Coupling ratios	—	V_B/V_A = 0·29	V_C/V_B = 0·72

***FIG. 8.10** Experimental design and a summary of the results of Spanswick (1972). The cells were part of an intact leaf of* **Elodea canadensis** *(Canadian pond weed).*
I *and* I_0 *are the internal and external current electrodes*
$E_A \ldots E_C$ *are voltage electrodes inserted in vacuoles* $V_A \ldots V_C$.
This scheme shows direct linkage of cytoplasmic phases of the cells through plasmodesmata; in fact the channels are probably not open in this way although plasmodesmata are abundant (see text).

The ultrastructure of plasmodesmata

When viewed in transverse section (Fig. 8.11) a plasmodesma clearly has several components. The outer pair of dark concentric rings are the inner and outer surfaces of plasmalemma which is continuous through the channel in the cell wall and thus lines the plasmodesma. Within this lining of membrane a second circle can be seen which does not have a distinct trilaminar appearance; there is some suggestion that it is composed of sub-units arranged in a spiral fashion. Robards (1968) has referred to this structure as the *desmotubule*. The nature of the desmotubule is still a matter of some controversy. Robards points out that it is only 20 nm in diameter and that, according to the calculation of Robertson (1964) (see chapter 2, page 36), the unit membrane would not be expected to assume a stable configuration when bent into a ring of such small circumference. Indeed, the smallest vesicle which can be enclosed by a unit membrane alone is estimated as 30 nm. The walls of the desmotubule appear to be 4·0–4·5 nm in thickness and this is only half of the thickness of the unit membrane seen in the endoplasmic reticulum. Robards (1971) points out that lipoprotein micelles (see chapter 2, page 33) can form sub-units of about 4·0 nm in diameter and on

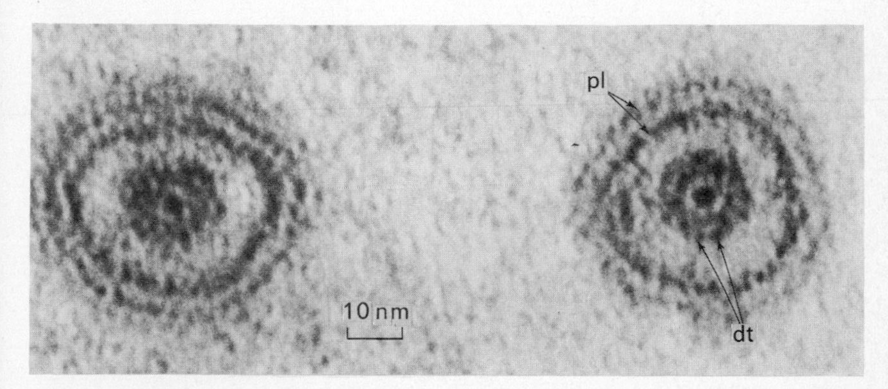

FIG. 8.11 Transverse section of two plasmodesmata from differentiating ray cells in Salix *(Willow). This is greatly magnified, about × 2 000 000, and shows the plasmalemma (pl) lining the pore and a central structure, the desmotubule (dt) which appears to be constructed of sub-units (compare the unit membrane of the plasmalemma). The central dark core is believed to be an artefact (see Robards, 1971). (By courtesy of Dr A. W. Robards.)*

this fact he has proposed a tentative model for the structure of the desmotubule in relation to the plasmodesmatal pore and the endoplasmic reticulum, see Fig. 8.12.

At the centre of the plasmodesma there frequently appears to be a dark rod or core, but electron microscopists suggest that this might be an artefact created by negative staining (see Robards, 1971).

In Fig. 8.12 the longitudinal section shows a desmotubule running the length of the pore enclosed by the plasmalemma. The pore itself is frequently flask shaped, being widest towards the centre of the wall and narrowest at the ends of the pore when the plasmalemma lining and the desmotubule appear to be in intimate contact. Plasmodesmata are frequently branched, there being more entrances to a pore through one wall than exits through the other; this is frequently seen in the plasmodesmatal connections between companion cells and sieve tubes in the phloem. In these cases, the desmotubules coalesce at a median nodule in the central cavity.

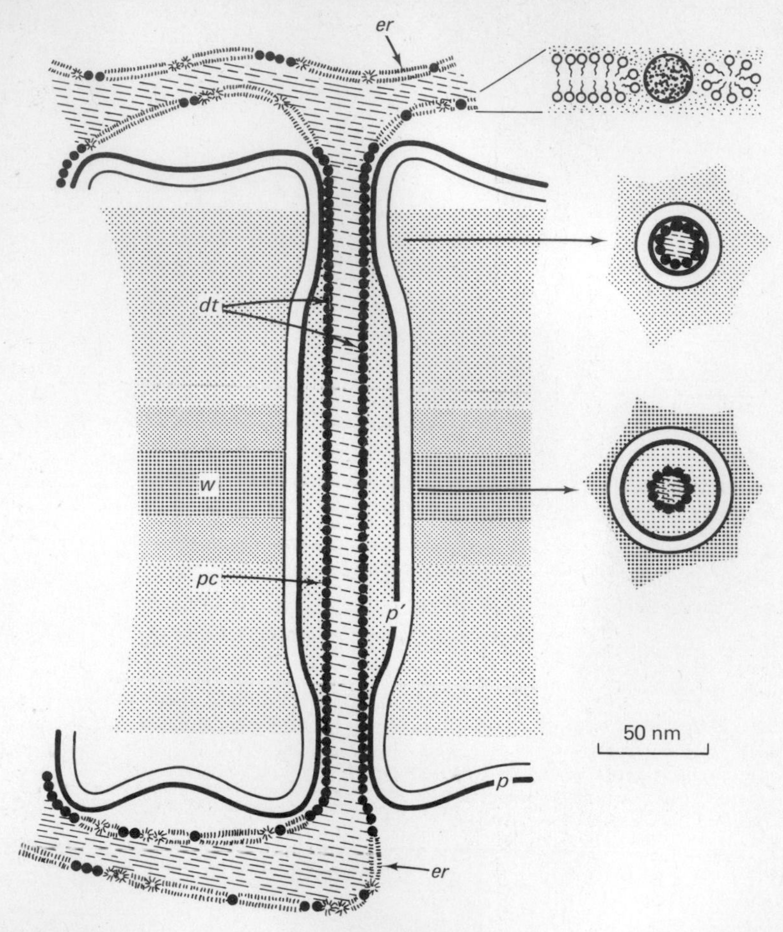

FIG. 8.12 An interpretation of plasmodesmatal ultrastructure. The endoplasmic reticulum, Er, is shown to be continuous with the desmotubule, dt, but in a modified form. The Er membrane is shown as a bimolecular leaflet with micellar inclusions (see chapter 2, page 218). The plasmalemma, p, lines the cavity, pc, and is in close contact with either end of the desmotubule. The various layers of the wall, w, are represented by stippling. (From Robards, 1971, with kind permission of the author and publisher.)

Relationship with endoplasmic reticulum. In a great many electron micrographs, profiles of endoplasmic reticulum seem to be directed towards, and associated with, the mouth of the plasmodesmatal pore and the desmotubule. While precise details of this association are rarely seen it seems to be the general view that the two structures are formally associated so that the plasmodesmata ensure the continuity of the endoplasmic reticulum from cell to cell. If this view proves to be correct then the plasmodesmata cannot be regarded as open pores of cytoplasm between the adjacent cells. The estimates of electrical resistance between cell junctions in *Elodea* and *Nitella*, which we described earlier, also support this view, since the values are appreciably greater than would be the case if the cytoplasmic phase were continuous. It may be that Spanswick's (1972) estimate of 0·051 kΩ cm^2 for the specific resistance of cellular junctions in *Elodea* really represents the specific resistance of the membrane of the endoplasmic reticulum.

It should be stressed that in some plasmodesmata, e.g., the filamentous green alga *Bulbochaete hiloensis*, there is no indication at all of an association between plasmodesmata and the endoplasmic reticulum (Frazer and Gunning, 1969). We should not exclude the possibility of there being plasmodesmata of several types, either in different organisms or perhaps within the same organism (see Burgess, 1971).

The distribution and frequency of plasmodesmata. Plasmodesmata appear to have their origin in parts of the endoplasmic reticulum which traverse the cell plate which forms between the daughter nuclei in a cell which has recently completed mitosis. It might be expected, therefore, that the number of plasmodesmata on a given wall of a mature cell will reflect the predominant planes of cell division which produced the cell. Thus, cells in a file which divides principally in a plane transverse to the long axis of the root will have more plasmodemata on their transverse walls than on longitudinal walls and *vice versa*. During cell enlargement the walls may expand to different extents; the extension of the longitudinal walls is usually greater than the transverse ones. This stretching in the absence of any secondary formation of plasmodesmata will reduce the frequency of plasmodesmata per unit area of wall. Clowes and Juniper (1969) measured the frequency of plasmodesmata in files of root cap cells in *Zea mays* and found values of 45 per µm^2 in the meristematic cells and 8 per µm^2 in the peripheral cells. This decreased density corresponded closely with the increase in the wall area of the cells, the actual number of plasmodesmata per cell remaining constant. Clowes and Juniper (1969) estimate that even small meristematic cells are joined to their neighbours by 1000–100 000 plasmodesmata. Some other estimates of plasmodesmatal frequency are given in Table 8.3.

The distribution of plasmodesmata in a cell is largely determined by its pattern of division and growth and, if plasmodesmata do represent the main intercellular connections of the symplast, this pattern will also play a part in determining the principal directions of symplast transport through the cell.

Table 8.3 ***Some estimates of the frequency of plasmodesmata between cells in various tissues***

Species	Tissue	Plasmodesmatal frequency (per $mm^2 \times 10^{-6}$)	Reference
Avena sativa (oat)	Cortical cells of mature coleoptile	3·6	Tyree (1970)
Metasequoia glyptostroboides	Phloem parenchyma—radial walls—in pit membrane	30–40	Kollman and Schumacher (1962)
	Sieve tubes—tapering radial end walls—sieve area	7	Kollman and Schumacher (1962)
Tamarix aphylla	Walls between stalk and mesophyll cells (see chapter 9, page 269	17	Thomson and Liu (1967)
Zea mays (corn)	Root cap cells:		
	Meristematic		
	Transverse	14·87	
	Longitudinal	5·30	
	Peripheral		Juniper and French (1970)
	Transverse	5·13	
	Longitudinal	0·45	
Hordeum vulgare (barley)	Endodermal cells: inner tangential wall		
	2 mm from root tip	1·19	
	4 mm from root tip	0·69	Robards *et al.* (1973)
	1 cm from root tip	0·65	
	30–40 cm from root tip	0·67	Clarkson *et al.* (1971)
Triticum aestivum (wheat)	Endodermal cells: outer tangential wall		
	5 mm from root tip	0·29	Tanton and Crowdy (1972)
Dryopteris felix-mas (male fern)	Root meristem; primary pit fields	140	
	Primary walls	10–20	Burgess (1971)

Choice of pathway for radial movement of ions

Having established that there can be two pathways we must now consider what factors predispose ions to move along them. In this discussion we shall consider only young root tissue within a few centimetres of the root apex; in the next section we shall consider how secondary anatomical development can affect movement through the two pathways.

The two main pathways for the movement of solutes across the root should not necessarily be thought of as exclusive. We can envisage a situation in which an ion travels a certain distance in the free space and then, because it finds itself in the vicinity of a vacant absorption site, enters the protoplast across the plasmalemma and continues its journey in the symplast. The structural design of the endodermis does seem to indicate that at some point every solute which eventually reaches the vascular tissue must enter the protoplast, at least briefly, so that the Casparian band is circumvented.

It has been asked on more than one occasion what purpose the root cortex actually serves. Surely if the endodermis represents some kind of a Rubicon which all solutes must cross, why not place it at the surface of the root? One can only give speculative answers to these questions.

It is clear that the cortex considerably increases the root diameter and thus the surface area of the root. As we shall see in chapter 11, this is of considerable significance for plants growing in soil where diffusion of ions to the root surface may become the rate limiting step in ion absorption by the root. However, even within the root itself it is obvious that a cortex—the walls of which can be freely permeated by the external solution—presents a much larger surface of plasmalemma to incoming ions than would be the case if a structure like the Casparian band occurred at the periphery. Thus, the cortex presents a large number of opportunities for ions to enter the symplast and for certain ions, e.g., $H_2PO_4^-$, Cl^-, NO_3^-, K^+, we might regard it as a collecting tissue for the symplast. In other instances, e.g., Ca^{2+}, which will be discussed below, this does not always seem to be the case.

For ions which are transported *via* the symplast we can imagine that, where they are present at low concentrations in the external medium, most of the incoming ions can be handled by the ion pumping sites or carrier molecules or pores in the root periphery, i.e., root hairs and epidermal cells (if not cuticularized). As the external concentration is increased the handling capacity of the superficial absorption sites is exceeded so that some of the ions are carried along with the water moving into the free space before they encounter a vacant absorption site. Thus, the extent of movement in the free space will be determined by the number and the turnover rate of specific absorption sites, the external solution concentration and the rate of water movement through the cell walls. It will also depend on the charge carried by the ion. Anions will move reluctantly through the free space since the fixed carboxyl groups in the matrix and the middle lamella are negatively charged. For this reason, cations tend to be attracted to the negatively charged cell walls. The interactions between a cation and the free space depend on the number of positive charges it carries and the hydrated ionic radius. Thus, from equivalent solutions the free space binds more calcium than potassium and more calcium than the more bulky, hydrated ions of magnesium (13 molecules of H_2O in its hydration shell: see Table 4.1, page 99). Aluminium (Al^{3+}) which is trivalent, is more firmly bound than calcium which it replaces on exchange sites in the free space (Clarkson and Sanderson, 1971).

While we can think of the cortex as a collecting tissue for anions and monovalent cations *en route* for the stele, the evidence suggests that, in some plants, this is not true for calcium. A simple tracer experiment combined with some micro-autoradiography provides evidence on this point.

Radial calcium movement. A barley plant is labelled for some time with ^{45}Ca, the roots washed and then replaced in a solution of the same calcium concentration but lacking tracer. For a certain period we can measure an increase in the radioactivity of the shoot after the supply of tracer to the root has been discontinued (Fig. 8.13). After six hours the increase in the shoot radioactivity with time becomes trivial and not significant statistically. Over the same period of time the radioactivity of the root first of all decreases very rapidly due to exchange from the Donnan free space (see chapter 7, page 178) and then declines more slowly for six hours. Beyond six hours the radioactivity of the root, unlike that of the shoot, continues to decline slowly. The experiment shows that only a certain fraction of the calcium present in the root when labelling was discontinued is translocated to the shoot, and that after six hours the compartment containing this calcium has been emptied by incoming non-radioactive calcium. The question we should now ask is where the calcium remaining in the root is located. Here micro-autoradiography can help us. The histogram in Fig. 8.14 shows that, once the calcium in the DFS is removed by exchange, the ^{45}Ca in the cortex declines very slowly in comparison with that in the cells in the stele. This ^{45}Ca is probably within the protoplasts of the cortical cells but for some reason it is transported with great reluctance, if at all, to the stele *via* the symplast. The results point strongly to the conclusion that calcium reaching the stele does not equilibrate with the bulk of the cortical calcium.

The work by Jarvis and House quoted earlier (page 213) indicates that, in maize, calcium can move *via* the symplast. Other evidence supports the view that the mechanism of calcium transport in maize may differ from that in barley and other species.

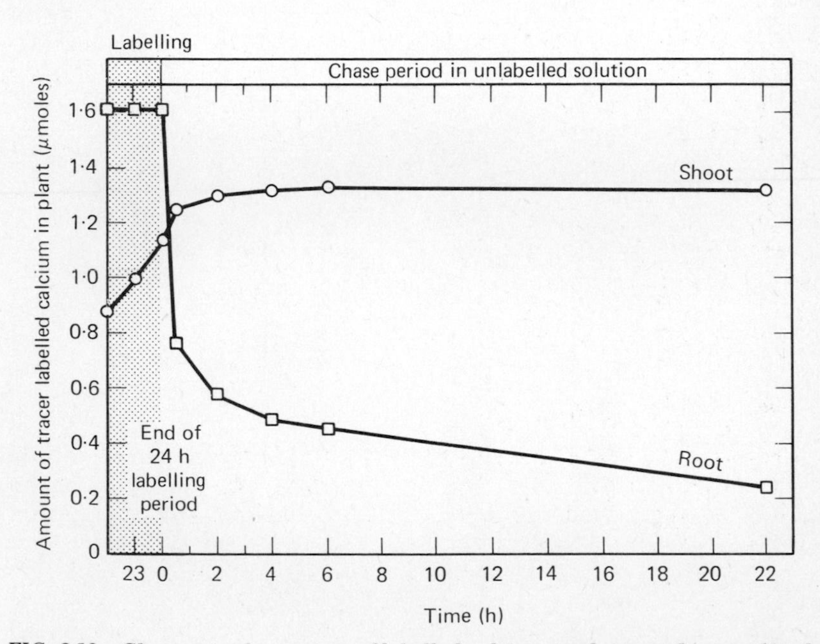

FIG. 8.13 Changes in the amount of labelled calcium in shoots and roots of barley plants. The plants were labelled in a solution containing 0·1 mM $CaCl_2$ for 24 hours and then transferred to unlabelled solution. Notice that the roots continue to lose their labelled calcium to the external medium rather than to the shoot which does not significantly gain labelled calcium after two or four hours. (Data from Clarkson and Sanderson, 1973, in press.)

Radial anion movement. Chloride and nitrate are rapidly transported from compartment to compartment in the cells *via* the symplast. Anderson, Goodwin and Hay (in press) have shown that both of these ions are released into the xylem exudate of excised maize roots for 22 hours and longer after the external supply of the respective ion has been discontinued. The quantity of chloride entering the xylem was such that it must have come largely from the vacuoles of the cortical cells. If all the exuded chloride or nitrate had been derived from the living cells of the stele or solely from the cytoplasm of the cortical cells it would be necessary to propose improbably large decreases in the concentration of the ions in these compartments. These results indicate that the chloride and nitrate in the various tissues and compartments of the root are in equilibrium and that ions can move freely between the vacuoles and the symplast.

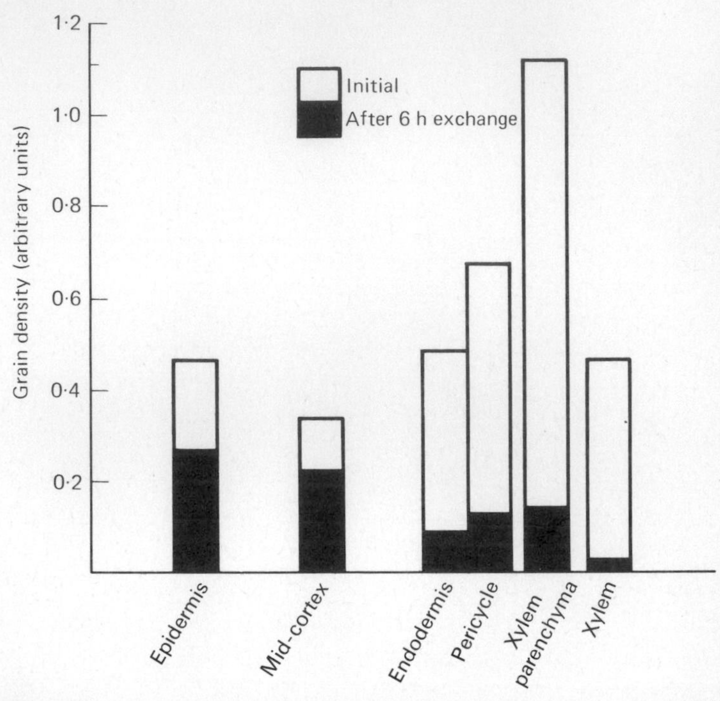

FIG. 8.14 ***Changes in the distribution of labelled calcium in root tissues of barley during a 'chase' period. The initial distribution was that found in the tissues after ten minutes exchange. The changes during the subsequent six hour 'chase' in unlabelled calcium chloride solution are much greater in the vascular tissue than in the cortex and epidermis. Data obtained by microautoradiography of sections cut from unfixed, frozen root. (From Clarkson and Sanderson, 1973, in press.)***

Anatomical development and radial ion movement

We saw, in earlier parts of this chapter, that certain tissues in the root undergo major changes in cell wall structure as they grow old. In this section I will review some evidence which suggests that these changes have a profound influence on the entry of certain ions into the vascular tissue, which move largely in the free space, but surprisingly little influence on those ions which move *via* the symplast. In this differentiation, the endodermis again seems to be a key structure.

In my own laboratory a piece of simple equipment has been designed which allows us to examine the absorption of labelled ions at various points along the length of roots of intact plants (Fig. 8.15). A selected portion of root is isolated from the bulk of the root system by placing it across the diameter of a polythene tube in which an incision has been made and sealing the slit with silicone rubber.

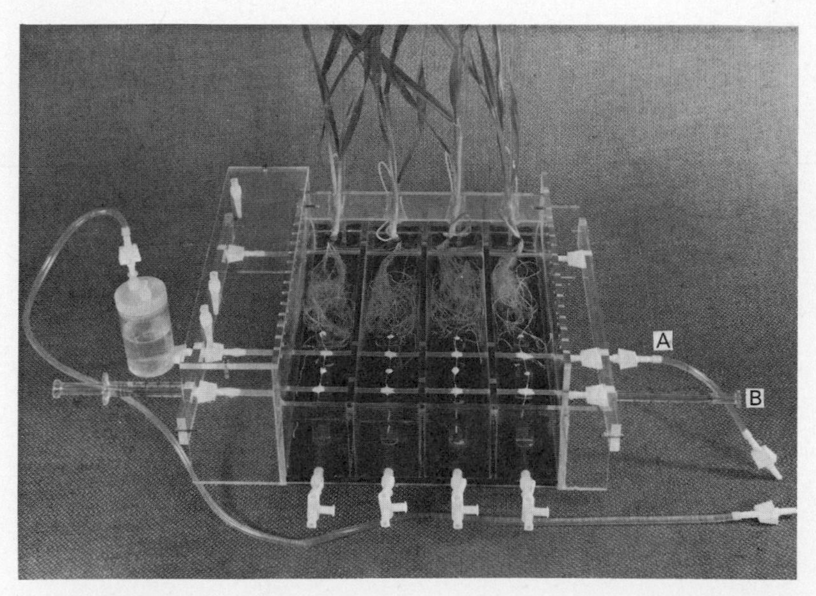

FIG. 8.15 Apparatus for studying uptake of ions and water by different parts of intact root systems. The segments under study are sealed into slits cut in plastic tubing. Tube A is fitted to measure ion uptake, the solution being circulated by a peristaltic pump (not shown). Tube B is fitted to measure water uptake; micropotometer on right, syringe for filling on left.

Radioactive tracers can be supplied to the tube enclosing the segments; in this way we can distinguish the uptake and translocation of ions by this small part of the root system while the remaining major portion of it is treated with unlabelled culture. Radioactivity detected in the shoot will have come principally from the segment exposed to the labelled culture solution.

Using an apparatus of this type we can observe directly the influence which developmental changes in root anatomy and physiology have on ion uptake by the root and the capability for radial transport into the xylem vessels. By labelling the uptake solution with several ions at the same time, we can see whether these developmental changes in any way discriminate between the ions. We have

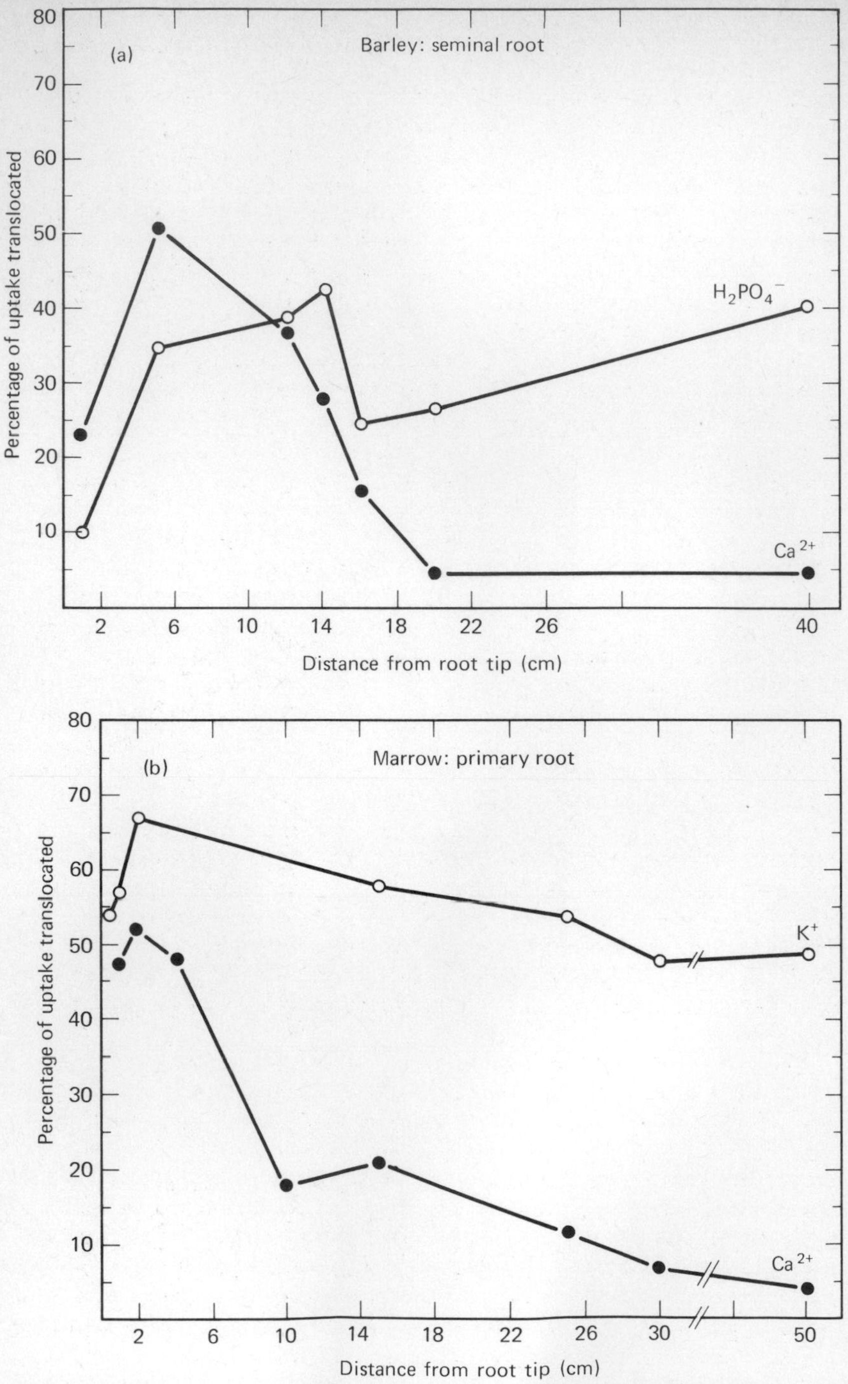

FIG. 8.16 Changes in the ability of roots to translocate ions absorbed by segments located at increasing distances from the root tip. (a) Ca^{2+} and $H_2PO_4^-$ translocation in seminal axis of barley (**Hordeum*****). (b) K^+ and Ca^{2+} translocation in primary roots of marrow (*****Cucurbita*****). (Harrison-Murray and Clarkson, 1973.)***

performed numerous experiments with the roots of barley and vegetable marrow (*Cucurbita pepo*) and the conclusion in both cases is that discrimination against the radial transport of calcium does occur at a certain point along the axis. In the same zone, the translocation of phosphate or potassium to the shoot is unaltered (Fig. 8.16). At a distance of approximately 14 cm from the root tip of both barley seminal axes and marrow primary axes translocation of labelled calcium from the treated segment virtually ceases. What anatomical changes occur in this zone?

The electron micrograph in Fig. 8.17 shows an early stage in the development of the tertiary wall of the endodermis in a barley seminal axis. This section was cut at 8 cm from the root tip and it can be seen that thin lamellae of suberin have been deposited on the surface of the secondary cellulose wall and that more cellulose is being deposited on top of this on the inner tangential wall. The completion of the suberin deposition is marked by a loss of the intimate attachment between the plasmalemma and the Casparian band, so that when the cells are plasmolysed the protoplasts shrink from all walls. It is as if the original function of the attachment between the plasmalemma and the Casparian band has been supplanted by the suberin formation on the wall. This interposes a layer of water repellent material between the porous cell wall and the plasmalemma of the endodermal cell across which the transfer of water and solutes must be quite slow. The only ready access point to the endodermis from the outside at this point would appear to be through the plasmodesmata in the outer tangential wall, i.e., *via* the symplast. In marrow, histochemical staining tests with Sudan IV show that suberization of the endodermis is evident at about the same point where the ability of the root to translocate calcium to the shoot is drastically reduced. This suggests that, once the endodermal plasmalemma becomes inaccessible from the free space, calcium transfer to the stele is effectively stopped because of its negligible movement in the symplast.

The developmental changes in the endodermis have almost no effect on the movement into the stele of potassium (in marrow) or phosphate (in barley) and we have grounds, therefore, for assuming that they are delivered to the endodermal cells *via* the symplast.

In barley, phosphate moves into the stele from dilute solutions in the zone 40–50 cm from the root tip just as readily as in the zone 1 cm from the tip (Clarkson, Sanderson and Russell, 1968), although the root tip is still described by many as the principal absorbing zone. At a distance of 50 cm from the root tip, the walls of the endodermis are massively thickened (Fig. 8.5). The fact that phosphate can move across this tissue as freely as when the cells are in a primary condition is probably explained by the fact that the plasmodesmatal

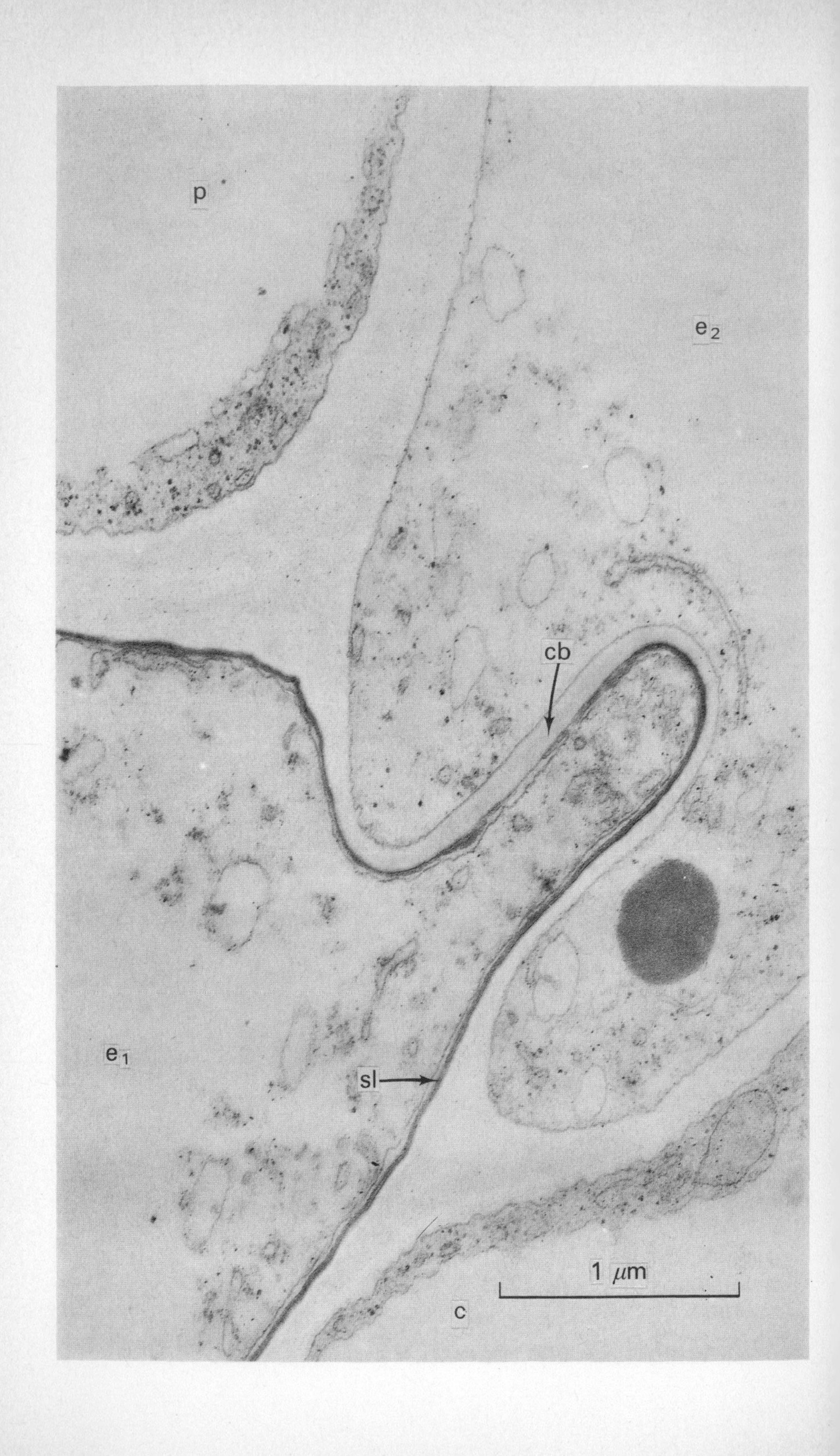
p
e2
cb
sl
e1
1 μm
c

FIG. 8.17 An electronmicrograph showing the early development of suberin lamellae in an endodermal cell. This section was cut 8 cm from the tip of a seminal axis of barley. Note that, in e_1, suberin lamella, sl, is much less distinct at this stage, in the vicinity of the Casparian band, cb, although later it is continuous. The second endodermal cell, e_2, in the picture shows no sign of lamella deposition. Other symbols; p = pericycle; c = cortex. (By courtesy of Dr A. W. Robards.)

connections with the surrounding cells remain intact. They can be seen at the base of the very deep pits in the wall (Figs. 8.18 and 8.19).

Thus, the secondary and subsequent development of the endodermis does not create a general barrier to ion movement into the stele and for neither phosphate nor potassium can barley or marrow roots be said to have an 'absorbing zone'. With both species, however, there does appear to be a correlation between the later stages of endodermal development and the failure of calcium to enter the stele.

Work by others indicates that considerable lengths of root axes absorb and translocate ions (e.g., Brouwer, 1954; Wiebe and Kramer, 1954). Short term uptake followed by 'chasing' with unlabelled phosphate showed that absorption and translocation occurred from all parts of the seminal root system of wheat seedlings

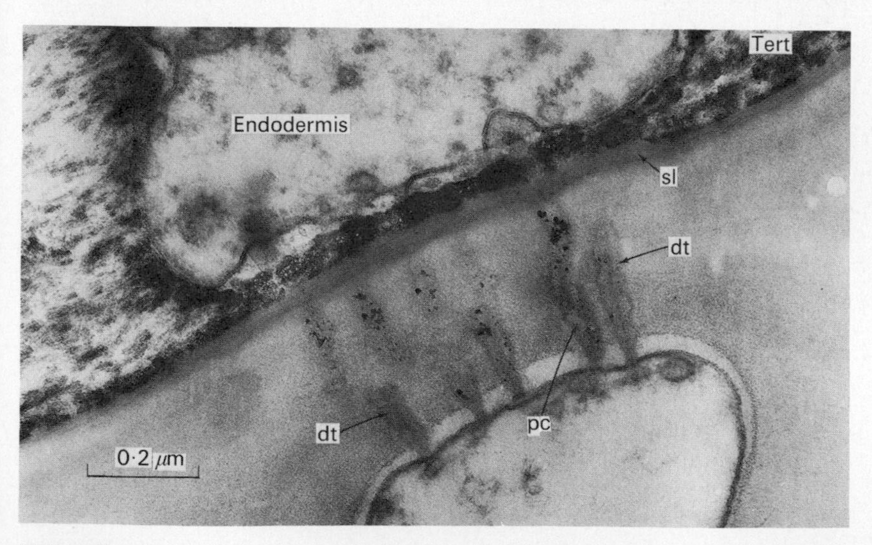

FIG. 8.18 An electronmicrograph showing plasmodesmata in a tertiary wall of the endodermis of barley seminal root axis. The plasmodesmata are cut longitudinally (obliquely) but the flask-like shape of the plasmodesmatal cavity (pc) and the desmotubule (dt) are evident. Note the lamellar deposition of suberin (sl) at the base of the tertiary endodermal wall (tert). Magnification about ×50 000. (Courtesy of Dr A. W. Robards.)

(Rovira and Bowen, 1970) and pine seedlings (Bowen, 1969, 1970). Burley *et al.* (1970) have shown that basal parts of seminal roots of *Zea mays* more than 20 cm from the tip absorb phosphate and translocate it to the shoot very effectively.

Drew, Nye and Vaidyanathan (1969) and Drew and Nye (1970) found that rates of P and K absorption were relatively constant from a segment located in a band of labelled soil at the basal end of an elongating onion root (*Allium cepa*) for 16 days. When observations were concluded the segment was 12 cm from the root tip. In water

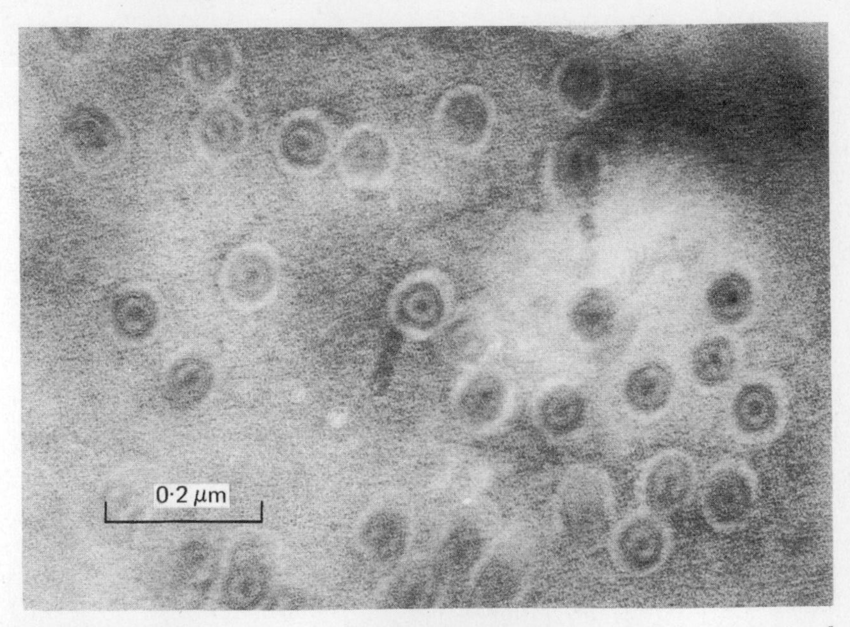

FIG. 8.19 Transverse section of plasmodesmata in a pit field linking pericycle and endodermis. Magnification about ×100 000. (Courtesy of Dr A. W. Robards.)

culture, the uptake and translocation of P and K by the roots of leek (*Allium porrum*) was constant over the entire length (>50 cm) of the seminal axis (Clarkson and Brewster, unpublished data).

Most direct observations on isolated parts of intact roots indicate a widespread capability for ion uptake. There are, however, reports in the literature pointing to the opposite conclusion but these are frequently based on indirect methods. Bar-Yosef (1971) observed the uptake of phosphate and calcium by whole roots of intact maize plants over a period of 10 days. Noting carefully the increments in root surface area and the total uptake of the ions at various sampling periods he showed that the fluxes of both ions declined markedly four days from germination. There are at least two ways in which this decline may be explained: first, the observed decline in rate was some consequence of plant development and the flux over all parts of the root surface was reduced to the same extent; second, that older parts of the axis ceased to function while the newly developed parts absorbed normally. Thus, in the latter explanation, the proportion of 'active' root surface to total root surface declined as the root elongated. Bar-Yosef (1971) chose the latter explanation and supported this choice with the assumption that the formation of suberized lamella in the endodermis prevented the uptake and translocation of both ions. Direct observations on maize by Burley *et al.* (1970) and other work cited above (see Fig. 8.14) indicate that this assumption is incorrect with respect to phosphate.

Physical driving forces in radial ion movement

If the shoot is removed from a root system and a piece of tubing placed over the cut end of the root there will be a slow accumulation of fluid in the tube. In appropriate conditions, fluid will be exuded at a more or less constant rate for several days. Periodic analysis can indicate the rate at which various ions are being released into the exudate. The ionic composition of the exudate frequently differs very markedly from that of the external solution; some of the ions are much more concentrated, while others tend to be excluded, particularly polyvalent ions. A glass electrode, similar to that described on page 83 can be placed in the exudate and connected through a salt bridge to a voltmeter, the other half cell being connected to an electrode immersed in the external solution. Between these two electrodes there is recorded invariably an electrical potential difference, the exudate being negative relative to the outside. The size of the electric potential varies within the same range as that given for the potential difference between cell vacuoles and the outside solution in Table 7.1 (page 182).

In some work, an attempt has been made to analyse the concentration gradient and the electrical gradient along which a given ion moves to the xylem so as to determine whether this movement is uphill or downhill thermodynamically. The situation is, however, somewhat more complicated than a similar analysis applied to a single cell. There has been a temptation to regard the xylem exudate as analogous to the cell vacuole, but a moment's thought shows this to be unsound. Transport into the vacuole of the cell involves movement first across the plasmalemma and then across the tonoplast. These membranes are demonstrably different from one another; e.g., the electrical conductance of the latter may be ten times that of the former. Transport into the xylem vessels, at very least, involves inward movement across one plasmalemma and efflux across another whether the ion moves *via* the symplast or the free space (see page 210). At a slightly more intricate level we might envisage the movement in the symplast involving, in addition, influx and efflux across the membrane of the endoplasmic reticulum or some other cytoplasmic compartment.

In short, without knowledge of the electrochemical potential of an ion in the several compartments which intervene between the sap and the outside solution we cannot be sure which stages of its journey across the root involve active or passive transport. The problems of interpreting electrochemical data in the xylem sap in terms of active and passive transport have been discussed with great clarity by Shone (1969).

In a series of papers Dunlop and Bowling (1971a, b, c) have made a careful study of the relationship between the electrical potentials and the electrochemical potentials of ions in the xylem exudate, the cell vacuoles in the cortex and epidermis and in the external solution. Using glass micro-electrodes they found the same electrical potential difference between the outside solution and the cell vacuole in all of the cells of the root. We illustrated these data earlier in chapter 7, page 184. The insertion of a potassium-selective glass electrode also showed them that the activity of potassium in the vacuoles was more or less uniform from epidermis to xylem parenchyma. When the values for the xylem exudate were compared with those from the vacuoles (Fig. 8.20) it can be seen that over a wide range of external potassium concentration the exudate is less electrically negative with respect to the external solution than the vacuole. The electrochemical potential differences for potassium are, however, the same in both exudate and vacuole.

With chloride there are marked differences between the compartments, thus thermodynamically the movement of chloride from vacuole to exudate is a downhill process. The active transport for both ions appears to be across the plasmalemma of the epidermal or cortical cells.

In another paper Dunlop and Bowling (1971c) measured simultaneously in the same maize root the electrical potential difference between the xylem sap and the outside (which they refer to as the 'trans-root' potential) and between the vacuole of an epidermal cell and the outside (membrane potential). The apparatus used (see Fig. 8.21) allowed them to change rapidly the composition of the solution bathing the roots. When the concentration of potassium was increased tenfold there was a rapid depolarization of both the membrane and trans-root potentials. The initial phase was completed within 20 seconds of changing the solution and the time course in both compartments appeared to be identical. While the membrane potential difference settled to a steady value there was a further slow depolarization of the trans-root potential for 25 minutes. This was followed by a small secondary rise, as can be seen in Fig. 8.22. These secondary movements were small in relation to the initial depolarization.

The identical kinetics of depolarization of the two compartments indicates that a major component of the trans-root potential resides at the plasmalemma of the epidermal cells. The rapid communication of electrical effects at the epidermis to the xylem exudate compartment supports the idea of continuous protoplasm linked by

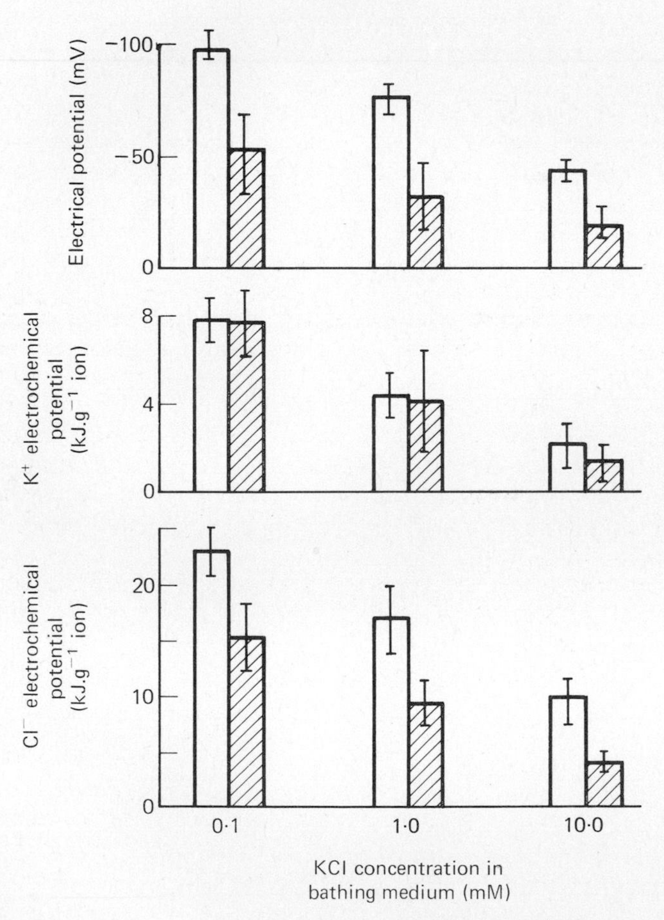

***FIG. 8.20** Comparison of electrochemical potentials of K^+ and Cl^- and the electric p.d. of the xylem exudate (shaded portions) and cell vacuoles (unshaded) in excised maize roots. Vertical bars indicate 95 per cent confidence limits. (From Dunlop and Bowling, 1971c, with the authors' kind permission.)*

channels of low resistance. The authors consider that the slow changes in the value of the trans-root potential may have been associated with the plasmalemma of the xylem parenchyma.

Figure 8.23, which summarizes the forces acting on potassium and chloride ions moving to xylem exudate in excised maize roots, shows that active transport is involved in the movement of both ions across the plasmalemma of the epidermal or cortical cells but that chloride movement into the xylem is passive. The potassium in the

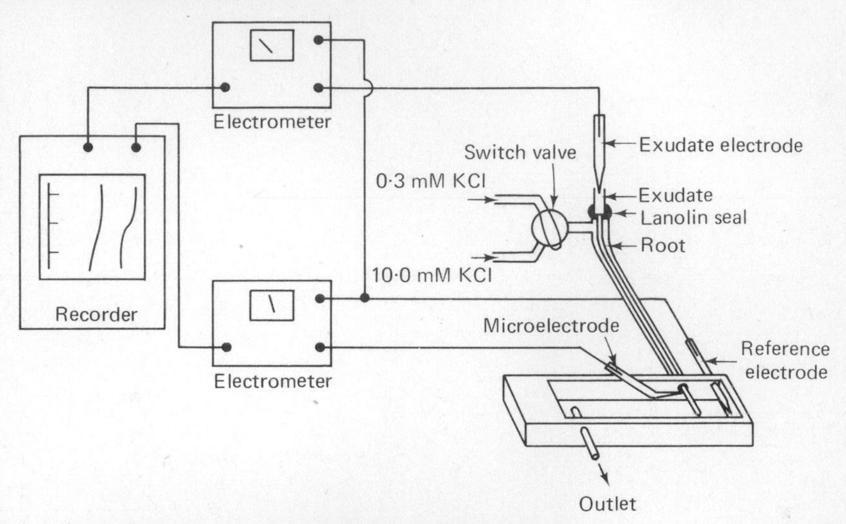

***FIG. 8.21** Apparatus for recording simultaneously the trans-root and epidermal cell membrane potentials in excised maize roots. The solution flowing over the root can be changed within a few seconds using the switch valve. (From Dunlop and Bowling, 1971c, with the authors' kind permission.)*

exudate is apparently in electrochemical equilibrium with that in the vacuoles of the cells. The results do not support the idea that there is some kind of metabolic gradient across the root which might drive or pump ions into the symplast. By contrast, inhibitor experiments by Pitman (1972) suggest that movement of chloride through the root of barley seedlings involves two active steps, at the plasmalemma and between the cytoplasm and the xylem exudate.

Root exudation

The foregoing discussion has rather begged the question of why it is that excised roots exude fluid from the cut ends of the xylem. The phenomenon is very commonly observed and formerly was said, rather mysteriously, to be a manifestation of root pressure.

The accumulation of ions by an excised root seems to be an almost

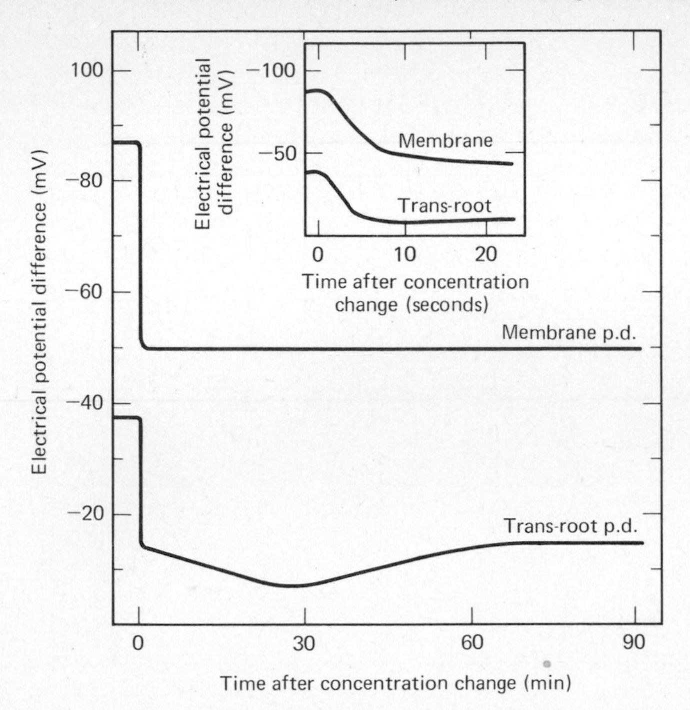

FIG. 8.22 Time course for the depolarization of membrane and trans-root potential difference when the concentration of KCl in the external solution increased from 0·3 to 10·0 mM. Inset shows the changes in the first 20 s. (From Dunlop and Bowling, 1971c, with the authors' kind permission.)

obligatory consequence of metabolic activity. Some of the absorbed ions are released into the xylem vessels where they establish an osmotic gradient. The water in which the ions are dissolved will be at a lower chemical potential than the dilute salt solution in the external medium (see page 6, chapter 1). Thus, water will move across the root and into the xylem down a gradient of chemical potential. This is what would happen to all of the cells if their volume were not rigidly confined by the cell wall. The xylem capillaries are, however, open ended and thus the expansion of the volume of sap is not subject to very large constraints. Water flow into the xylem vessels eventually causes the sap to exude from their ends. Clearly, the rate at which the root will exude will be principally determined by the salt flux into the xylem vessels. Since virtually all of the anion flux is driven by active transport across the plasmalemma (see page 191), exudation from the root is a manifestation of active transport. It is sensitive to a variety of metabolic inhibitors to the same extent as other active transport processes.

Doubt is sometimes expressed whether the excised root is a suitable model for studying radial ion movement into the xylem. Quite apart from subtle interactions between shoots and roots (e.g., hormonal), there is obviously a significant lack of rapid water movement

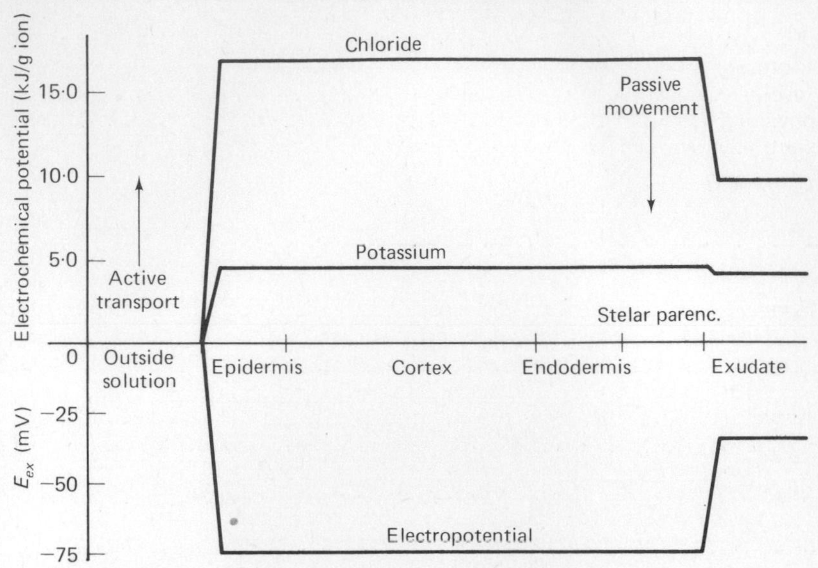

FIG. 8.23 Summary of the forces acting upon K^+ and Cl^- moving to the xylem exudate of excised maize roots bathed in 1·0 mM KCl + 0·1 mM $CaCl_2$. (From Dunlop and Bowling, 1971c, with the authors' kind permission.)

through the root which is caused, in intact plants, by evaporation from the shoot system. Where the movement of ions into the xylem is determined largely by the active transport, there seems to be no reason why mass flow of water should greatly influence this process. The question which remains, therefore, is whether the water moving across the root carries with it any dissolved salt in addition to that which is actively transported. As we shall see, in chapter 10, this depends to a large extent on the concentration of salt in the external solution.

Bibliography

a Further reading

Kramer, P. J. (1969) *Plant and Soil Water Relationship: A Modern Synthesis.* McGraw-Hill, N.Y.

This excellent book covers some of the same ground as this chapter but contains a great deal more information on water movement, an important matter not dealt with in this book (see also Slatyer in bibliography to chapter 1).

Clowes, F. A. L. and Juniper, B. E. (1969) *Plant Cells*. Blackwell, Oxford. Has an excellent survey of cell types and cell wall structure.

Arisz, W. H. (1956) Significance of the symplasm theory for transport across the root. *Protoplasma* **46**, 5–62.
(1969) Intercellular polar transport and the role of plasmodesmata in coleoptiles and *Vallisneria* leaves. *Acta Bot. Neerl.* **18**, 14–38.

Professor Arisz summarizes much of his own work in these stimulating, but discursive, papers, and predicts in a remarkable way the direction which physiological research will take. It is always instructive to read the questions such men ask as well as to see which ones they can answer.

b References in the text

Anderson, W. P., Goodin, L. and Hay, R. K. (in press) Evidence for vacuole involvement in xylem ion supply in excised primary roots of two species, *Zea mays* and *Allium cepa*. In *The Structure and Function of Primary Root Tissues* (Ed. J. Kolek). Slovak Academy of Sciences.

Appleton, T. C. (1972) Autoradiography of diffusible substances. In: Gahan, P. B. (ed.) *Autoradiography for Biologists*, 51–64. Academic Press, London.

Arisz, W. H. (1954) Transport of chloride in the 'symplasm' of *Vallisneria* leaves. *Nature* **174**, 233.

Arisz, W. H. (1964) Translocation of labelled chloride ions in the symplasm of *Vallisneria* leaves. *Proc. Kon. Nederl. Akad. Wetensch, Amsterdam* **C67**, 128–137.

Bar-Yosef, B. (1971) Fluxes of P and Ca into intact corn roots and their dependence on solution concentration and root age. *Pl. Soil* **35**, 589–600.

Böhm-Tüchy, E. (1960) Plasmalemma und Aluminiumsalz. *Protoplasma* **52**, 108–142.

Bowen, G. D. (1969) The uptake of orthophosphate and its incorporation into organic phosphates along roots of *Pinus radiata*. *Aust. J. Biol. Sci.* **22**, 1125–1135.

Bowen, G. D. (1970) Effects of soil temperature on root growth and on phosphate uptake along *Pinus radiata* roots. *Aust. J. Soil Res.* **8**, 31–42.

Briggs, G. E. (1967) *Movement of Water in Plants*. p. 45. Blackwell, Oxford.

Brouwer, R. (1954) The regulating influence of transpiration and suction tension on water and salt uptake by the roots of intact *Vicia faba* plants. *Acta. Bot. Neerl.* **3**, 264–312.

Bryant, A. E. (1934) A demonstration of the connection of the protoplasts of the endodermal cells with the casparian strips in the roots of barley. *New Phytol.* **33**, 231.

Burgess, J. (1971) Observations on structure and differentiation in plasmodesmata. *Protoplasma* **73**, 83–95.

Burley, W. J., Nwoke, F. I. O., Leister, G. L. and Popham, R. A. (1970) The relationship of xylem maturation to the absorption and translocation of ^{32}P. *Am. J. Bot.* **57**, 504–511.

Clarkson, D. T., Robards, A. W. and Sanderson, J. (1971) The tertiary endodermis in barley roots: Fine structure in relation to radial transport of ions and water. *Planta* (Berlin) **96**, 296–305.

Clarkson, D. T. and Sanderson, J. (1971) Inhibition of the uptake and long-distance transport of calcium by aluminium and other polyvalent cations. *J. Exp. Bot.* **23**, 837–851.

Clarkson, D. T. and Sanderson, J. (In press *a*) The endodermis and its development in barley roots as related to radial migration of ions and water. In: *The Structure and Function of Primary Root Tissues* (Edited by J. Kolek). Slovak Academy of Sciences.

Clarkson, D. T. and Sanderson J. (In press *b*) The accumulation of calcium in the tissues of barley roots.

Clarkson, D. T., Sanderson, J. and Russell, R. S. (1968) Ion uptake and root age. *Nature* **220**, 805–806.

Clowes, F. A. L. and Juniper, B. E. (1969) *Plant Cells*. Blackwell, Oxford.

Crafts, A. S. and Crisp, C. H. (1971) *Phloem Transport in Plants*, Freeman (San Francisco).

Drew, M. C., Nye, P. H. and Vaidyanathan, L. V. (1969) The supply of nutrient ions by diffusion to plant roots in soil. I. Absorption of potassium by cylindrical roots of onion and leek. *Pl. Soil* **30**, 252–270.

Drew, M. C. and Nye, P. H. (1970) The supply of nutrient ions by diffusion to plant roots in soil. III. Uptake of phosphate by roots of onion, leek and rye-grass. *Pl. Soil* **33**, 545–563.

Dunlop, J. and Bowling, D. J. F. (1970a) The movement of ions to the xylem exudate of maize roots. I. Profiles of membrane potential and vacuolar potassium across the root. *J. Exp. Bot.* **22**, 434–444.

Dunlop, J. and Bowling, D. J. F. (1970b) The movement of ions to the xylem exudate of maize roots. II. A comparison of the electrical potential and electrochemical potentials of ions in the exudate and in the root cells. *J. Exp. Bot.* **22**, 445–452.

Dunlop, J. and Bowling, D. J. F. (1970c) The movement of ions to the xylem exudate of maize roots. III. The location of the electrical and electrochemical potential differences between the exudate and the medium. *J. Exp. Bot.* **22**, 453–464.

Frazer, T. W. and Gunning, B. E. S. (1969) The ultrastructure of plasmodesmata in the filamentous green alga *Bulbochaete hiloensis* (Nordst.) Tiffany. *Planta* (Berlin) **88**, 244–254.

Goldsmith, M. H. M., Fernandez, H. R. and Goldsmith, T. H. Electrical properperties of parenchymal cell membranes in the oat coleoptile. *Planta* (Berlin) **102**, 302–323.

Harrison-Murray, R. S. and Clarkson, D. T. (1973) Relationships between structural development and absorption of ions by the root system of *Cucurbita pepo*. *Planta* (Berlin) **114**, 1–16.

Jarvis, P. and House, C. R. (1970) Evidence for symplastic ion transport in maize roots. *J. Exp. Bot.* **21**, 83–90.

Juniper, B. E. and French, A. (1970) The fine structure of the cells that perceive gravity in the root tip of maize. *Planta* (Berlin) **95**, 314–330.

Kollmann, R. and Schumacher, W. (1962) Über die Feinstruktur des Phloems von *Metasequoia glyptostroboides* und seine jahreszeitlichen Veränderungen. II. Vergleichende Untersuchungen der plasmatischen Verbindungsbrücken im Phloemparenchymzellen und siebzellen. *Planta* (Berlin) **58**, 366–386.

McElgunn, J. and Harrison, C. M. (1969) Formation, elongation and longevity of barley root hairs. *Agron. J.* **61**, 79–81.

MacRobbie, E. A. C. (1971) Phloem translocation. Facts and mechanisms: A comparative survey. *Biol. Rev.* **46**, 429–481.

Pitman, M. G. (1972) Uptake and transport of ions in barley seedlings. II. Evidence for two active stages in transport to the shoot. *Aust. J. Biol. Sci.* **25**, 243–257.

Preston, R. D. (1965) Interdisciplinary approaches to wood structure. In: Cote (ed.) *Cellular Ultrastructure of Woody Plants*, 1–31. Syracuse U.P.

Robertson, J. D. (1964) Unit membranes: A review with recent new studies of experimental alterations and new sub-unit structure in synaptic membranes. In: M. Locke (ed.) *Cellular Membranes and Development*, 1–81. Academic Press, N.Y.

Robards, A. W. (1968) A new interpretation of plasmodesmatal ultrastructure. *Planta* (Berlin) **82**, 200–210.

Robards, A. W. (1971) The ultrastructure of plasmodesmata. *Protoplasma* **72**, 315–323.

Robards, A. W., Jackson, S. M., Clarkson, D. T. and Sanderson, J. (1973) The structure of barley roots in relation to the transport of ions into the stele. *Protoplasma.* **77**, 291–312.

Rovira, A. D. and Bowen, G. D. (1970) Translocation and loss of phosphate along roots of wheat seedlings. *Planta* (Berlin) **93**, 18–25.

Sanderson, J. (1972) Micro-autoradiography of diffusible ions in plant tissues—problems and methods. *J. Micros.* **96**, 245–254.

Scott, F. M. (1963) The root hair zone of soil-grown plants. *Nature* **199**, 1009–1010.

Shone, M. G. T. (1969) Origins of the electrical potential difference between xylem sap of maize roots and the external solution. *J. Exp. Bot.* **20**, 698–716.

Spanswick, R. M. (1971) Electrical coupling between cells of higher plants: A direct demonstration of intercellular communication. *Planta* (Berlin) **102**, 215–227.

Spanswick, R. M. and Costerton, J. W. F. (1967) Plasmodesmata in *Nitella translucens*: structure and electrical resistance. *J. Cell Sci.* **2**, 451–464.

Tanton, T. W. and Crowdy, S. H. (1972) Water pathways in higher plants. II. Water pathways in roots. *J. Exp. Bot.* **23**, 600–618.

Thomson, W. W. and Liu, L. L. (1967) Ultrastructural features of the salt gland of *Tamarix aphylla*, L. *Planta* (Berlin) **73**, 201–220.

Tyree, M. T. (1970) The symplast concept. A general theory of symplast transport according to the thermodynamics of irreversible processes. *J. Theor. Biol.* **26**, 181–214.

Weaver, J. E. (1925) Investigation of root habits of plants. *Am. J. Bot.* **12**, 502–509.

Wiebe, H. H. and Kramer, P. J. (1954) Translocation of radioactive isotopes from various regions of roots of barley seedlings. *Plant Physiol.* **29**, 342–348.

nine

Ion movement between tissues within the plant

Once ions have entered the xylem sap they are carried in a stream of rapidly moving water, in a transpiring plant, to the shoot system where they are reabsorbed by the cells of the leaf and the stem. In this chapter, as in the previous one, I shall describe first the structure and some of the properties of the vascular system which supplies shoots and leaves with ions and other solutes. There are similarities between the movement of ions *out of* the xylem into leaves and the movement across the root cortex *into* the xylem and the same kind of anatomical barriers have to be negotiated.

Having examined the design of the system I shall consider how ions are circulated and redistributed once they reach the leaves; the redistribution can occur between widely separated organs or adjacent tissues. In the latter instances, solute movement may be an important aspect of physiological processes which are not usually associated with ion transport such as stomatal behaviour, leaf movements and the phytochrome system.

Finally, I shall outline the provisions made by specialized plants to cope with large quantities of salt delivered to the shoot by roots growing in saline habitats.

Movement of ions in the xylem

Components of the xylem sap. The xylem sap is a relatively simple, mainly inorganic solution; its organic components vary both with the species and with the nature of the ions present in the external medium. If the root is treated with a salt solution containing cations which are more readily absorbed than anions, e.g., a solution of potassium sulphate, there may be an excess of cation relative to anion uptake. This imbalance is compensated for by the synthesis of organic anions, usually carboxylic acids. Collins and Riley (1968) found that if the solution bathing excised maize roots was changed from 10 mM potassium chloride to an equivalent concentration of potassium sulphate, cation influx exceeded anion influx into the xylem by 13·5 meq/l; this excess charge was neutralized by, in order of abundance, succinate, malate, citrate, aspartate, glutamate, lysine and serine. In KCl-treated roots these anions were either absent or present in trace amounts only.

In conditions where the nitrate reductase activity of roots is high, amino acids may be prominent components of the xylem sap. Ivanko and Ingversen (1971) found that half of the total nitrogen in the xylem sap from the roots of maize plants was organic, mostly as glutamine, asparagine and lysine. The remaining nitrogen was present as nitrate anion.

There seems to be good evidence to support the view that low concentrations of iron and some other trace elements in the xylem sap are complexed with carboxylic acids and amino acids. Tiffin (1970) has shown strict proportionality between the iron and citrate concentration in the xylem sap of soybean plants.

Ion exchange in the xylem walls. The walls of the xylem contain numerous sites with fixed negative charges which can selectively bind divalent cations and thus retard their movement through the xylem. If we imagine that the binding sites are fully saturated with unlabelled calcium throughout the length of the xylem, we can see that if we were to introduce a small amount of ^{45}Ca-labelled calcium into one end of this column, the activity of the $^{45}Ca^{2+}$ ions will fall off rapidly with distance from the point of introduction because of exchange with the unlabelled calcium ions already present in the xylem walls. In the leaf, calcium ions move from the xylem into the mesophyll cells leaving some vacant sites on the column. These are filled from calcium on the sites below and from calcium ions in the sap. In this way, Bell and Biddulph (1963) envisage that movement of any given calcium ion in the xylem resembles the ascent of an ion exchange column.

If roots absorb radioactive calcium and radioactive phosphorus from a dilute, doubly labelled solution, phosphate is first detected in the shoot within a few minutes while there is an appreciable delay (more than one hour) in the arrival of the labelled calcium. The mobility of the calcium is greatly increased if it is supplied to the root as a chelate or a complex. Isermann (1971) showed that even though the uptake of calcium from calcium citrate and a chelate, calcium ethylenediaminetetraacetic acid (Ca EDTA) was slower than from calcium chloride, the calcium absorbed was much more mobile, suggesting that some Ca-citrate and Ca EDTA entered the plant unchanged, and was not held up in the usual way on the exchange sites in the xylem.

In the presence of adequate amounts of calcium, the ascent of potassium does not seem to be affected by ion exchange in the xylem although there is some suggestion in the work of Jacoby (1965) and of Shone *et al.* (1969) that there is some extraction of sodium from the xylem sap as it passes up the root. This matter will be discussed below.

Extraction of ions from the xylem sap

Once they are within the xylem, there is some evidence to suggest that ions may be withdrawn from the xylem sap by energy dependent processes before they reach the stem or leaves.

Roots. In roots there is not a great deal of information on the interchange of ions between the xylem sap and the surrounding tissues. In *Zea* it has been shown that sodium ions, which freely enter the xylem sap of younger parts of the root axis, are progressively withdrawn from the sap as they ascend towards the shoot, see Table 9.1. Shone *et al.* (1969) showed that the sodium concentration in the exudate from excised roots decreased as a greater length of root was immersed in the radioactive sodium solution (Table 9.1, column 3). At the same time, however, there was no decrease in the sodium concentration in the root tissue itself. If only the apical part of the root of an intact plant was labelled, and the remaining part of the root was suspended in damp air, there was a strong tendency for labelled sodium to become accumulated in the root tissue some distance above the labelled zone. Progress in an upward direction beyond this point was very slow. The results suggest that, as the vascular tissue aged, some mechanism extracting sodium from the xylem became more active. The mechanism for this extraction was thought to be more than a simple ion exchange or sequestering by some other passive process, since less sodium was extracted and more moved to the shoot at low temperatures.

Table 9.1 ***Absorption and translocation of sodium by excised maize roots differing in length after 18 hours' labelling with ^{22}Na****

Length of root (cm)	Exudate volume (ml)	Labelled Na concentration in exudate (mM)	Labelled Na concentration in root tissue (mM kg^{-1} fresh weight)
4	0·026	0·618	0·195
8	0·071	0·261	0·144
16	0·092	0·088	0·173
21	0·179	0·006	0·288

* *External solution contained* 0·2 mM Na^+. *(Data from Shone* et al., *1969.)*

Stems. Sodium extraction is found also in stem tissue (e.g., Jacoby, 1965 and Rains, 1969) and may be a fairly general mechanism accounting for the low concentrations of sodium which are found in the leaves of many species.

One of the earliest plant physiological experiments using radioactive potassium showed that there was an appreciable lateral movement of ^{42}K from the xylem (the wood) into the bark in the stems of willow saplings. Stout and Hoagland (1939) were able to separate these tissues in an intact stem by making a longitudinal incision in the bark and pushing a slip of waxed paper around between the bark and the wood, thus eliminating direct contact between the xylem and the phloem (Fig. 9.1). The plants were supplied with a solution containing ^{42}K for five hours, then the stem was divided into a number of sections and the radioactivity in the bark and the wood was determined. The results, in Fig. 9.1, show that, in the normal course of events, a considerable amount of ^{42}K moving up the stem passes into the bark. The results also show that there is some longitudinal movement of potassium in the bark itself, especially in the downward direction, which was presumably in the phloem. The rate of movement in this direction is, however, quite slow in comparison with that of longitudinal movement in the xylem.

The xylem sap in large herbaceous plants and trees has to pass through great lengths of stem and leaf stalks (petioles) before it actually reaches the leaves and during this passage its composition may be altered by the extraction of ions and water by cells surrounding the xylem (the xylem parenchyma). We shall see in our discussion of transfer cells in a later section that the xylem parenchyma in stems and petioles seems to be structurally adapted for this extraction.

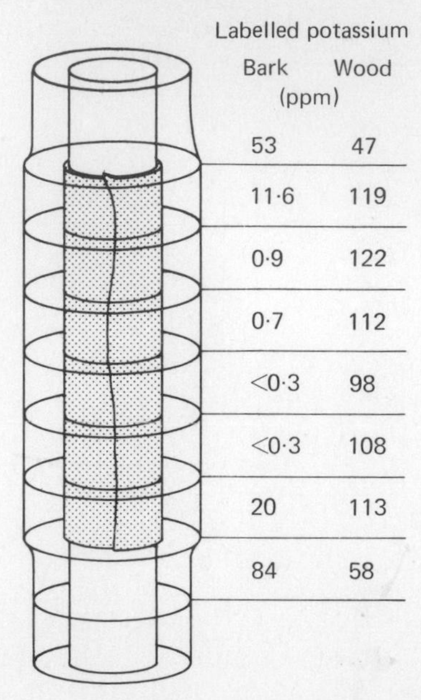

Labelled potassium
Bark Wood
(ppm)
53 47
11·6 119
0·9 122
0·7 112
<0·3 98
<0·3 108
20 113
84 58

FIG. 9.1 The effect of a physical blockage on the lateral movement of ^{42}K from the wood into the bark of the stem of Salix *(willow). The piece of waxed paper making the barrier is shown as the stippled part of the central core of xylem. Based on data from Stout and Hoagland (1939).*

Guttation fluid. When an intact plant is kept in a water saturated atmosphere drops of fluid frequently appear in places on the leaf surface. This fluid is produced by the process known as guttation. The drops appear close to modified vein endings; the solution in them has passed through the entire length of the xylem system from root to leaf and emerges because of root exudation pressure (see page 234) in the absence of evapotranspiration.

The use of a guttating plant as an experimental system allows us to see how much of the salt present in the xylem sap from the roots is extracted during its movement through the shoot system. The data in Table 9.2 show that the total solute concentration of the guttation fluid is much lower than that of the exudate from decapitated root systems of tomato. The reduction in solute concentration was much greater in plants which had been starved of mineral nutrients before the experiment began (low salt plants) even though the concentrations in the root exudate were comparable.

The composition of the fluid is variable but usually inorganic cations are greatly in excess of inorganic anions; the balance is kept by organic anions, of which glutamate is a frequent component,

Table 9.2 Osmotic pressure of guttation fluid and the exudate derived from stumps of excised roots of tomato plants

	Normal salt plants	Low salt plants
	Osmotic pressure (bar)	
Exudate from stumps	−2·40	−2·21
Guttation fluid (leaves)	−0·76	−0·13
Ratio: Exudate/Guttation	3·2	17·0

Data taken from Eaton (1943).

which may make up more than half of the total weight of dissolved material.

The results in Table 9.3 are from an experiment with marrow (*Cucurbita pepo*) seedlings in which the exudate from roots and petioles and guttating leaves are compared. The solution exuded from petioles was collected from approximately their midpoint. The concentration of all of the ions increased to about the same extent between the root stump and the petiole; this may reflect the fact that water was being extracted from the sap rather more quickly than salts by the stem and cotyledons. The concentrations of calcium, potassium and nitrate in the guttation fluid were very substantially reduced with a somewhat less marked decline in phosphate concentration. The only ion to be withdrawn from the xylem less readily than water (the leaf was growing, therefore some water from the exudate would have been incorporated into the expanding cells) was sodium, the concentration of which was more than twice as great in the guttation fluid as in the root exudate. Thus, discrimination against sodium absorption in the marrow plant occurs at two stages—during the initial uptake and transport in the root and again in the reabsorption of ions from the xylem sap in the leaf. This latter stage seems to

Table 9.3 Changes in the concentration of ions during passage through the xylem of stems and leaves of marrow seedlings

	Ca^{2+}	K^+	Na^+	H_2PO_4'	NO_3'
	Concentration (mM)				
Xylem exudate	5·1	12·8	0·3	0·2	7·1
Petiole exudate	8·0	26·2	0·6	0·5	12·9
Guttation fluid	0·9	0·5	0·7	0·2	0·4
External solution	0·15	0·5	0·2	0·01	1·0

(Unpublished data from Mrs C. A. Basham, Letcombe Laboratory.)

contrast with the situation in other parts of the plant body, viz., maize roots and bean stems, where sodium seems to be withdrawn preferentially from the xylem sap (see page 242).

Delivery of ions to the leaf

There is notably much less detailed information available on the movement of ions from the transpiration stream into the protoplasts of the cells in the leaf than there is for the movement of ions from the external solution into the cells of the root. There are a few interesting analogies with the situation in roots but, before we consider these, we should examine the design of the vascular tissue system in leaves and the organization of the tissues which it supplies.

Some aspects of leaf anatomy

Despite their different appearances most leaves have certain common features.

Epidermis. The two leaf surfaces are composed of epidermal cells, which are usually coated by cutin (see chapter 8, page 199) in land plants. This coating greatly increases the resistance to water permeation and therefore reduces evaporation from the leaf surface; this function may be supplemented by the deposition of waxy materials, often in characteristic patterns. The lower epidermis (abaxial) is perforated by numerous small pores, *stomata*, which permit diffusive gas exchange between the photosynthetic mesophyll cells of the leaf interior and the outside air.

Mesophyll. There are large air spaces between the mesophyll cells so that the internal surface area exposed to the intercellular air may be 10 to 30 times that of the surface area of the outer boundary of the leaf. The walls of mesophyll cells are usually free from lignin and suberin so that resistance to water movement through the microcapillaries is generally low (see chapter 8, page 198). In the cells bordering on the sub-stomatal cavity a layer of cutin is often found. The cells contain numerous chloroplasts and the cytoplasm of adjoining cells is linked by abundant plasmodesmata.

Vascular tissue. Large xylem traces entering the leaf either divide into successively finer branches in leaves of dicotyledons or run the entire length of the leaf as parallel veins (monocotyledons) interconnected by very fine lateral anastomoses which separate the leaf up into a multitude of 'blocks', the centres of which are only a few cells away from the nearest xylem delivery point. In both dicotyledons and monocotyledons, the veins are distinctly separated from the mesophyll by a collar of cells, known as the bundle sheath, which is a layer of modified parenchyma. This structure is of interest because, in monocotyledons especially, it resembles the endodermis of the root.

In dicotyledons the long axis of the bundle sheath cells runs parallel to the vein; the cells have thin walls and may, or may not, contain chloroplasts. The bundle sheath extends to and encloses the

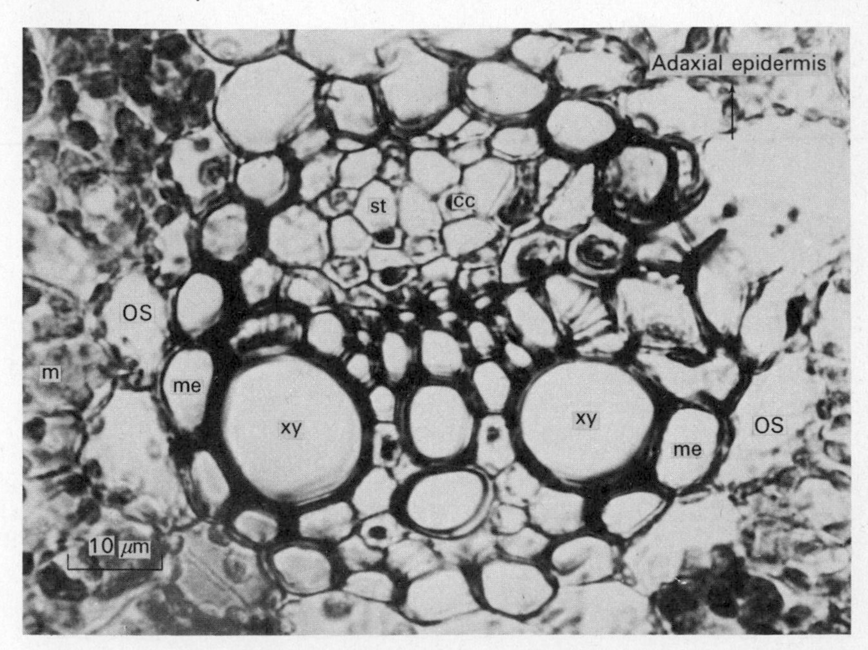

FIG. 9.2 Transverse section of a vein from the leaf of **Hordeum vulgare** *(barley). The section was cut towards the tip of a mature leaf and stained with saffranin and light green. The vein is enclosed by the outer bundle sheath cells (o.s.) which have a few chloroplasts and the inner bundle sheath or mestome (me) which lacks chloroplasts and has distinctly thickened and lignified radial and inner tangential walls. The thickening is most prominent adjacent to xylem tracheids (xy). Other tissues; mesophyll (m), phloem sieve tube (st), phloem companion cell (cc).*

termination of the xylem tracheids at the extreme ends of the veins. The thin walls of the bundle sheath seem to be highly permeable and dyes injected into one of the major veins of the leaf quickly move into the minor veins, across the bundle sheath and into the mesophyll wall. In certain families a structure resembling the Casparian band is found in the radial walls of the bundle sheath cells.

In certain monocotyledons, e.g., *Zea*, *Sorghum* and *Saccharum*, the design of the bundle sheath superficially resembles that seen in dicotyledons; it consists of a single rank of cells which frequently contain very large chloroplasts, and in *Zea* they appear to be the main centre for starch formation (hence the term *'starch sheath'* sometimes found in the older literature). In other grasses and cereals, e.g., *Avena, Triticum, Hordeum* and *Lolium*, the sheath is two layered, the outer layer being unexceptional parenchyma and the inner one being relatively thick walled; this inner layer is termed the

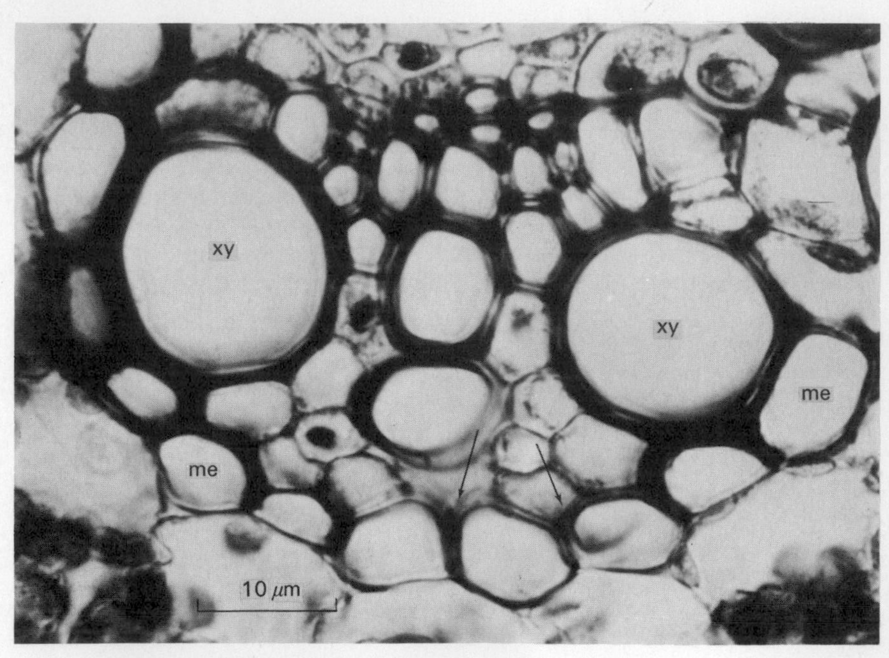

FIG. 9.3 ***Transverse section of part of the bundle sheath from a vein in the leaf of*** **Hordeum vulgare** ***(barley). The micrograph shows a highly magnified part of the mestome (me) in which the deposition of material stainable with saffranin is particularly marked in the outer layers of the radial walls (arrows); this material can also be detected using Sudan IV and may, therefore, be suberin.***

mestome. The thick wall contains deep pits, with numerous plasmodesmata at their bases, which connect the protoplasts of the mestome with those of the outer bundle sheath and the parenchyma of the vein itself. O'Brien and Carr (1969) have shown that suberin lamellae are laid down over the entire wall surface in both the single layered bundle sheath of *Zea* and in the mestome of the bilayered bundle sheath in *Triticum* and *Avena*. This layer of suberin resembles very closely that seen in the older endodermis in the root of *Hordeum* (see Fig. 8.17, page 228). There is also a strong suggestion of an especially heavy deposition of suberin (O'Brien and Carr, 1969) in the thickened radial walls (Fig. 9.3) in a position analogous to that occupied by the Casparian band. Structurally, therefore, the bundle sheath seems to be the counterpart of the endodermis of the root.

Movement of water and solutes within leaves

When xylem sap reaches the ultimate branches of minor veins water and solutes move into the walls of the mesophyll cells. Most of the water will evaporate, the vapour escaping from the leaf through the stomata, the number and aperture of which control the rate of water loss.

Transpiration from leaves in bright light and low humidity can be high; in barley plants mature leaves transpire their own fresh weight of water every 2–3 hours at a temperature of 20 °C. Evaporation of water from the mesophyll will tend to increase the salt concentration in the fluid in the pores of the wall. This will create a minor gradient for back diffusion towards the xylem against the mass flow. If solutes were not reabsorbed across the plasmalemma of mesophyll cells there would be a rapid accumulation of salts in the free space of the leaf. As we saw from the guttation data in Table 9.3, resorption occurs at a rate comparable with delivery.

The solution presented to the plasmalemmata of mesophyll cells will be much more concentrated than that bathing the cortical cells of the root (see Table 9.3). The actual concentration will depend on the rate of transpiration. When it is low, the solution may be as concentrated as that exuding from the ends of excised roots (see Table 9.3), e.g., potassium may be 10–100 times greater than in the external solution. When transpiration is high the concentration will be reduced, but even so in barley we find that potassium is usually 10–20 times that in the solution bathing the roots. The relatively concentrated solution delivered to leaves by the xylem justifies studies of ion fluxes into leaf cells from high salt solutions (see chapter 10, page 290) which have sometimes been dismissed as 'unphysiological'.

Restrictions on water and solute movement out of the veins. There seems to be some division of opinion about the restriction which is placed on movement into and out of the veins by the endodermis-like cells of the bundle sheath. Briefly, evidence on the movement of certain dyes out of veins suggests there is little restriction (see Weatherley, 1963, for a review), while other experiments and the suberized walls of the bundle sheath strongly suggest that there is restriction or, at least, should be.

The weaknesses of experiments with dyes were stressed by O'Brien and Carr (1969) who pointed out that contradictory results with the leaves of grasses are easily obtained. Thus, an injection of acid fuchsin (1 per cent) into the leaf of wheat indicated only a very slow movement of dye from the xylem to the mesophyll, while basic fuchsin (0·5 per cent) spread rapidly from the veins. With other non-vital dyes movement out of the veins was clearly restricted until the leaf lost turgor or was poisoned by the dye itself.

In an attempt to resolve the path of water from the xylem to the mesophyll in wheat leaves, Crowdy and Tanton (1970) introduced a relatively non-toxic chelate of lead into the vascular tissues of the leaves and after the leaves had transpired for 12 hours they were suspended in an atmosphere of hydrogen sulphide (H_2S) gas which precipitated *in situ* lead sulphide in the leaf. Examination of both fixed and unfixed, freeze-dried tissue in the electron microscope revealed that the bulk of the lead sulphide (which is electron dense and can thus be visualized directly) was confined to the cell walls and was present largely in the primary wall layer. Regrettably, these experiments were not conducted on a time course but there is an indication in one of their micrographs of a considerable accumulation of lead sulphide in the radial walls of the mestome cells in the position of the suberin deposition noted by O'Brien and Carr (1969). It is also noteworthy that, 12 hours after the application of the lead chelate to the leaves, the deposition of lead sulphide in the xylem and other vascular tissues was much more prominent than in the mesophyll and the epidermis. This result suggests that the movement of the chelate out of the major veins was considerably restricted.

The reluctance with which some solutes move out of the veins in monocot leaves may well be correlated with the suberized layer and Casparian band in the wall of the cells of the bundle sheath (see page 210). If, in addition to looking like an endodermis, the bundle sheath mimics its functions then we might suggest that material crossing the sheath must travel for a short distance in the symplast, or the protoplasts of the sheath cells, before being released into the free space of the mesophyll. This progress would, therefore, be exactly the reverse of that taken by materials entering the vascular tissues in the root. If it possesses no other virtue the suggestion has the attraction of being neatly consistent.

The further implications of a barrier to free diffusion between the mesophyll and the veins are discussed by O'Brien and Carr (1969).

In leaves of dicots there seem to be few reports on the fine structure of the bundle sheath cells and they are usually thought of as being highly permeable. There is evidence to suggest, however, that there can be restrictions on the movement of solutes from the veins into the mesophyll. Shone and Wood (1972) have shown that the uptake and translocation of a non-polar herbicide molecule (simazine) by the roots of plants is strongly correlated with the uptake of water, but in the leaves of blackcurrant (Shone and Wood, 1973) which are resistant to the toxic effects of the herbicide, simazine was shown to be largely restricted to the veins. Other cases of herbicide restriction in veins can be found in a review by Crafts (1964). Detailed anatomical investigations of the bundle sheath cells in instances of this kind might well repay study.

Absorption of ions by cells in the leaf

If the impermeable walls of cells of the bundle sheath direct most of the solutes leaving the xylem into the protoplasts then the rate of ion transport across the plasmalemmata of these cells may regulate the entry of ions both into the symplast and into the free space of the leaf. There is no direct evidence on this point and most studies of ion uptake by leaf tissue have been conducted in such a way that ions are presented to the mesophyll directly from the free space. Because of its highly convoluted structure, diffusion of ions in the free space of the leaf can introduce considerable errors in the ion flux experiments of the type discussed in chapter 7. For this reason, very thin slices or strips of leaves are used to minimize the diffusion pathways.

The basic mechanisms of ion transport in leaves, green algae and non-photosynthetic tissue are probably the same. I shall not discuss these under this heading, but reference should be made to earlier chapters and chapter 10. The one point which should be stressed is that in leaf tissue, as in algae, ion fluxes are usually stimulated by light. In leaf tissue from *Vallisneria* (van Lookeren Campagne, 1957) and *Elodea* (Jeschke and Simonis, 1969), active chloride influxes are up to tenfold greater in light than in darkness but both dark- and light-stimulated fluxes depend on the coupling of electron transport to ATP synthesis. Light stimulation is most simply explained, therefore, by the increased supply of 'fuel' (ATP) for the operation of a common pump, and not by the operation of a separate one. Similar conclusions were reached by Rains (1968) on the dependence, in leaves of *Zea*, of both dark- and light-stimulated influxes of potassium on ATP.

Smith and Epstein (1964) found many closely similar features in the uptake of potassium by *Zea* leaf tissue and the roots of *Hordeum*, viz., uptake kinetics, K-selectivity, anion effects, low temperature effects.

The continuous delivery of ions to the mesophyll cells ensures that a growing leaf is well supplied with inorganic salts and nitrogenous compounds essential for its growth and the maintenance of the osmotic potential of its increasing volume. In the mature leaf these continued supplies can clearly lead to an embarrassment of riches. If we recall that the volume of water transpired in a day may exceed by 5–10 times the volume of the leaf cells, some kind of procedure is necessary to deal with the incoming salt if the osmotic pressure of the cells is not to become alarmingly high. Briefly, there are three main courses of action available:

(a) Salt can be removed from solution by precipitation.
(b) Salt may be exported from the leaf at a rate comparable with import.
(c) Salt may be accumulated in special cells adapted for this purpose and which are able to withstand colossal osmotic pressures. These are known as *salt glands* (see page 269).

In most herbaceous plants in non-saline habitats a strategy based on the first two procedures is usually adopted. Both are necessary because certain ions are not exportable to any practical extent from leaves. This is because they do not seem to move, in most circumstances, along with sugars and other ions in the phloem.

Precipitation of salts in leaves

As leaves age their calcium content continues to rise; microscopical examination of the cells commonly reveals that there are large crystals of calcium oxalate in the vacuoles. The crystals increase in abundance as the tissues mature, being most prominent in deciduous plants just before leaf fall. In certain species, e.g., *Mercurialis perennis*, most of the cell volume seems to be occupied by crystals. Calcium may also be precipitated, both internally and in the cell wall as carbonate, and within mitochondria as calcium phosphate in certain circumstances.

Calcium is not appreciably mobile in the phloem although most other ions are. The failure to move in phloem may be caused by failure to enter the sieve tubes but Fritsch and Salisbury (1957) showed crystals of calcium oxalate in the phloem of horse chestnut leaves (*Aesculus*), and in a recent report Hall and Baker (1972) showed that phloem exudate from *Ricinus* contained 0·5–2·3 mM calcium.

Boron is not mobile in phloem but its contribution to the osmotic potential is very small. There do not seem to be any reports of precipitation of this element, but such a drastic procedure is perhaps unnecessary when such small quantities are involved.

Export of ions from leaves

With the exceptions mentioned above, most elements or ions can be exported from leaves which have matured and which are approaching senescence. This movement is dependent upon entry into, and transport within, the phloem and generally the pattern of redistribution closely resembles that of the exported products of photosynthesis.

Mobility of ions in the phloem. The autoradiograph in Fig. 9.4 shows the distribution of ^{32}P in a seedling of barley after a small segment (*ca.* 3·5 mm) of a seminal root had been treated with radioactive phosphorus for 20 hours. The remaining parts of the root system were kept in unlabelled phosphate solution, but they were strongly labelled by the end of the experiment, especially the root tips. To have reached them ^{32}P must have moved up the treated root in the xylem and down to the tips of untreated ones in the phloem. Whether the transfer between xylem and phloem occurred in the leaf or at the stem where the vascular supplies of various seminal roots come into proximity is not known, but the results would have been similar if the ^{32}P had been applied as an injection to one of the mature leaves. The picture betrays a vigorous circulation of ^{32}P in the xylem and phloem.

Most of the other major plant nutrients can be transported in the phloem and thus circulation patterns for the following ions can also be observed: potassium, sodium, sulphur, chloride, magnesium and nitrogenous compounds.

Since calcium and boron are not phloem-mobile, the developing plant must receive a steady supply of calcium and boron throughout its growth; the older and senescing tissues cannot contribute to the developing leaves and meristems. It is interesting to note that calcium is transported effectively in neither the symplast (page 222) nor phloem, even though its entry into those cells forming part of the transport system does not seem to be prevented.

Magnesium is the only divalent cation to be appreciably mobile in phloem and it seems possible that it may become complexed or chelated in order to do so. Many of the minor elements move to some extent from mature to immature tissue in the phloem, e.g., iron, manganese, zinc and molybdenum. In view of the insoluble nature of iron at intracellular pH values it is difficult to see how it could move about the plant in any way other than in a non-ionic or chelated form; there is widespread evidence for complex formation by the other elements mentioned in biochemical systems.

Transfer cells and the circulation of nutrients in the plant. In dicotyledonous leaves there is an interesting development of cells in the minor veins at the point in time where the leaf has expanded fully and the export of carbohydrate commences. The walls of parenchyma cells in the vicinity of sieve elements and xylem elements begin to thicken by the rapid deposition of cellulose at certain loci forming ridges on the wall directed into the cytoplasm. The building process

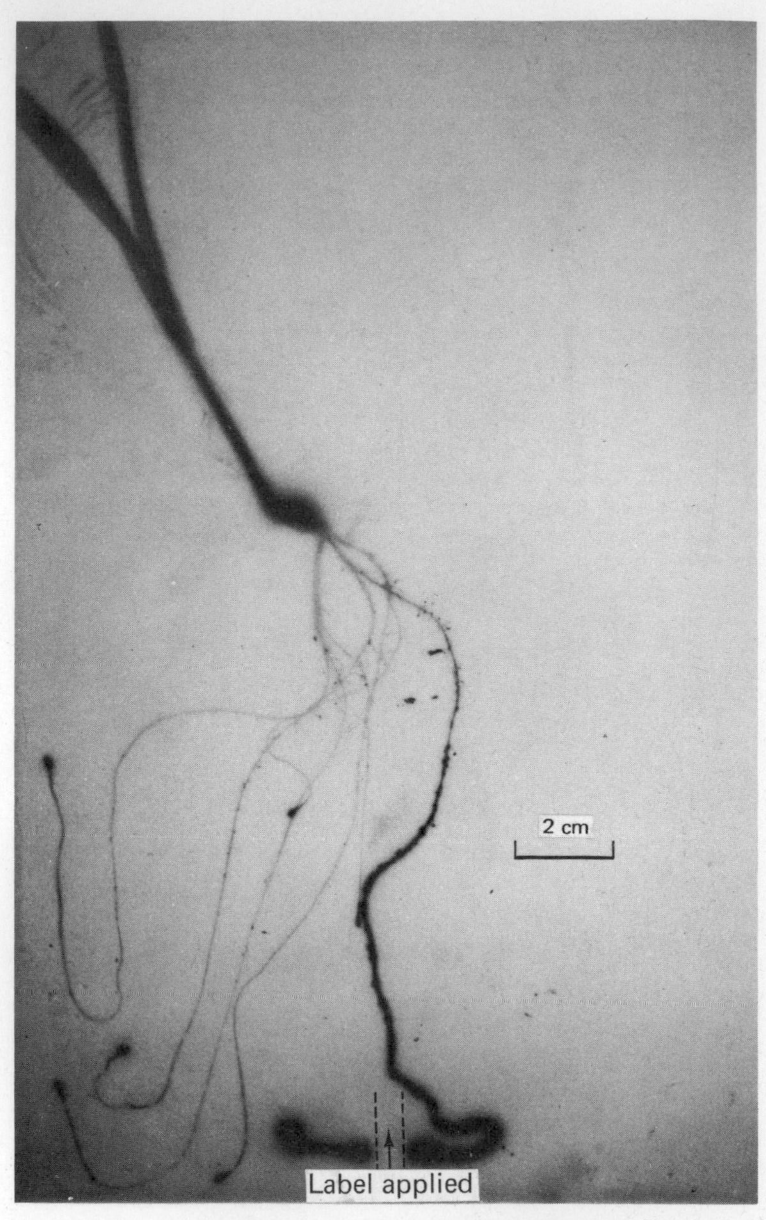

FIG. 9.4 X-ray photograph of a young barley plant labelled with ^{32}P near the tip of one of the seminal root axes. The radio-activity was supplied directly to the treated zone (which has been cut out to prevent fogging of the film) using the apparatus described in chapter 8 and illustrated in Fig. 8.15. Note the heavy labelling of root tips of main axes remote from the labelling point and the prominent labelling of meristems of lateral roots.

can continue to produce a very elaborate wall with projections and branching ingrowths which form what appears, in transverse section, to be a labyrinth of great complexity. In sections, like the one in Fig. 9.5, it is rather difficult to appreciate how the apparently isolated outliers of wall are attached to the main structure. Cells developing to this condition are known as *transfer cells*; their structure, occurrence and possible functions have been reviewed in a series of most readable papers by Gunning and Pate (1969, 1972), Pate and Gunning (1969), Gunning, Pate and Briarty (1968), Pate, Gunning and Briarty (1969).

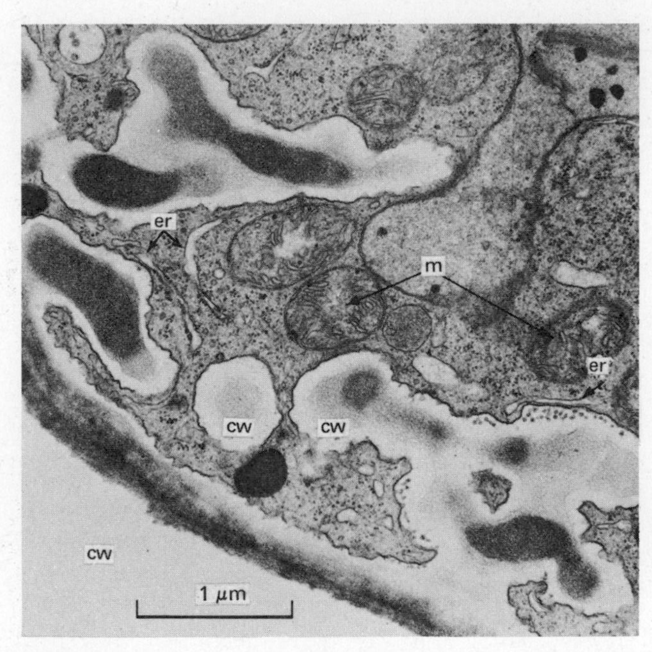

FIG. 9.5 Electronmicrograph of part of a transfer cell from a leaf vein of **Vicia faba.** ***The lighter patches intruding into the cytoplasm are ingrowths of the cell wall (cw). Note that the plasmalemma follows the ingrowths faithfully and that mitochondria (m) profiles of endoplasmic reticulum (Er) and vesicles are seen close to the expanded wall. (I am indebted to Dr Gunning for this micrograph.)***

There seem to be two consequences of the peculiar wall development in transfer cells which we might relate at once to transport functions. The first is that the surface area of the plasmalemma, which faithfully follows the labyrinth, is enormously expanded. The second is that because the wall ingrowths penetrate deeply into the cytoplasm no part of the plasmalemma is far removed from the intracellular organelles. Indeed, Gunning and Pate (1969) draw particular attention to the fact that profiles of endoplasmic reticulum follow the contours of the irregular wall in close proximity to the plasmalemma. These authors also note that mitochondria approach the wall ingrowths and that they are frequently seen, in transverse section, to be completely enclosed in chambers of the highly reticulate wall. Thus, the ER, which may be associated with symplast transport, and the mitochondria, a source of 'fuel' for active ion transport, are dominant features of the transfer cell.

Transfer cells are relatively constant features of the vascular tissue of herbaceous dicotyledons in which there is a vigorous circulation of nutrients and carbohydrate. They are of several different kinds and may be associated with both xylem and sieve elements. The functional association between transfer cells and those surrounding it may be judged by the walls on which ingrowths occur. They can be highly polarized. This is emphasized in the diagram (Fig. 9.6) which

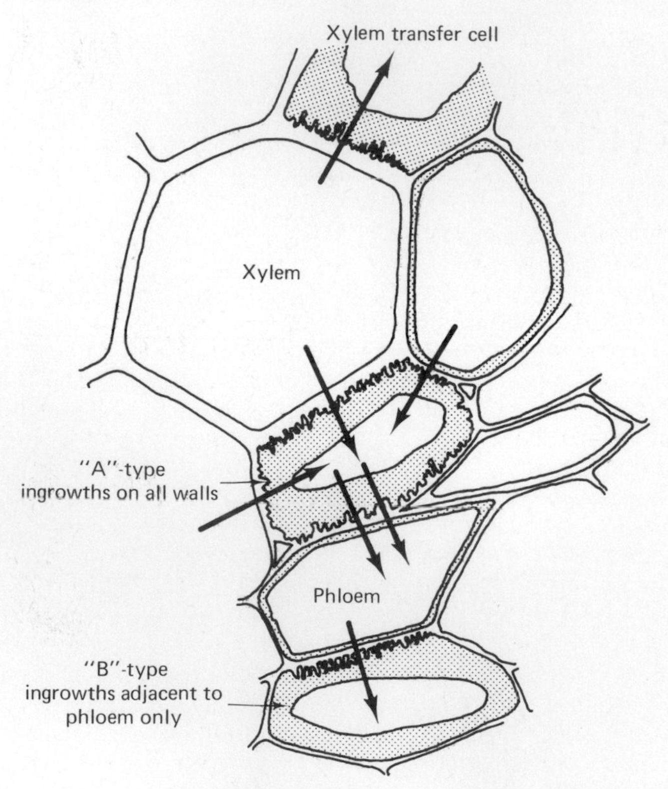

FIG. 9.6 Diagram to illustrate some of the most common types of transfer cells and the probable major directions of transfer they effect.

illustrates the various types distinguished by Gunning and Pate (1968). Xylem transfer cells are uncommon in leaf veins but are prominent in petioles and stems; it is tempting to relate their presence with the extraction of ions from the xylem sap into the symplast of the xylem parenchyma (see page 243). The 'A' type cell can have ingrowths on walls adjacent to both phloem and xylem. Such cells seem well adapted to act as a 'bridge' between xylem and phloem so that, perhaps, the salt carried to the leaf can be exported largely without its ever leaving the vein. The fact that these cells are not developed until the leaf is ready to export carbohydrate (and, indirectly, ions) supports this idea. They become especially prominent in the finer veins where the loading of the sieve elements with photosynthate occurs.

In variegated leaves there is a nice correlation between the prominence of transfer cells in minor veins and the expected pattern of loading of the sieve elements. In *Impatiens*, *Sedum* and *Ligularia* Gunning and Pate (1969) show that ingrowths in the walls of parenchyma are small and sparse in the chlorotic patches of the leaves but abundant in the green patches.

Pate and Gunning (1969) suggest that the 'B' type cell, which is found in the veins of expanding leaves, is concerned with unloading substrates, imported *via* the phloem, into the symplast of parenchyma cells or the bundle sheath.

Redistribution patterns of ions from leaves. In earlier research literature, Mason and Maskell (1931) appear to have been the first to implicitly recognize the roles of xylem and phloem in the circulation of ions in the leaf. Subsequently, the use of radioactive tracers in both qualitative and quantitative experiments has greatly assisted in turnover studies. Biddulph *et al.* (1958) allowed plants of *Phaseolus vulgaris* (red bean) to absorb ^{32}P for some hours before replacing the plants in unlabelled phosphate solution. Immediately following the labelling period and at intervals thereafter, plants were placed against X-ray film and autoradiographs were produced. During the labelling period the ^{32}P was directed mainly into the young leaves. Subsequently, as these leaves matured, most of their ^{32}P was exported to the leaves which had differentiated after the supply of ^{32}P to the roots had been discontinued. Thus, the ^{32}P gradually moved up the plant as it grew. The same general features were seen when the plants were treated with $^{35}SO_4$ solution even though the proportion of export to import in the maturing leaves was lower (presumably due to the slow turnover of sulphur containing proteins).

More quantitative information comes from experiments by Greenway and Pitman (1965) who supplied ^{42}K to the roots of intact plants of *Hordeum vulgare* (barley) which had reached the third leaf stage. By measuring both the stable potassium content of the leaves before the application of ^{42}K and both the stable and radio-potassium afterwards they were able to make a simple balance sheet for the import and export processes. Table 9.4 shows that the influx of tracer in the older leaves was almost the same as the rate of export. The potassium leaving the oldest leaf was recovered in the younger leaves. In leaf 3, 40 per cent of the incoming potassium had been contributed by leaf 1 and other parts of the plant.

Table 9.4 Potassium turnover in the shoot tissue of a young barley plant

	Oldest leaf (L_1)	Second leaf (L_2)	Youngest emerged leaf (L_3)
Initial dry weight (mg)	20	13	1·8
Initial K content (μmoles)	46	26	2·9
Intake of K (μmoles day^{-1})			
^{42}K from roots	1·9	2·7	2·0
Stable K from other tissues	−1·6	0·7	1·3

Based on data from Greenway and Pitman (1965).

Redistribution in relation to physiological processes

The circulation patterns of ions and metabolites are strongly influenced by growth and metabolic activity. The growth of a particular plant organ is said to create a 'demand' for solutes to sustain its metabolic activity and to maintain the osmotic potential of its increasing volume. Thus, the rate of growth of a plant will influence both the rate of net uptake of ions and the turnover of ions in a given part of the plant. Of course, the simple analogy of a growth zone or meristem as a kind of bottomless pit attracting nutrients to itself is by no means an accurate picture of the relationship between supply and demand in the plant. Plant growth is controlled by numerous interacting factors; it is the interactions rather than the factors themselves which are poorly understood and which seem to produce confusion and circular arguments in the literature. We shall see that some authors have attempted to distinguish between the effects which hormones have on ion transport and those which they have on growth. In some cases, but not all, this distinction is difficult and to emphasize the caution necessary in such an exercise we might point out that growth affects ion transport, that hormones are a product of growing cells and that ion supply and transport can affect growth. The student should not be discouraged if he finds that such a ring is hard to penetrate, for his dilemma is a common one.

Hormone directed ion transport

For the reasons outlined above it is often difficult to decide whether the movement of an ion to a zone where a hormone is being synthesized, or where it has been deliberately placed by the experimenter, is due to a direct effect of the hormone on polar transport or due to the growth caused by the hormone. The distinction is of considerable importance if we are to understand fully such matters as apical dominance but we should say, at the outset, that direct effects of hormones and other growth regulators on ion transport have not been demonstrated to everyone's satisfaction.

Auxins. The familiar auxin, indole acetic acid (IAA), is synthesized in the meristems of shoots and is transported away from the meristem. Its distribution is matched by a gradient of metabolic activity in a developing organ.

In attempts to see whether auxins can direct transport, a favourite experiment has been to take a piece of tissue lacking an active meristem, apply auxin to part of it and see whether solutes, either ^{14}C-labelled substrates or radioactive ions, move towards the auxin treated zone in a polar fashion. They have been found frequently to do so. On the other hand observable changes in the growth of cells in the auxin treated zone often occur over a similar period of time. It becomes a matter of interpretation, therefore, which factor, growth or hormone gradient, established the polarity of solute movement.

A slightly more convincing experiment of this kind, in which growth complications were apparently minimal, was reported by Davies and Wareing (1965) working with disbudded shoots of poplar (*Populus robusta*). Figure 9.7 illustrates the system used. The excised bud bases were smeared with lanolin containing IAA or with plain lanolin and the ^{32}P was applied to the stem surface in drops from a syringe. The ^{32}P in the bud bases was analysed after 48 h. The IAA increased the movement of ^{32}P away from the point of application; the effect was highly significant in an upward (acropetal) direction but much less so when the IAA was applied basipetally. No major growth of cells at the bud bases occurred as a result of auxin treatment but no other metabolic consequences were examined. If the phloem above or below the ^{32}P application was disrupted by banding, little movement occurred, suggesting that the ^{32}P moved in the sieve tubes. Davies and Wareing (1965) suggested that, in view of the known stimulatory effect of IAA on protoplasmic streaming (Thimann and Sweeney, 1937), auxin may control directly the movement of material in the sieve tubes of the phloem. Such an interpretation is, however, not consistent with explanations of phloem transport based on mass flow (see MacRobbie, 1971, for review of phloem transport).

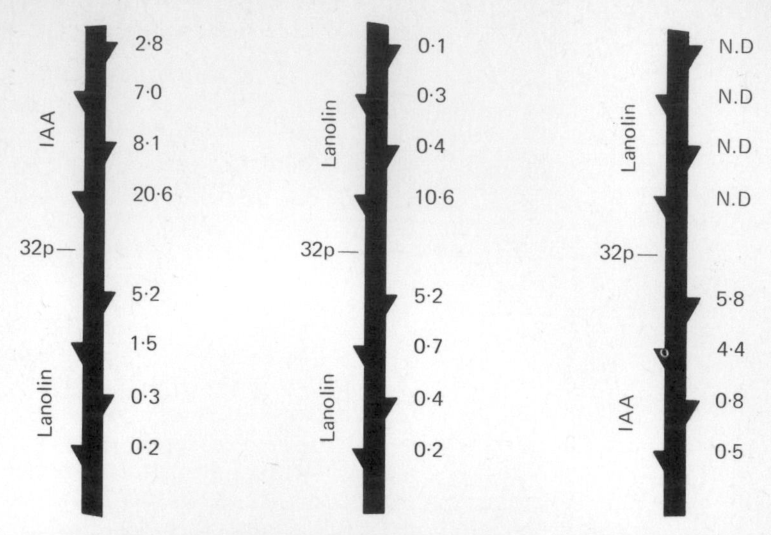

***FIG. 9.7** The effect of indole acetic acid on the movement of ^{32}P from a point of application on debudded stems of* Salix *(willow). Note that IAA above the application zone (left) was more effective in mobilizing the ^{32}P than below (right) the zone. The control to which no IAA was applied is represented by the central 'twig'. The numerical values are counts per minute $\times 10^{-2}$ in the basis of the excised buds. (Redrawn from Davies and Wareing, 1965.)*

Etherton (1970) showed that 2–4 hours after presenting IAA to excised segments of oat (*Avena*) coleoptile, the transmembrane potential increased by up to 30 per cent. Over a range of concentrations (10^{-9} to 10^{-6} M) of IAA the pattern of the response of membrane potential and extension growth were similar (Fig. 9.8).

Observations by Weigl (1969) go further than most in an attempt to provide some rational basis for effects of auxins on transport processes. He found that IAA and other synthetic auxins, e.g., 2,4-dichlorophenoxyacetic acid (2,4-D), indole propionic acid (IPA) and indolebutyric acid (IBA), entered into a strong physical association with the phospholipid, lecithin (see page 28). One mole of lecithin bound 0·8 mole IAA and other phospholipids somewhat less. While other molecules with auxin activity competitively inhibited the binding IAA to lecithin, a range of related substituted fatty acids lacking auxin activity did not. The association between the auxins and the polar regions of the phospholipids seemed, therefore, to be relatively specific.

Weigl (1969) also notes that the efflux of both ^{86}Rb and ^{36}Cl tracers from mesocotyl segments of oat (*Avena*) was markedly stimulated by 10^{-9} to 10^{-7} M IAA present in the uptake medium. The effect was most marked in the first hour of treatment.

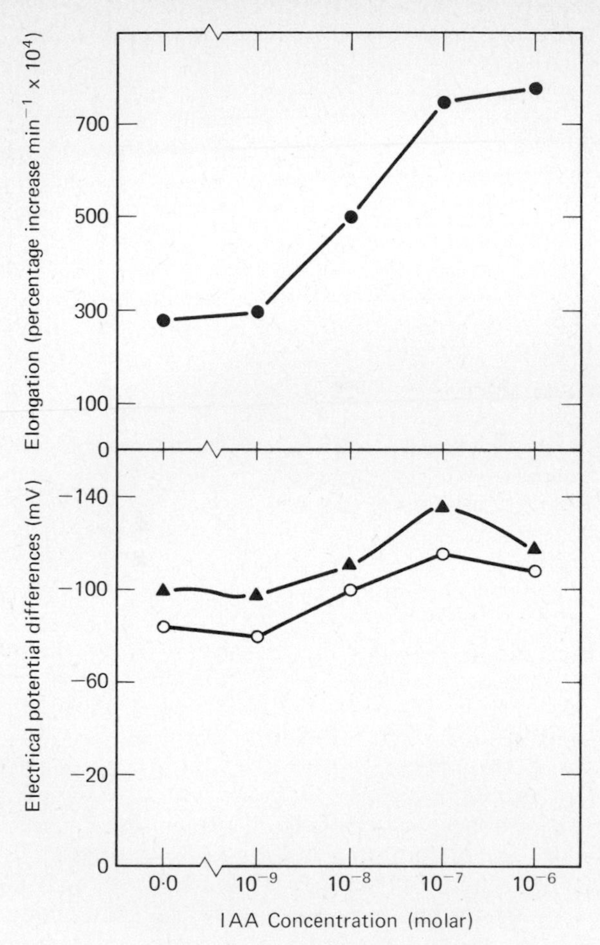

FIG. 9.8 The effect of indole acetic acid (IAA) on the growth and cell electropotentials of excised segments of* Avena *(oat) coleoptile. Note that the change in potential occurs over the same range of auxin concentration which promotes the elongation of the coleoptile segments. Solid circles = growth rate; open circles = p.d. across plasmalemma; triangles = p.d. across plasmalemma + tonoplast. (Redrawn from Etherton, 1970.)

Cytokinins. Kinetin and other cytokinins have the capacity to delay the senescence of tissues and to maintain their rates of protein synthesis and other vital metabolic functions. Senescence commences quickly when leaves are detached from plants but the application of kinetin to the leaf greatly reduces the rate of senescence. The effect is restricted to the points of application so that if kinetin is applied locally, a gradient of metabolic activity is set up in the detached leaf.

If, before excision, the leaf had been uniformly labelled with ^{32}P or some ^{14}C-labelled metabolite we would find that, some hours after excision and local kinetin application, tracer movement towards the kinetin-treated spots had occurred. Eventually, nearly all of the radioactivity would be found in the kinetin-treated zone (e.g., Müller and Leopold, 1966; Atkin and Wain, 1967).

The principal effect of the growth regulator is concerned with establishing a gradient of metabolic activity, not necessarily accompanied by growth, within the leaf and the results show how powerfully this can influence the pattern of solute migration. Similar effects are frequently noted in lesions caused by pathogenic fungi in leaves (see Wood, 1967).

Ilan (1971) suggests that cytokinins play some part in the selective absorption of K^+ versus Na^+ in disks of leaves from sunflower. Since it has been shown that the K^+ selectivity of damaged or metabolically impaired tissue is much reduced the results could be explained by the prevention of senescence in the kinetin-treated cotyledons thus maintaining their normal K-selectivity, while in the senescing controls the selectivity declined. Both K^+ influx and Na^+ efflux from cells are dependent to some extent on active transport.

Abscisic acid. This substance may be regarded as a naturally occurring growth retardant. It is prominent in senescing tissue and in plants whose growth is limited by environmental stress. In several respects its effects on metabolism are opposite to those of kinetin.

Collins and Kerrigan (1973) showed that after four to five hours pretreatment with 2 μM abscisic acid (ABA) the exudate flux from excised maize roots was increased relative to the control without any change in the concentration of the major inorganic ions in the sap (Table 9.5). Kinetin at the same concentration had the reverse effect and there was some reduction in the K^+ concentration in the exudate. Since the treatments had small effects on the ionic concentration of the sap they conclude that ABA increased the apparent hydraulic conductivity of the root and kinetin decreased it. It should be remembered, however, that the water flux is determined by the osmotic gradient provided by active salt transport into the xylem. The primary effect of the ABA and kinetin must, therefore, have been on this active transport. The effect of ABA was measurable after only an hour of pretreatment.

Further results by Collins and Kerrigan (1973) are consistent with an effect of ABA on active transport since the K^+ content of ABA-treated root tissue was appreciably greater than the controls. This report is at odds with experiments with stomata where the opening response in the light, which seems to be dependent on active K^+-pumping into guard cells (see below), is almost completely inhibited by ABA. The changes in water and ion transport in shoot and root brought about by ABA are, then, in opposite directions (see Tal and Imber, 1971).

Table 9.5 Volume flows and ionic concentrations in exudate from maize roots treated with abscissic acid (ABA) and kinetin for 4–5 hours

	Volume flow $\mu l\ cm^{-2}\ h^{-1}$ ± S.E.	Exudate concentration (mM ± S.E.)		
		K^+	Ca^{2+}	Cl
ABA	2·56 ± 0·26	20·5 ± 0·7	1·10 ± 0·03	17·2 ± 0·7
Control	1·79 ± 0·42	22·1 ± 0·7	1·84 ± 0·09	23·1 ± 1·7
Kinetin	1·13 ± 0·17	16·1 ± 1·1	2·90 ± 0·27	15·4 ± 1·2
Control	1·73 ± 0·27	22·1 ± 0·7	1·84 ± 0·09	23·1 ± 1·7

Both growth regulators supplied at 2 μM solubilized in a small quantity of dimethyl sulphoxide; this substance added to the controls.

Ion and solute accumulation in stomatal guard cells

Ion and solute movements in response to local gradients of metabolic activity are involved in a number of important physiological processes. Stomatal aperture and its regulation is a matter of great significance in plant growth since it controls both the loss of water by the plant and the influx of CO_2 into the photosynthetic tissue.

Stomatal aperture is determined by the osmotic potential of the two guard cells relative to the other cells in the epidermis. In bright light and where water supply is not limiting, the aperture and the difference in osmotic potential is greatest (Table 9.6). In darkness

Table 9.6 Osmotic pressures of guard and surrounding epidermal cells in leaves with open and closed stomata

Species	Stomata open (light)		Stomata closed (dark)	
	Guard cells	Epidermis	Guard cells	Epidermis
	Osmotic pressure of vacuolar sap (bar)			
Vicia faba (bean)	15	6	6	4
Beta vulgaris (sugar beet)	32	12	22	12

Data from Meidner and Mansfield (1968).

the solute concentration, and hence the osmotic potential, of the guard cells drops and the aperture is reduced.

Plasmodesmatal connections, and hence symplastic continuity, between guard cells and the surrounding tissues are rare in dicotyledons and reportedly absent in the guard cells of grasses (Brown and Johnson, 1962). Thus, if the changes in osmotic potential depend on substantial movement of solute into and out of the guard cells, considerable fluxes must occur across their plasmalemmata.

There are two ways in which a guard cell might increase its osmotic potential. Solutes can be imported or some insoluble material within the cell must be brought into solution; two major theories are based on these possibilities. In the first, potassium ions are believed to be the specific solute imported and in the second the products of starch breakdown have been implicated. In giving more attention to the former in this book on ionic relations the student should not assume that the evidence necessarily favours this interpretation; there remains much support for the starch breakdown hypothesis. An attempt to review the common ground can be seen in an excellent paper by Mansfield and Jones (1971).

Potassium transport and stomatal opening

Perhaps the first clear evidence that stomatal opening involved the transport of a solute into the guard cell was given by Fischer (1968a) and was subsequently elaborated in a series of papers (Fischer, 1968b, 1970; Fischer and Hsiao, 1968). Fischer made use of isolated strips of epidermal cells from leaves of *Vicia faba*. He found that normal responses of the stomata to illumination could be obtained only if the strips were floated on a medium containing potassium ions. Both Fischer (1968b) and Thomas (1970) working with stomata from tobacco leaves found that the response was not markedly influenced by the anion accompanying the K^+. The initial observations were based on the absorption of ^{86}Rb, a tracer whose behaviour resembles that of potassium. More recently, however, more direct observations using electron probe microanalysis by Humble and Raschke (1971) have confirmed that the increase in osmolarity of guard cells of *Vicia faba* is almost wholly accounted for by the increase in cellular potassium. The analysis also showed that neither Cl, P nor S accompanied the K^+ and it was concluded that endogenous organic acid anions maintained electroneutrality (Fig. 9.9).

The results of Fischer, Thomas, Humble and Raschke all suggest that the opening response is highly specific to the potassium (or rubidium) ion and that sodium is ineffective. In the stomata of *Commelina*, however, Mansfield has found that sodium was more

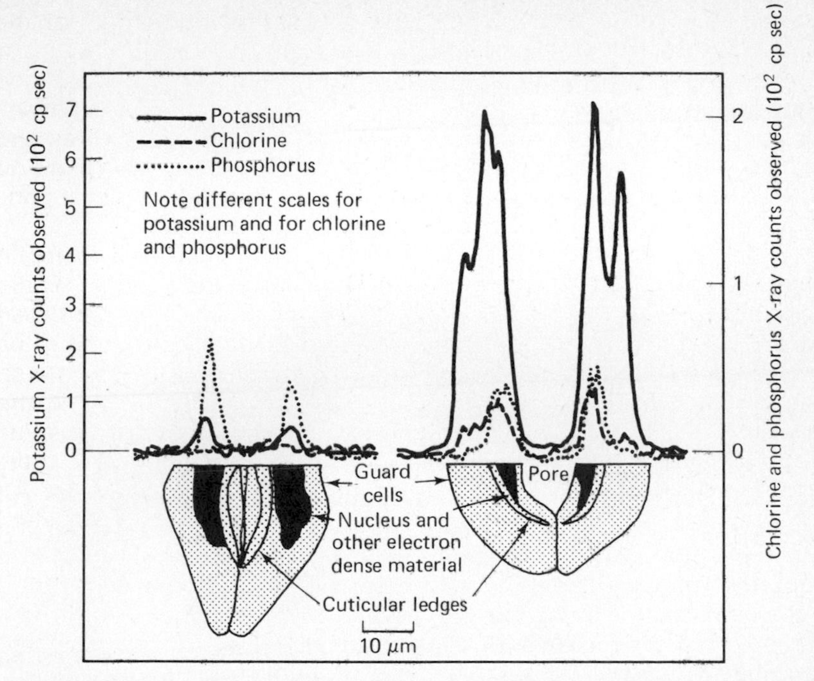

FIG. 9.9 Profiles of the relative amounts of K, Cl, and P across an open and a closed stoma. The analytical data was obtained by the electron-probe micro-analysis of whole, freeze-dried cells. Note that the increase in potassium in the guard cells of the open stoma is very much greater than the increase of chlorine (mostly chloride) and that the phosphorus content does not change. (From Humble and Raschke (1971) by kind permission of the authors and publisher.)

effective than K^+ in stomatal opening. Another instance of sodium transport and stomatal activity was reported by Thomas (1970) for the succulent plant, *Kalenchoe marmorata*. The stomata of succulent plants are unusual in that they open in the dark and close in the light; closure in the light was found to be due to a light-stimulated efflux of Na^+ which was not affected by K^+ in the medium.

In epidermal strips of tobacco (*Nicotiana tobaccum*) Thomas (1970c) found that the addition of ATP to the medium enhanced the light-stimulated opening and promoted opening in the dark but only if K^+ was present. This result suggests that the K^+ transport into the cell is dependent on ATP which would be produced in the normal course of events by photophosphorylation. The K^+ pump involved is probably a K^+-stimulated ATPase since the effects of both K^+ and ATP on stomatal opening were severely inhibited by ouabain (see chapter 5, page 147) (Thomas, 1970b, c).

Starch–sugar conversion and stomatal opening

For many years it has been known that the size of starch grains in guard cells is reduced on illumination and as the stomata open. The simple conversion of starch into molecules of glucose-1-phosphate in solution cannot, in itself, increase the osmotic pressure of the cell since for each mole so formed one mole of phosphate ion H_2PO_4' is taken from the solution: see eq. (9.1) (Heath, 1959).

$$\text{Starch} + P_i \rightleftharpoons \text{glucose-1-phosphate} \qquad (9.1)$$

Further breakdown of the carbon skeleton into C_3 and C_4 pieces leads to the formation of organic acids, and increases in osmolarity of the cell sap.

Fischer and Hsiao (1968) and Mansfield and Jones (1971) have noted that the breakdown of starch and the increase in potassium content of guard cells are simultaneous occurrences and that the kinetics do not indicate whether one of the processes is directly dependent on the other. Certainly both processes require ATP but it is difficult to believe that the failure of stomata to open in the presence of ouabain could be due to interference of ATP utilization in starch breakdown into C_3 or C_4 pieces. There is, however, a strong indication that the intracellular pH exerts a strong influence on starch hydrolysis and it seems possible that H^+ ions from dissociated organic acids may exert a kind of end product inhibition on the process. Should this be the case starch hydrolysis might proceed only when endogenously produced protons can be exchanged for exogenous K^+ ions through an ATP-dependent pump at the plasmalemma of the guard cell.

Potassium pumping in leaf movements

Flowering plants in certain families can alter the orientation of their leaves in light and darkness. We shall consider an example of this behaviour, known as *nyctinasty*, in the pinnules of Acaciod leaves, the opposite pairs of which are spread open in the light but which tend to close at night, with their adaxial surfaces innermost (Fig. 9.10). The rate at which the pinnules close in darkness is controlled by the wavelength of light they receive at the end of the light period. Thus, if plants were last illuminated by red light the pinnules close after 30–90 minutes in the dark, but they remain open for several hours if the red light is followed by a brief period of illumination with far red light ($>$730 nm). The closure response can be repeatedly reversed by red/far red light sequences; this is usually taken as evidence that the pigment phytochrome is the photoreceptor in this response to light. Koukkari and Hillman (1968) established that the small stalks of the pinnules, known as pulvinules, are the most photosensitive area.

Satter, Sabnis and Galston (1970) have shown that the closure of the pairs of pinnules is due to a reduction in volume, or loss of turgor, in the ventral cells of the pulvinules and an increase in volume of the dorsal cells (Fig. 9.11). Electron probe microanalysis of freeze-dried sections showed that these changes in osmotic potential were accompanied by a substantial loss of K from the ventral cells and a corresponding gain of K by the dorsal cells (Satter, Marinoff and

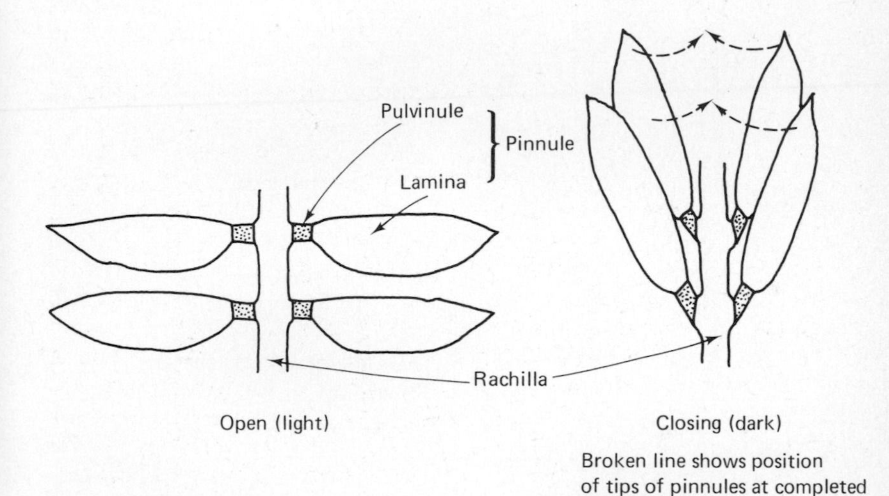

***FIG. 9.10** Schematic view of the ventral surface of open and closing pairs of pinnules from a leaflet of* **Albizzia julibrissin.**

Galston, 1970). There was no net loss of K from the pulvinule and closure was, therefore, associated with redistribution. The specificity of K movement was indicated by barely significant changes in the distribution of Ca and P. Further evidence on this matter came from an experiment where excised pairs of pinnules were transferred to darkness while floating on solutions of a number of salts of K, Na, Mg and Ca. All K-salts examined at concentrations greater than 200 mM inhibited closure (presumably by eliminating the K concentration gradient in the tissue) while similar concentrations of sodium, calcium and magnesium salts had little or no inhibitory effect on closure. The involvement of metabolism in the K-redistribution was suggested by experiments with metabolic inhibitors.

As well as providing another example of the way in which K-pumping is used in physiological processes these results have been

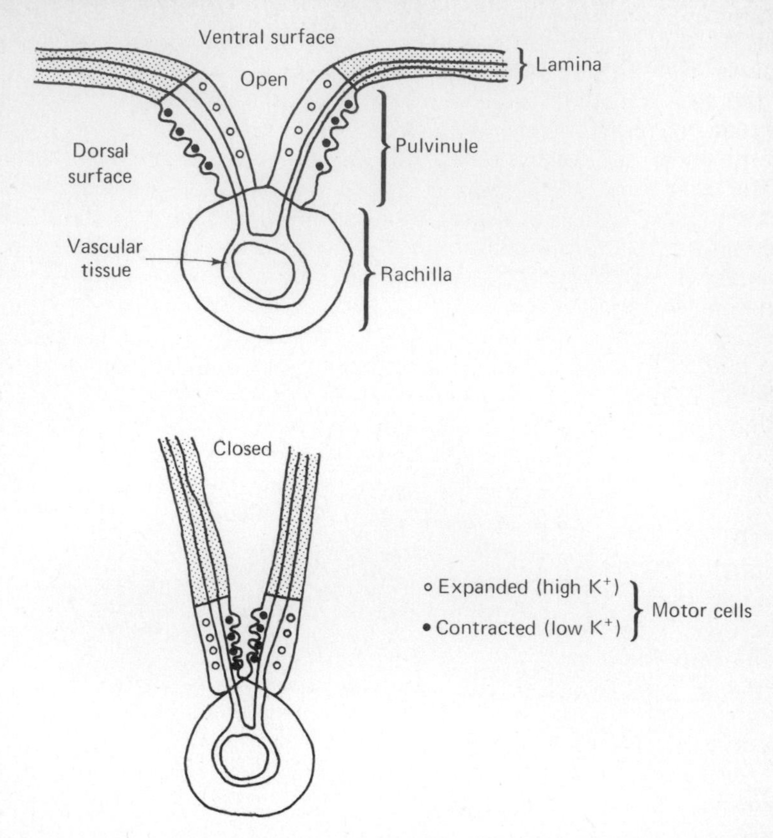

FIG. 9.11 ***Schematic transverse sections of open and closed pairs of pinnules of*** **Albizzia julibrissin.** ***Note the location and changes in volume of the motor cells in the pulvinule. Potassium pumping from ventral to dorsal surface probably causes closure. (Based on Satter, Marinoff, and Galston, 1970.)***

of great interest to research workers in photomorphogenesis. The immediate reactions of the phytochrome system after it has been 'activated' by red light are not known, but an increasing body of evidence suggests that phytochrome is associated with cell membranes in some way and that it can control either general or selective aspects of their permeability (see Hillman, 1967; Jaffe, 1970).

Disposal of excessive amounts of salt

The delivery of large amounts of salt to the leaves of plants growing in saline habitats requires some special processes beyond the normal ones of dilution by growth, to prevent the salt concentration of the cells reaching toxic levels. Several types of salt gland have been evolved into which salt can be diverted. Some of these glands store up the salt in bladder like cells, while others steadily excrete it onto the leaf surface to be removed by wind and rain (the latter process resembling guttation).

The concentration of salt found in the vacuoles of gland cells can be very impressive but depends on the salinity of the environment and the metabolic activity of the leaf, for salt secretion is an energy consuming process. Table 9.7 shows the total electrolyte concentration (measured as electrical conductivity) and the concentration of the major ions in gland cells of *Atriplex halimus* which was grown in two solutions which differed greatly in salinity. Comparison of the gland cells with the leaf sap shows that, while the bulk of the leaf is buffered against changes in external salt concentration, the gland cell conductivity increased by 450 per cent when the NaCl + KCl concentration was raised from 12 to 200 mM (Mozafar and Goodin, 1970). The osmotic potential within the glands was calculated to be −110 and −559 bar in the lower and higher salt solutions respectively. The gland cells need to have very strongly constructed walls to withstand the enormous hydrostatic pressures (turgor) which may develop in this way.

The difference in osmotic potential between the mesophyll and the glands may be more than −400 bar in the high salt regime. Since the glands may be densely distributed in leaf surfaces it is clear that they form a very important diversionary 'trap' for excess salt.

Physiological anatomy of salt glands. The salt glands of a variety of species consist of two cells, a stalk and a bladder. The former contains dense cytoplasm in which prominent small vacuoles or vesicles are often seen, while most of the volume of the latter is taken up by a large vacuole which contains the high concentrations of salt characteristic of the gland (see Table 9.7). Direct access to the protoplast of the bladder cell by way of the cell walls, or leaf free-space, is usually prevented because the walls of both stalk and bladder cells are impregnated with suberin, cutin and, occasionally, lignin. This limitation of the free space is in some respects analogous to the endodermis in roots, a point emphasized by Lüttge (1971) and illustrated in Fig. 9.12 by one of his diagrams. The bladder and stalk in *Atriplex* spp., *Tamarix* and *Limonium* are, in this way, isolated from the mesophyll except for cytoplasmic connections through abundant plasmodesmata. Observations by Thomson and Liu (1967) suggest that there may be as many as 17×10^6 plasmodesmata per mm^2 in this zone (in the endodermal walls of barley the density is *ca*. $0{\cdot}7 \times 10^6\ mm^{-2}$). The principal path into salt glands of this kind would seem to be *via* the symplast. Lüttge and Pallaghy (1969) have shown

Table 9.7 Effect of salt treatment on the electrolyte concentration of salt glands and leaf sap in **Atriplex halimus**

	External solution		Total electrolyte conductivity		Ionic concentration: gland		
	Concentration (M)	Conductivity (m mhos)	Leaf sap (m mhos)	Gland cell (m mhos)	Na^+	K^+ (M)	Cl^-
NaCl + KCl	0·012	4	14	277	1·33	0·98	1·7
NaCl + KCl	0·200	29	20	1204	5·48	6·17	9·2

Note that concentrations are measured in terms of *Molar* as opposed to our usual convention of millimolar.

(Data taken from Tables I and II and Fig. 2 of Mozafar and Goodin, 1970.)

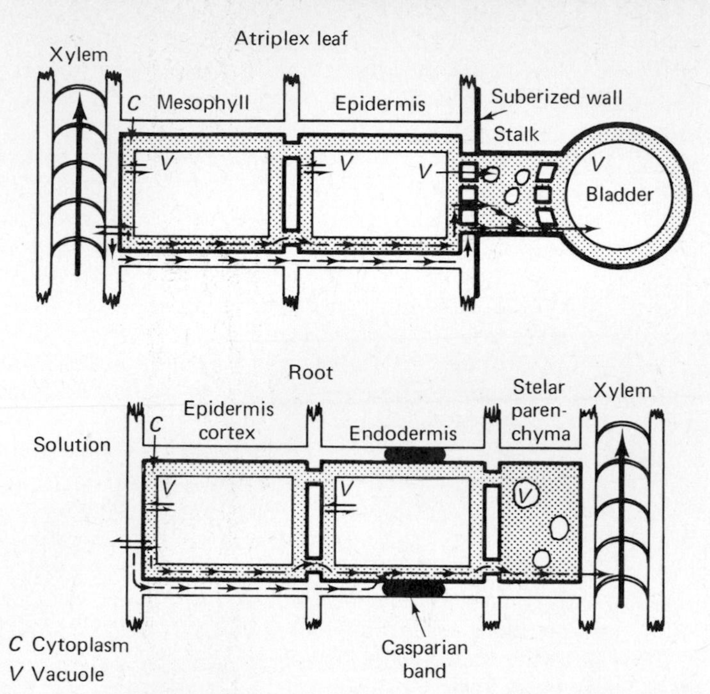

FIG. 9.12 Simplified structural models of a leaf salt gland and a root showing certain analogies with respect to the long distance transport of solutes. In both models movement in the cell wall pathway is severely restricted by the suberin deposited in cell walls, obliging solutes to cross the plasmalemma (shown as a shaded portion inside the wall). The symplast pathway through the cells and linking plasmodesmata is also shown. (Adapted from Lüttge, 1971, by kind permission of the author and Annual Reviews, Inc.)

that in *Atriplex spongiosa* there is a pathway of very low electrical resistance between the protoplasts of the mesophyll and the salt gland; this pathway is probably analogous with the symplast (see chapter 8, page 215). They found that changes in the transmembrane potential in mesophyll cells caused by illumination were almost immediately reflected by similar changes in the bladder cells, even though they contain no chloroplasts. This rapid propagation of changes in membrane potential resembles that found in maize roots by Dunlop and Bowling (see chapter 8, page 235).

Location of ion pumps. Although measurements of the electrochemical potentials of ions in salt bladders and mesophyll cells do not seem to have been made, it is almost certain that accumulation of salt is an active process. The enormous concentration gradients and the observed sensitivity of the process to low temperatures strongly suggest that this is so. Lüttge and Osmond (1970) found that chloride fluxes into glands of *Atriplex spongiosa* were partly light-dependent although the glands have no chloroplasts. This suggests two possibilities for the location of the chloride pump, one, at the tonoplast of the bladder cell, in which case there would have to be a very efficient communication between the pump and the chloroplasts in the mesophyll, and two, within the mesophyll itself where salt might be secreted into a compartment linked *via* the symplast to the bladder cell. In either case the numerous vesicles in the stalk cell, referred to on page 269, may play an important part.

In the former scheme some product of photosynthetic electron transport might move rapidly *via* the symplast and by vesicular transport to the tonoplast of the bladder cell. The vesicles would therefore be carrying 'fuel' for the tonoplast pump.

In the second scheme the vesicles themselves would be carrying salt. In *Tamarix*, Thomson and Liu (1967) suggest that salt is secreted actively into these vesicles. Kinetic studies with other species also suggest that tracer chloride and sodium passing to the gland cells equilibrate only with the cytoplasmic phase of the mesophyll cells. This suggests that in the mesophyll cells there is a system, which diverts the salt crossing the plasmalemma and prevents it from entering the vacuole.

It is not possible to arrive at any firm decision on the mechanism of salt secretion; there seem to be precedents for both systems and it is possible that both may operate. Research workers have appreciated only recently the interesting properties of salt gland as an experimental system. It seems likely that its thorough investigation may eventually help us to understand how the rapid interactions between intracellular organelles and the main permeability barriers of cells are brought about.

Bibliography

b References in the text

Atkin, R. K. and Wain, R. L. (1967) Studies on plant growth regulating substances XXIV. Factors influencing kinetin-induced phosphorus mobilization in detached radish cotyledonary leaves. *Ann. Appl. Biol.* **60**, 321–331.

Bell, C. W. and Biddulph, O. (1963) Translocation of calcium: Exchange versus mass flow. *Plant Physiol.* **38**, 610–614.

Biddulph, O., Biddulph, S., Cory, R. and Koontz, H. (1958) Circulation patterns for phosphorus, sulfur and calcium in the bean plant. *Plant Physiol.* **33**, 293–300.

Brown, W. V. and Johnson, S. C. (1962) The fine structure of the grass guard cell. *Am. J. Bot.* **49**, 110–115.

Collins, J. C. and Kerrigan, A. P. (1973) Hormonal control of ion movements in the plant root. In: *Ion Transport in Plants* (W. P. Anderson, ed.), 589–594. Academic Press, London.

Collins, J. C. and Reilly, E. J. (1968) Chemical composition of the exudate from excised maize roots. *Planta* (Berlin) **83**, 218–222.

Crafts, A. R. (1964) Herbicide behaviour in the plant. In: *Physiology and Biochemistry of Herbicides* (L. J. Audus, ed.), 75–110. Academic Press, London.

Crowdy, S. H. and Tanton, T. W. (1970) Water pathways in higher plants. 1. Free space in wheat leaves. *J. Exp. Bot.* **21**, 102–111.

Davies, C. R. and Wareing, P. F. (1965) Auxin directed transport of radiophosphorus in stems. *Planta* (Berlin) **65**, 139–156.

Eaton, J. M. (1943) Osmotic and vitalistic interpretations of exudation. *Am. J. Bot.* **30**, 663–674.

Etherton, B. (1970) Effect of indole-3-acetic acid in membrane potentials of oat coleoptile cells. *Plant Physiol.* **45**, 527–528.

Fischer, R. A. (1968a) Stomatal opening in isolated epidermal strips of *Vicia faba*. I. Response to light and CO_2-free air. *Plant Physiol.* **43**, 1947–1952.

Fischer, R. A. (1968b) Stomatal opening: role of potassium uptake by guard cells. *Science* **160**, 784–785.

Fischer, R. A. and Hsiao, T. C. (1968) Stomatal opening in isolated epidermal strips of *Vicia faba*. II. Responses to KCl concentration and the role of potassium absorption. *Plant Physiol.* **43**, 1953–1958.

Fritsch, F. E. and Salisbury, E. (1953) *Plant Form and Function*. Bell, London.

Greenway, H. and Pitman, M. (1965) Potassium retranslocation in seedlings of *Hordeum vulgare*. *Aust. J. Biol. Sci.* **18**, 235–247.

Gunning, B. E. S. and Pate, J. S. (1969) 'Transfer cells'. Plant cells with wall ingrowths, specialized in relation to short distance transport of solutes—their occurrence, structure and development. *Protoplasma* **68**, 107–133.

Gunning, B. E. S., Pate, J. S. and Briarty, L. G. (1968) Specialized 'Transfer cells' in minor veins of leaves and their possible significance in phloem translocation. *J. Cell. Biol.* **37**, C_7–C_{12}.

Hall, S. M. and Baker, D. A. (1972) The chemical composition of *Ricinus* phloem exudate. *Planta* (Berlin) **106**, 131–140.

Heath, O. V. S. (1959) Light and carbon dioxide in stomatal movements. *Handbuch der Planzenphysiologie 17/1*, 415–464. Springer-Verlag, Berlin.

Hillman, W. S. (1967) The physiology of phytochrome. *Ann. Rev. Pl. Physiol.* **18**, 301–324.

Humble, G. D. and Raschke, K. (1971) Stomatal opening quantitatively related to potassium transport. Evidence from electron-probe analysis. *Plant Physiol.* **48**, 447–453.

Ilan, I. (1971) Evidence for hormonal regulation of the selectivity of ion uptake by plant cells. *Physiol. Plant.* **25**, 230–233.

Isermann, K. (1971) Der Einfluss von chelatoren auf die Ca-aufnahme und Ca-Verlagerung bei höheren Pflanze. *Z. Pflanzenr. Bodenkunde* **128**, 195–207.

Ivanko, S. and Inguerson, J. (1971) *Physiol. Plant.* **24**, 355–362.

Jacoby, B. (1965) Sodium retention in excised bean stems. *Physiol. Plant.* **18**, 730–739.

Jaffe, M. J. (1970) Evidence for the regulation of phytochrome-mediated processes in bean roots by the neurohumor, Acetylcholine. *Plant Physiol.* **46**, 768–777.

Jeschke, W. D. and Simonis, W. (1969) Über die Wirkung von CO_2 auf die lichtabhängige Cl^- aufnahme bei *Elodea densa*. Regulation zwischen nichtcyclischer und cyclischer Photophosphorylierung. *Planta* (Berlin) **88**, 157–171.

Koukkari, W. L. and Hillman, W. S. (1968) Pulvini as photoreceptors in the phytochrome effect on nyctinasty in *Albizzia julibrissin. Plant Physiol.* **43**, 698–704.

Lüttge, U. (1971) Structure and function of plant glands. *Ann. Rev. Plant Physiol.* **22**, 23–44.

Lüttge, U. and Osmond, C. B. (1970) Ion absorption in *Atriplex* leaf tissue. III. Site of metabolic control of light-dependent chloride secretion to epidermal bladders. *Aust. J. Biol. Sci.* **23**, 17–25.

Lüttge, U. and Pallaghy, O. K. (1969) Light-triggered transient changes of membrane potentials in green cells in relation to photosynthetic electron transport. *Z. Pflanzenphysiol.* **61**, 58–67.

MacRobbie, E. A. C. (1971) Phloem translocation. Facts and mechanisms: A comparative survey. *Biol. Rev.* **46**, 429–481.

Mansfield, T. A. and Jones, R. J. Effects of abscisic acid on potassium uptake and starch content of stomatal guard cells. *Planta* (Berlin) **101**, 147–158.

Mason, T. G. and Maskell, E. J. (1931) Further studies on transport in the cotton plant. I. Preliminary observations of the transport of phosphorus, potassium and calcium. *Ann. Bot.* **45**, 125–173.

Meidner, H. and Mansfield, T. A. (1968) *Physiology of Stomata*. McGraw-Hill, London.

Mozafar, A. and Goodin, J. R. (1970) Vesiculated hairs: A mechanism for salt tolerance in *Atriplex halimus* L. *Plant Physiol.* **45**, 62–65.

Müller, K. and Leopard, A. C. (1966) The mechanism of kinetin-induced transport in corn leaves. *Planta* (Berlin) **68**, 186–205.

O'Brien, T. P. and Carr, D. J. (1970) A suberized layer in the cell walls of the bundle sheath of grasses. *Aust. J. Biol. Sci.* **23**, 275–287.

Pate, J. S. and Gunning, B. E. S. (1969) Vascular transfer cells in angiosperm leaves: A taxonomic and morphological survey. *Protoplasma* **68**, 135–156.

Pate, J. S., Gunning, B. E. S. and Briarty, L. G. (1969) Ultrastructure and functioning of the transport system of the leguminous root nodule. *Planta* (Berlin) **85**, 11–34.

Rains, D. W. (1968) Kinetics and energetics of light-enhanced potassium absorption by corn leaf tissue. *Plant Physiol.* **43**, 394–400.

Rains, D. W. (1969) Cation absorption by slices of stem tissue of bean and cotton. *Experientia* **25**, 215–216.

Satter, R. L., Sabnis, D. D. and Galston, A. W. (1970) Phytochrome controlled nyctinasty in *Albizzia julibrissin*. I. Anatomy and fine structure of the pulvini. *Am. J. Bot.* **57**, 374–381.

Satter, R. L., Marinoff, P. and Galston, A. (1970) Phytochrome controlled nyctinasty in *Albizzia julibrissin*. II. Potassium flux as a basis of leaflet movement. *Am. J. Bot.* **57**, 916–926.

Shone, M. G. T., Clarkson, D. T. and Sanderson, J. (1969) The absorption and translocation of sodium by maize seedlings. *Planta* (Berlin) **86**, 301–314.

Shone, M. G. T. and Wood, A. V. (1972) Factors affecting absorption and translocation of simazine by barley. *J. Exp. Bot.* **23**, 141–151.

Shone, M. G. T. and Wood, A. V. (1973) Factors responsible for the tolerance of blackcurrants to simazine. *Weed Research* **12**, 337–347.

Smith, R. C. and Epstein, E. (1964) Ion absorption by shoot tissue: kinetics of potassium and rubidium absorption by corn leaf tissue. *Plant Physiol.* **39**, 992–996.

Stout, P. R. and Hoagland, D. R. (1939) Upward and lateral movement of salt in certain plants as indicated by radioactive isotopes of potassium, sodium and phosphorus absorbed by roots. *Am. J. Bot.* **26**, 320–324.

Tal, M. and Imber, D. (1971) Abnormal stomatal behaviour and hormonal imbalance in *Flacca*, a wilty mutant of tomato. III. Hormonal effects on the water status in the plant. *Plant Physiol.* **47**, 849–850.

Thimann, K. V. and Sweeny, B. M. (1937) The effect of auxins on protoplasmic streaming. *J. Gen. Physiol.* **21**, 123–135.

Thomas, D. A. (1970a) The regulation of stomatal aperture in tobacco leaf epidermal strips. I. The effect of ions. *Aust. J. Biol. Sci.* **23**, 961–979.

Thomas, D. A. (1970b) The regulation of stomatal aperture in tobacco leaf epidermal strips. II. The effect of ouabain. *Aust. J. Biol. Sci.* **23**, 981–989.

Thomas, D. A. (1970c) The regulation of stomatal aperture in tobacco leaf epidermal strips. III. The effect of ATP. *Aust. J. Biol. Sci.* **24**, 689–707.

Thomson, W. W. and Liu L. L. (1967) Ultrastructural features of the salt gland of *Tamarix aphylla* L. *Planta* (Berlin) **73**, 201–220.

Tiffin, L. O. (1966) Iron translocation. II. Citrate/Iron ratios in plant stem exudates. *Plant. Physiol.* **41**, 510–518.

van Lookeren Campagne, R. N. (1957) Light-dependent chloride absorption in *Vallisneria* leaves. *Acta. Bot. Neerl.* **6**, 543–582.

Weatherley, P. E. (1963) The pathway of water movement across the root cortex and leaf mesophyll in transpiring plants. In Rutter, A. J. and Whitehead, F. H. (eds.), *The Water Relations in Plants*. Wiley, N.Y.

Weigl, J. (1969) Wechsewirkung pflanzlicher Wachstumhormone mit Membranen. *Z. Naturforschg.* **24b**, 1046–1052.

Wood, R. K. S. (1967) *Physiological Plant Pathology*. Blackwell, Oxford.

ten

Some factors affecting the rate of ion uptake by higher plants

The complete understanding of all the interactions which determine the rate of uptake by plant roots depends on the analysis of a complex supply and demand system. It is clear that rates of ion uptake depend not only on the inherent capability of the transport system in a given set of environmental conditions but on other more subtle factors such as growth and the 'demand' which this creates. This latter subject is poorly understood although there is unequivocal data to suggest that rates of uptake can be adjusted within fairly wide limits to suit the nutrient status of the plant and the stage of its development.

We will begin the chapter, however, with simpler matters. The response of transport processes to the ionic concentration of the external solution is of physiological as well as of ecological interest. We shall see that, in many plant cells, there is provision to carry ions across the membrane from very dilute solutions and up steep gradients of electrochemical potential, while at high external salt concentrations there are changes both in the properties of the cell membrane and the dependence of transport processes on metabolism.

Although water and ions are conducted through the plant in the same vascular system, the relationship between the rate of water and ion uptake is by no means simple and is dependent on the ionic concentration both in the plants and in the medium surrounding their roots.

In the closing section I shall give some examples of the overall control exerted by growth and 'demand' on ion absorption.

Uptake of ions from dilute solutions

Except for saline and maritime habitats the roots of plants are bathed by a solution in which many inorganic ions may be dilute. Many of the ion species move into the cells of the root against a steep gradient of electrochemical potential; such active transport must be mediated by ion 'pumps'. Some ions may reach the cell interior by passive diffusion; in these cases there is evidence for the participation of other types of ion carrier. The operation of such carriers, like the ionophorous antibiotics monactin and valinomycin described in chapter 4, is not directly linked to metabolic processes. We can learn something of the association of an ion and its carrier (whichever type it is) by a special application of enzyme kinetics.

The carrier concept

We established earlier that all transport processes are dependent on metabolism, the 'active' ones directly and the 'passive' ones indirectly. Against this background it is hardly surprising that the kinetics describing the rates of metabolism and enzyme reactions should be applicable to ion transport.

For many years the kinetics of enzyme reactions have been studied because they describe the affinity of the enzyme, *E*, and its substrate, *S*, and the rate of dissociation of the enzyme–substrate complex into the end products, *P*. Enzyme kinetics can also be used to predict the maximum velocity, V_{max}, at which a reaction can proceed in a given set of physical conditions.

If we consider the very simple enzyme reaction:

$$S + E \underset{K_2}{\overset{K_1}{\rightleftharpoons}} SE \xrightarrow{K_3} E + P \qquad (10.1)$$

It is clear that the two rate constants K_2 and K_3 govern the dissociation of *SE* while K_1 governs its rate of formation. The ratio of these opposing processes is known as the Michaelis Constant, K_m (in recognition of the early work on enzyme kinetics by Michaelis and Menten). Thus,

$$K_m = \frac{K_2 + K_3}{K_1} \qquad (10.2)$$

If K_m is very small then the affinity between the enzyme and its substrate is high and $K_1 \gg K_2 + K_3$; the reverse is true as K_m approaches unity.

The maximum velocity at which a reaction can proceed is related to the amount of enzyme present and the rate at which it is turning over. Once the capacity of the enzyme system is saturated the overall rate of the process is quite independent of the amount of substrate available for catalysis. Thus, if we plot the velocity of the reaction, V, against the substrate concentration in the medium we obtain a familiar hyperbola (Fig. 10.1).

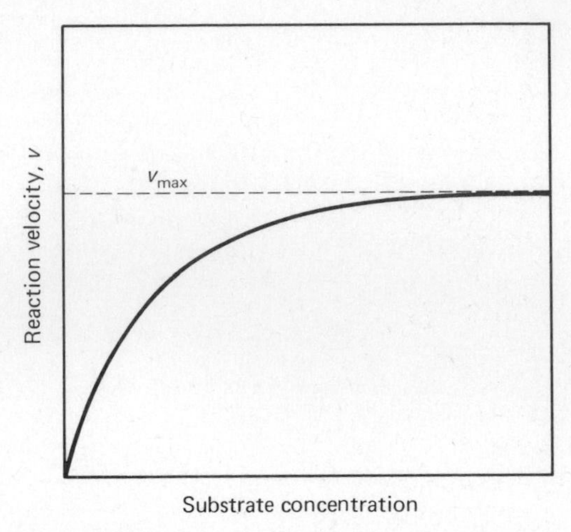

FIG. 10.1 The relationship between substrate concentration and the velocity of a reaction catalysed by an enzyme.

Ion fluxes into plant tissues show exactly similar 'saturation' as the external concentration is increased. This observation led Epstein and Hagen (1952) to apply the well established principles of enzyme kinetics to ion uptake by plant roots. In this analysis, summarized clearly in Epstein (1972), the velocity with which ion transport into the root, ϕ_j^{in}, proceeds is simply related to V_{max} and K_m. Thus:

$$\phi_j^{in} = \frac{V_{max} C_j}{K_m + C_j} \qquad (10.3)$$

where C_j is the concentration of the ion in the external solution. Using the special case where ϕ_j^{in} is half V_{max} we can rewrite eq. (10.3) and arrive at a practical definition of K_m in terms of solute concentration. Thus,

$$V_{max}/2 = \frac{V_{max}C_j}{K_m + C_j} \quad (10.4)$$

Rearranging and solving for K_m we get

$$K_m + C_j = 2\frac{(V_{max}C_j)}{V_{max}} \quad (10.5)$$

Cancelling V_{max} we get

$$K_m + C_j = 2C_j \quad (10.6)$$

Therefore at half maximal velocity

$$K_m = C_j \quad (10.7)$$

We can therefore define K_m as that concentration of an ion which produces a half maximal rate of influx. In common with the more formal description of K_m in eq. (10.2), a small value for K_m implies a high affinity between the ion and its carrier.

Practical determination of K_m *and* V_{max}

Most influxes of ions from dilute solutions appear to reach saturation in the range 0·2 to 1·0 mM. For our present purpose we will take this as the most concentrated solution to be considered and compare the rates of uptake in progressively more dilute solutions.

Samples of plant material are immersed in solutions where the ion being examined is labelled with a radioactive tracer. The incubation continues for a standard time (20 min to 2 h); at the end of this time tracer will be present in the free space of the tissue and within the protoplasts. The former fraction, since it in no way depends on membrane transport, is removed by washing and ion exchange; the residual radioactivity measures the absorption by the tissue and from this the rate of net influx is calculated (usually expressed as μmoles g^{-1} (dry weight) h^{-1}). To obtain K_m and V_{max} from these data two procedures can be followed.

Lineweaver-Burk Plot. If the reciprocals of both uptake rate, V, and ionic concentration in the medium are plotted a straight line relationship is usually obtained if the system displays the general saturation behaviour illustrated in Fig. 10.1. The extrapolation of the line through the experimental points to the ordinate gives the reciprocal of V_{max} (Fig. 10.2); K_m can easily be calculated since it is the ionic concentration producing half maximal velocity.

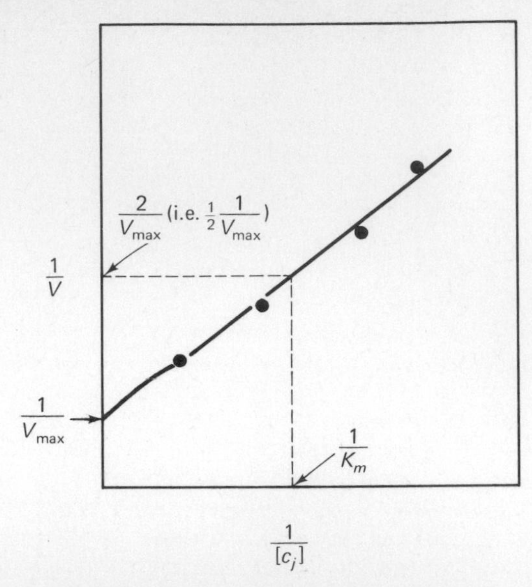

FIG. 10.2 ***Lineweaver-Burk double reciprocal plot of data from an absorption isotherm. Extrapolation of the line determines the value of V_{max} from which K_m can be deduced (see text).***

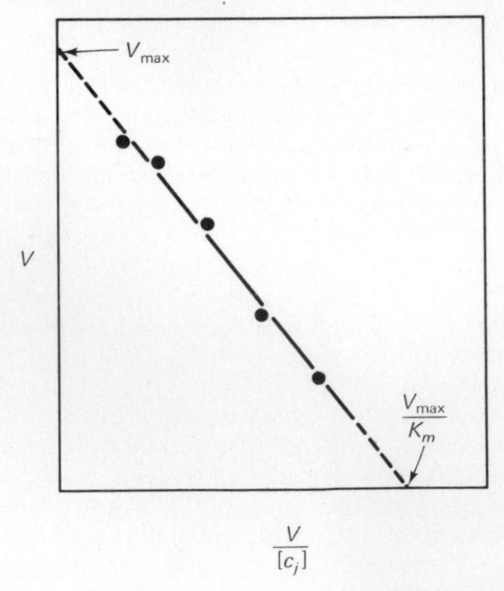

FIG. 10.3 ***Hofstee plot of data from an absorption isotherm. Extrapolation of the experimental data gives V_{max} and V_{max}/K_m.***

Hofstee Plot. An alternative procedure is to plot the rate of uptake, V, against V/C_j as illustrated in Fig. 10.3. V_{max} and V_{max}/K_m can be obtained by extrapolation of the experimentally determined slope to the ordinate and abscissa respectively.

Value for the apparent K_m *and* V_{max} *of various ion/tissue systems*

Values of K_m and V_{max} have been gathered assiduously for a number of years using a variety of tissues and most ions normally encountered by plants. Some representative values for K_m (Table 10.1) indicate a remarkable constancy for K_m for a given ion in a variety of experimental systems, e.g., K_m for K^+-absorption is similar for roots of barley and leaves of maize. The apparent K_m values for a variety of ions fall within a fairly narrow range, e.g., K_m values for Rb^+ are not unlike those for divalent cations, chloride and phosphate. In nearly every instance the carrier system or ion pump became saturated when the concentration of the ion in question was 0·2 to 0·5 mM.

This constant pattern of absorption has become known as *system I absorption* and will subsequently be referred to as such. System I is characterized by 'carriers' which have a high affinity for the ions they carry and which are not subject to much interference from other ions. Note that this definition tells us nothing about the nature of the carrier concerned.

Competition between ions

There is much evidence that the relationship between an ion-carrier and the ion-species it carries is highly specific. Interferences are usually found only with ions of similar size which are closely related in the Periodic Table. Frequently, these ions are rather uncommon in nature and we might regard them as 'analogues' of the ions which the carriers are designed to deal with in the normal course of events. Examples of these common–exotic ion pairs are, K^+ and Rb^+, Ca^{2+} and Sr^{2+}, Cl^- and Br^-, SO_4^{2-} and SeO_4^{2-}. There is usually marked discrimination against other ions when the concentrations in the solution are in the range where system I operates; thus, in barley roots, chloride uptake was unaffected by the other halide anions, F^- and I^-, even though it was inhibited competitively by Br^- (Elzam and Epstein, 1965). In a similar way, K^+ uptake by corn leaf tissue was inhibited by Rb^+ but was insensitive to Na^+, Li^+, Cs^+ and NH_4^+ (Smith and Epstein, 1964).

In sugar cane (*Saccharum officinarum*, L.) leaf tissue, the uptake of manganese had almost identical kinetics in the presence of zinc and copper even though for all three ions the values for K_m and V_{max} were similar (Table 10.2).

Table 10.1 Some examples of K_m and the 'saturating' concentration of ions in system I absorption by various tissues

Ion	Species	Organ/tissue*	Saturating concentration (mM)	K_m (mM)	Competing ions	Non-competing ions	References
K^+	*Hordeum vulgare* (Barley)	Excised roots	0·2	0·021	—	Na^+	(1)
K^+	*Zea mays* (Maize)	Leaf tissue	0·2	0·038	Rb^+	Na^+	(2)
K^+	*Elodea densa* (Water weed)	Leaf (whole)	0·2	0·099	Rb^+	—	(3)
Rb^+	*Hordeum vulgare* (Barley)	Excised roots	0·5	0·017	—	—	(4)
Rb^+	*Zea mays* (Maize)	Leaf tissue	0·2	0·015	K^+	Na^+, Li^+ Cs^+, NH_4^+	(2)
Cl^-	*Agropyron elongatum* (Tall wheat-grass)	Excised roots	0·2	0·013	—	—	(5)
NO_3^-	*Zea mays* (Maize)	Roots of intact plants	0·24	0·021	—	—	(6)
$H_2PO_4^-$	*Hordeum vulgare* (Barley)	Excised roots [Sterile]	0·2	0·050	—	—	(7)
$H_2PO_4^-$	*Beta vulgaris* (Beetroot)	Disks of storage tissue [Sterile]	0·2	0·050	—	—	(7)
$H_2BO_3^-$	*Saccharum officinarum* (Sugar Cane)	Leaf tissue	0·2	0·029	—	—	(8)
SO_4^{2-}	*Elodea densa* (Water weed)	Whole leaves	0·1	0·006	—	—	(9)
SO_4^{2-}	*Zea mays* (Maize)	Excised roots					
		Inbred lines	—	0·166, 0·179	—	—	(10)
		Hybrid	—	0·055	—	—	(10)

* *Unless stated otherwise the tissues were contaminated by micro-organisms.*

References: **(1)** ***Epstein, Rains and Elzam (1963)*****; (2)** ***Smith and Epstein (1964)*****; (3)** ***Jeschke (1970)*****; (4)** ***Jackman (1965)*****; (5)** ***Elzam and Epstein (1969)*****; (6)** ***Van der Honert and Hooymans (1955)*****; (7)** ***Barber (1972)*****; (8)** ***Bowen (1969)*****; (9)** ***Jeschke and Simonis (1965)*****; (10)** ***Ferrari and Renosto (1972)*****.**

Table 10.2 Apparent values for K_m and V_{max} for divalent cation uptake by sugar cane leaf tissue*

Ion	K_m (mM)	V_{max}
		μmoles g^{-1} (dry wt) h^{-1}
Cu^{2+}	0·0145	5·37
Zn^{2+}	0·0111	5·88
Mn^{2+}	0·0161	5·35

** Tissue treated for 1–2 h at 30 °C in a solution containing 0·5 mM $CaCl_2$ at pH 5·7. (Data from Bowen, 1969.)*

Table 10.3 shows that the rate of Mn^{2+} uptake was unaffected by the presence of 0·1 mM Zn^{2+} or Cu^{2+}. Similarly, the absorption of neither Zn^{2+} nor Cu^{2+} was affected by Mn^{2+}. There was, however, marked competition between Zn^{2+} and Cu^{2+}, Zn^{2+} uptake in the presence of an equimolar solution of Cu^{2+} being reduced by 20 per cent. This reduction is exactly that which would have been found had we simply reduced the Zn^{2+} concentration by one-half. These results suggest that Zn^{2+} and Cu^{2+} can be carried by a common carrier while Mn^{2+} is carried into the leaf tissue by a completely separate one.

Bowen (1969) also found that 0·1 mM solutions of a number of other cations (viz., Na^+, Li^+, Cs^+, Ag^+, Co^{2+}, Cr^{2+}, Al^{3+}, Mg^{2+}, Ba^{2+}, NH_4^+) were completely without effect on the absorption of Mn^{2+}, Cu^{2+} and Zn^{2+}.

The evidence from experiments on ion interactions seems to demand that most ions absorbed from dilute solutions should be carried by a selective carrier; we need to think, therefore, of a plasmalemma with as many as a dozen different carrying mechanisms if this interpretation of the evidence is correct.

Some exceptions to the usual pattern of 'system I' absorption

The characteristic 'saturation' kinetics for the uptake of ions from dilute solutions is not always found. A notable exception is provided by observations on uptake of Cl^-, Na^+ and K^+ by slices from mature leaves of orange, (*Citrus sinensis*) (L.) cv. Valencia, reported by Robinson and Smith (1970) and Smith and Robinson (1970).

The uptake of sodium was directly proportional to the concentration of Na^+ in the external medium over a 1000-fold concentration range, 0·03 to 30·00 mM. At higher concentrations the slope of the absorption isotherm declined but was far from saturation at 100 mM.

Table 10.3 Interactions in the absorption of divalent cations from mixed solutions by excised sugar cane leaf tissue*

Ions present	Absorption rate μmoles g^{-1} (dry wt) h^{-1}		
	Cu^{2+}	Zn^{2+}	Mn^{2+}
Cu^{2+}	5·14	—	—
Zn^{2+}	—	5·28	—
Mn^{2+}	—	—	5·91
Cu^{2+} + Zn^{2+}	[4·28]	[4·05]	—
Cu^{2+} + Mn^{2+}	5·10	—	5·87
Zn^{2+} + Mn^{2+}	—	5·22	5·84
Cu^{2+} + Zn^{2+} + Mn^{2+}	[4·34]	[3·98]	5·80

Each ion supplied at 0·1 mM either singly or in combination in a solution of 0·5 mM $CaCl_2$, pH 5·7, at 30 °C. Boxed values represent significant inhibitions.

(Data from Bowen, 1969.)

The absorption isotherm for potassium (Fig. 10.4) gives an indication of an approach to saturation between 3 to 10 mM. Using conventional procedures, i.e., the double reciprocal method of plotting the data (see Fig. 10.2), Smith and Robinson (1970) calculated a value of K_m for K^+ uptake of 1·3 mM. This is clearly very much greater than any of the values in Table 10.1 and might indicate a very low affinity of K^+ for the K-carrier. Where this is the case one would usually expect the uptake to be subject to much interference from other ions, since the selectivity of the carrier might be correspondingly low. In fact, this was not so, because there was virtually no interference from sodium ions until the Na: K concentration ratio reached 10:1.

There is not enough information available to decide whether these results from *Citrus* represent an isolated case or a fairly common one. In one respect at least they seem to be logical for leaf tissue in particular. In chapter 9 I pointed out that the concentration of K^+ and other ions reaching the plasmalemma of mesophyll cells from the free space of the leaf was a relatively concentrated solution in which 10–50 mM K^+ might be in no way exceptional. Our data on guttation from the marrow leaf (chapter 9, page 245) also showed that from these concentrated solutions K^+ was selectively extracted relative to sodium. It would seem to be a sensible thing for the leaf to arrange that its highly selective system I absorption should extend over a wider concentration range.

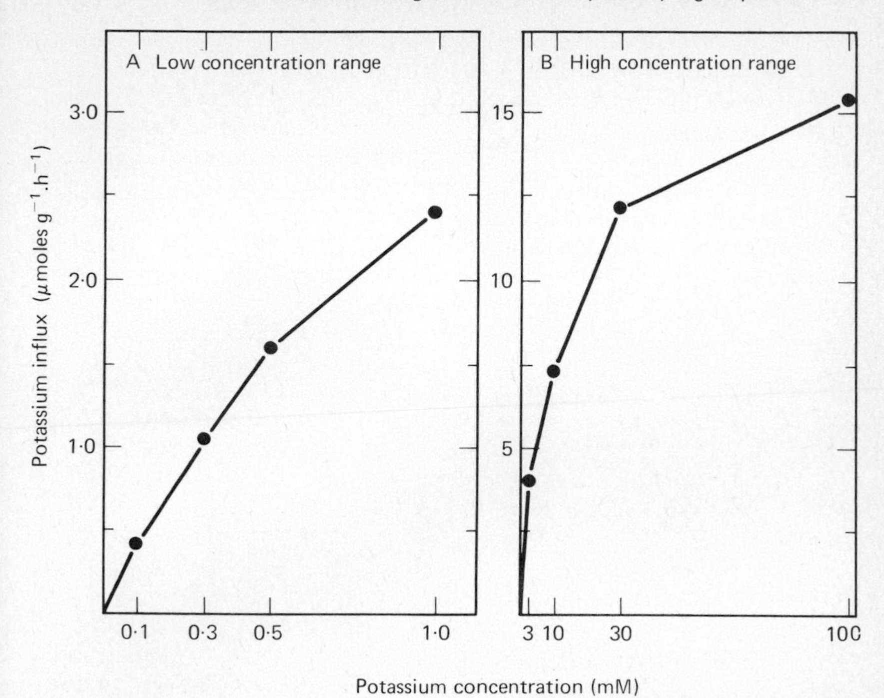

FIG. 10.4 The effect of a wide range of potassium concentrations on the potassium influx into slices of citrus leaf tissue. Note that in the low concentration, (A), the influx has not reached saturation in the range normally found for system I. In the upper concentration range, (B), an approach towards saturation is found at 100 mM K^+. (Taken from Figure 6 in Smith and Robinson, 1971.)

Smith and Robinson (1970) also make the observation that the shape of the absorption isotherm can be strongly influenced by factors which may not, in themselves, directly affect the binding of an ion to a selective carrier. They found that the Cl^- absorption isotherm was determined largely by the nature of the accompanying monovalent cation; Cl^- influxes being much greater from solutions of KCl than from solutions of NaCl. On the other hand, monovalent cation uptake by barley and maize roots was virtually independent of the accompanying anion, even when supplied as sulphate which is absorbed slowly relative to the cations (e.g., Epstein *et al.*, 1963; Lüttge and Laties, 1966).

Micro-organisms and 'system I' absorption

A recent paper by Barber (1972) has confirmed, what has often been assumed, that system I absorption kinetics in sterile material are quite similar to those from the non-sterile plant materials generally used in experiments (Table 10.1). In earlier work, however, Barber and Frankenburg (1971) showed that, at low concentrations, the uptake of several ions by excised barley roots was greater in non-sterile roots than in sterile ones. The presence of contaminating micro-organisms will clearly affect the shape of the absorption isotherm and may influence the apparent K_m values. For instance, Barber (1968) showed that in phosphate concentrations in the 1–10 μM range the presence of micro-organisms increased the apparent phosphate absorption by the roots by 100 to 300 per cent. Micro-autoradiographs made it quite clear that much of the ^{32}P associated with

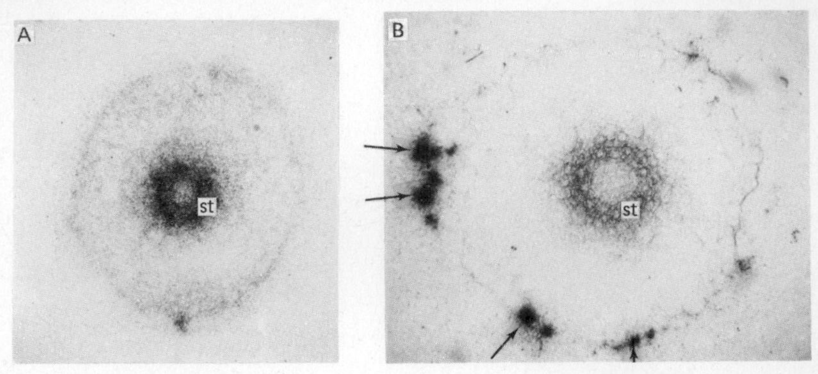

FIG. 10.5 Autoradiographs of transverse sections of barley root labelled with ^{32}P. (A) cut from a sterile root (B) non-sterile root contaminated by ambient laboratory micro-organisms. The dark spots at the periphery of (B) (arrows) probably represent ^{32}P accumulation in bacterial colonies; they are absent in the sterile root. Note the somewhat higher radio-activity of the stele, st, in the sterile root. (See also Barber, Sanderson, and Russell, 1968.)

non-sterile roots was located on the root surface in what seem to be bacterial colonies (Fig. 10.5). There seem to be similar interactions between micro-organisms and sulphate uptake and somewhat smaller ones in the uptake of rubidium and potassium (Barber 1972).

There can be little doubt that their high metabolic rate and their location at the root periphery place bacteria and other micro-organisms in a very favourable position to intercept incoming ions from dilute solutions at the expense of the host plant. With this in mind, it is obvious that reliable conclusions about rates of absorption by 'carrier' systems of the root at very low concentrations can be drawn only from experiments where the plant tissue is sterile.

When non-sterile tissue is examined it is sometimes possible to detect one or several (Nissen 1971) changes in slope of the absorption isotherm within what is normally regarded as the system I concentration range. The experiments of Carter and Lathwell (1967) illustrate this point. They measured phosphate uptake by excised maize roots at KH_2PO_4 concentrations ranging from 1 to 256 μM. On the basis of kinetic analysis they proposed that there were two sites (carriers) for phosphate uptake, one with a K_m of 0·136 mM which dominated the uptake kinetics at concentrations greater than 10 μM, and a second, with a K_m of 0·0061 mM which dominated at the lowest concentrations of phosphate in the external solution. Transport of phosphate at both of these absorption sites was linked to metabolism. In the light of what was said above, it seems that this experiment may have elegantly distinguished between the affinity of phosphate for carriers in the root cells and in the bacteria rather than, as the authors suggested, two distinct carriers in the membranes of the maize root.

Boundary layer effects and 'system I' absorption

If a plant root is immersed in a dilute solution, ions will be absorbed by the cells from the solution in the immediate vicinity of the root/solution boundary. In the absence of turbulent mixing, ions from the bulk solution will diffuse into the zone where this depletion has occurred. If the rate of absorption exceeds the rate at which ions can be supplied by diffusion, then the ionic concentration at the root surface will be consistently lower than the nominal concentration in the bulk solution. We would therefore underestimate the apparent influx at the nominal concentration. To a certain extent this situation can be remedied if the solution is stirred or turbulently mixed since this will reduce the length of the diffusion path which ions from the bulk solution must travel in order to reach the root/solution boundary. There will remain, however, an unstirred layer of solution which cannot be mixed with the bulk solution at velocities which are practicable in biological experiments. At very low concentrations diffusion across this layer may be the rate limiting step in ion uptake. As the concentration rises we know by experiment that between 0·2–0·5 mM the system I mechanisms become saturated and ion uptake becomes independent of external concentration. Thus, if we conduct an experiment over a range of concentrations, say 1 to 100 μM, we will introduce a variable into the experiment due to unstirred layers. This variable has seldom been considered but at low concentrations it may influence considerably the shape of the absorption isotherm.

Polle and Jenny (1971) found that rubidium uptake by barley roots from dilute solutions was much more rapid if the solution was stirred (Fig. 10.6) but that when the saturation of system I was approached (92 μM treatment) the effect was absent. The turbulent mixing would have reduced the diffusion pathlength and shows that, at low concentrations, diffusion was limiting the rate of uptake. If we take this data and construct absorption isotherms for Rb^+ uptake at zero stirring and at 15 rev/min, it is clear that the shapes of the curves are different (Fig. 10.7).

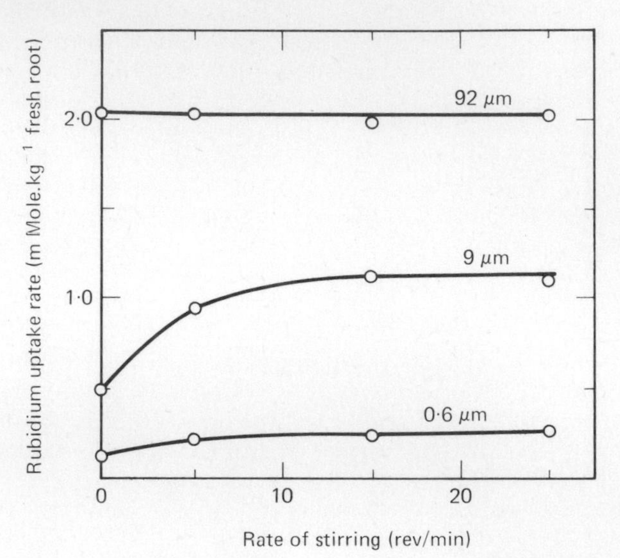

***FIG. 10.6** Effect of stirring on rubidium uptake by excised barley roots. The three concentrations are all within the system I range. (After Polle and Jenny, 1971.)*

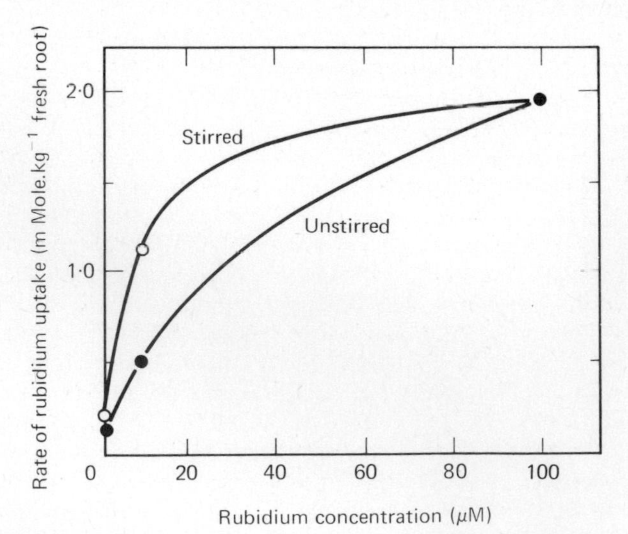

***FIG. 10.7** Effect of stirring on the shape of rubidium absorption isotherms. Constructed from data in Polle and Jenny (1971).*

In some circumstances effects of stirring can be measured at the upper end of the system I concentration range. Jacobson, Cooper and Volz (1971) showed that chloride absorption was reduced in unstirred solutions containing 5 mM KCl in comparison with stirred solutions. Analysis of the system showed that this was due to the accumulation of bicarbonate in the boundary layer with a consequent rise in pH which inhibited Cl^- absorption. The stirring produced by vigorous aeration reduced the pH differential between the boundary layer and the outer medium by reducing the length of the outward pathway for HCO_3^- diffusion.

Experimental procedures which restrict the rapid mixing of the boundary and bulk solutions should therefore be avoided when working at low ionic concentrations and the rate of mixing in various treatments should be constant.

A final word on the carrier concept

Although the application of Michaelis–Menten enzyme kinetics to ion transport has produced a convenient way of describing the association of ions with the uptake system, the results do not, and cannot, tell us how transport is effected or anything about the nature of the carriers themselves. The formal similarity between ion transport processes and enzyme reactions should not mislead the student into thinking that the carriers themselves have to be enzymes or even specific chemical entities. Almost any physiological process dependent on metabolism displays Michaelis–Menten kinetics. The transport of anions must be directly linked to metabolism at the range of concentrations where system I operates; the carriers detected in kinetic analysis must, in these instances, be ion pumps as we have come to understand them in the earlier part of the book. The nature of the other carriers is less certain. They may be molecules with a chemical nature similar to the ionophorous antibiotics (e.g., Monactin or Valinomycin) which carry ions in the direction determined by the gradient of electrochemical potential. The effects of metabolic inhibitors may extend to transport by such 'passive' carriers if, as is commonly found, the inhibitor alters the electrical driving forces across the membrane (e.g., depolarizes the cell). The fact that all transport processes are to some extent disturbed by metabolic inhibitors does not, as Epstein (1972) implies, indicate that all the carriers are 'active' ion pumps in the strict sense. We might illustrate the point further by suggesting, not unreasonably, that transient defects in the membrane structure provide a means whereby certain ions, with appropriate dimensions, can be carried across the membrane. Our carrier is, therefore, neither ion pump nor ionophore but a 'hole'. Metabolic activity clearly influences both the rate of synthesis and turnover as well as the mobility of the highly dynamic plasmalemma. Such factors can easily be envisaged as affecting both the number of ion-selective 'holes' and the rate at which ions pass through them (e.g., Träuble's theory of 'kink' movement, page 34). Such a system would still have a K_m and a V_{max}.

One of the disturbing features of interpreting K_m values as the association of a solute for a specific carrier molecule is the multiplicity of carriers required to equip the cell for life in its normal environment. The credibility of the specific carrier concept is further stretched when it is found that a herbicide, such as simazine, is absorbed by barley roots in a manner which indicates the operation of system I. When a double reciprocal plot of the data was made a good straight line was obtained from which V_{max}, and a very low value of 4 μM for the apparent K_m, were calculated (Shone, personal communication). The result suggests a high affinity between the herbicide and a carrier if it is interpreted in the conventional way. It is hard to believe that the plasmalemma is equipped for encounters with such foreign substances by the provision of specific carriers.

Absorption of ions from solutions of higher concentration

The influx of ions into plant tissues does not, in fact, reach saturation between 0·2 and 0·5 mM. If the concentration is increased well above this range the influx continues to increase rather slowly but eventually can be very much greater than the V_{max} of system I. At these higher concentrations the rate of absorption falls, approaching a second 'saturation' at concentrations usually greater than 50 mM; this second process is known as system II absorption.

It is rather difficult to illustrate graphically the results of an experiment where the ionic concentration may have ranged from 0·001 to 100 mM on a single diagram. I raise this point simply because one commonly used solution to this problem may give the student the unfortunate impression that, above 0·50 mM, a second uptake mechanism suddenly springs into life and that ions are absorbed by it at a rapid rate (Fig. 10.8). In Fig. 10.8 the change in scale beyond 0·5 mM should be properly noted. An alternative illustration of the same data is shown in Fig. 10.9 where high and low concentration ranges are plotted on separate scales. The first graph is the same as the lefthand portion of Fig. 10.8; on the second graph these data are compressed to a very short linear distance over which the rate of uptake increases very sharply with increasing concentration. This presentation does, perhaps, give a clearer picture of the relationship of system I and system II absorption but does not make the point, evident in Fig. 10.8, that the absorption isotherm appears to be hyperbolic in shape in both systems.

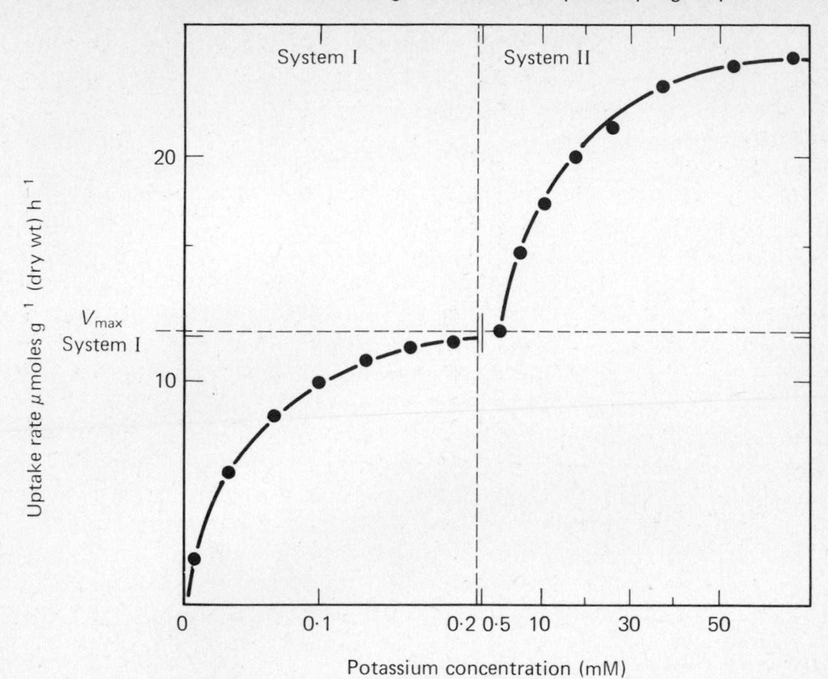

***FIG. 10.8** A method of plotting absorption data from a very wide range of concentrations. This procedure is widely used and emphasizes the dual 'mechanisms' involved in ion transport, both of which show saturation kinetics.*

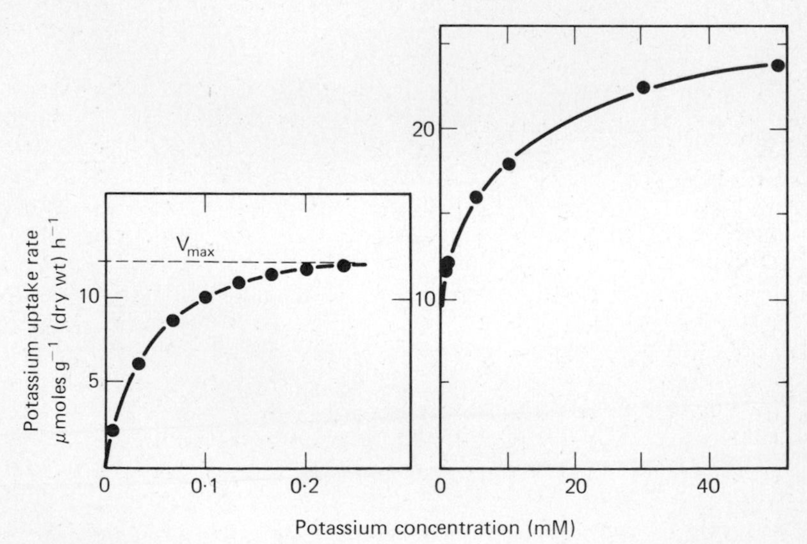

***FIG. 10.9** The data in Fig. 10.8 replotted on two separate scales (see text).*

The nature of ion absorption by system II

The absorption of ions at high concentrations continues well beyond the point where system I is saturated. The nature of the mechanism and its location in the cell remain very controversial and in what follows I shall try to give an account of the opposing views.

Metabolic dependence. Some evidence points to the conclusion that ion transport in the system II concentration range does not show a strong dependence on metabolism; there is no need, therefore, for ion carriers in system II, if they exist, to be active pumps.

Figure 10.10 shows some data from Barber (1972) gathered from phosphate uptake by sterile disks of beetroot tissue. Over the range of concentrations in which system I operates, phosphate uptake was severely depressed by low temperature but, as the concentration was increased above 10 mM, the temperature dependence declined so that at 50 mM there was little difference in the amount of phosphate absorbed by disks in the two treatments. The declining temperature sensitivity with increasing concentration is reflected in a continuous decline in the Q_{10} for the process. At the higher concentration the Q_{10} values are typical of a purely physical process as opposed to a metabolic one.

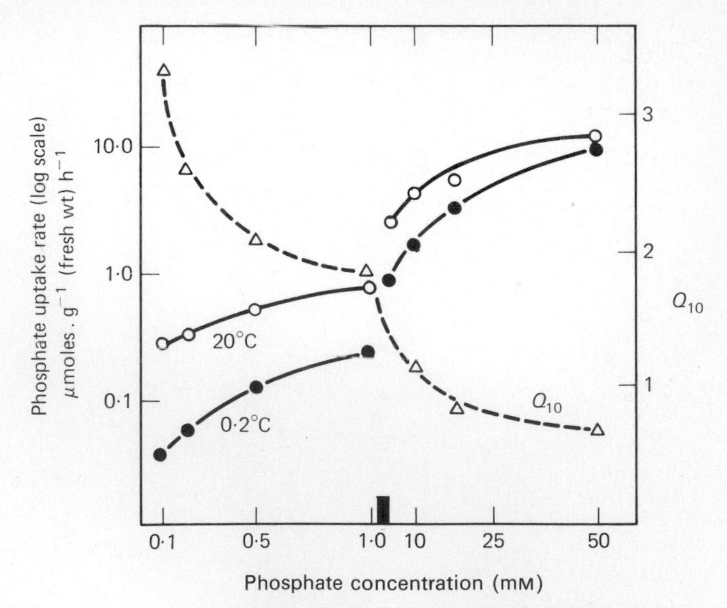

*FIG. 10.10 Phosphate absorption isotherms for sterile disks of beetroot (***Beta vulgaris***) at 0·2 °C and 20 °C, and the temperature coefficient, Q_{10}, for the process at various concentrations. (From Barber, 1972 with kind permission.)*

Barber (1972) obtained rather similar results for phosphate uptake by excised, sterile barley roots in the presence of the uncoupler DNP; severe inhibition was found at low phosphate concentrations but very little at 50 mM.

These experiments show that absorption of phosphate in the system II concentration range was largely a physical process, perhaps passive diffusion down an electrochemical potential gradient.

Effects on membrane properties. There are several lines of evidence which suggest that at high electrolyte concentration there are changes in the physical properties of biological membranes. In *Nitella* the electrical resistance fell, and the electrolyte conductance increased, when the KCl concentration in the medium was increased from 1 to 10 mM (Spanswick, 1970). Weigl (1970a) has shown that the efflux of chloride from excised maize roots increased eight-fold when the bathing medium was changed from 0·5 mM to 50 mM KCl (Fig. 10.11). A similar response of Cl^- efflux was produced by similar

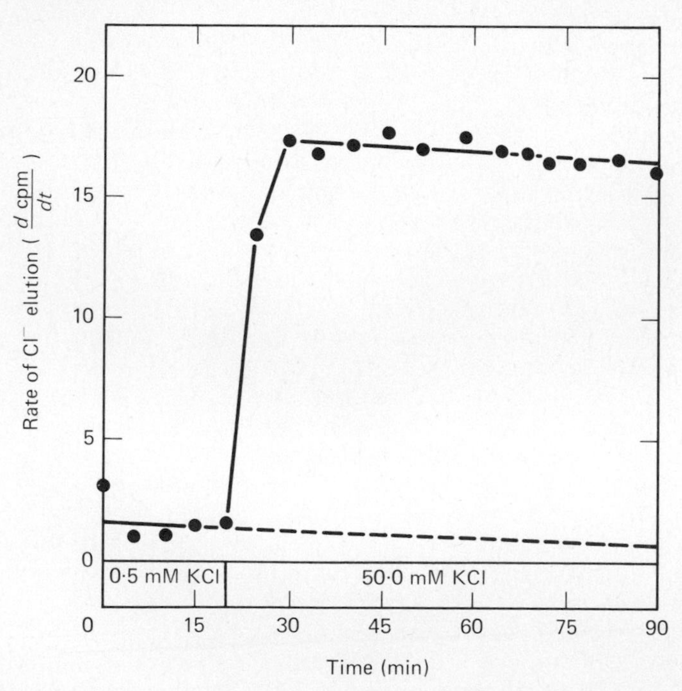

FIG. 10.11 Effect of a hundred-fold increase in the external KCl concentration on the elution of labelled chloride from excised roots of Zea mays. The elution rate changes abruptly when the concentrated KCl solution is added, but then reaches a new steady level. (Redrawn from Weigl, 1970a.)

concentrations of NaCl, K_2SO_4, $CaCl_2$ and choline chloride indicating the rather non-specific effect. The results strongly suggest that the passive permeability of the membrane to chloride and other anions, e.g., SO_4^{2-} (Weigl, 1970b), is increased in the high salt solutions. He also found that these treatments had no consistent effect on the efflux of ^{86}Rb. Weigl (1970a) thinks it probable that the increased passive anion permeation results from interactions between cations and the negatively charged head of phospholipid molecules, although there is no suggestion how this effect is achieved. One readily observable consequence of increased anion permeability will be to depolarize the membrane potential (see chapter 3, page 71) and, in common with the algae, increases in external salt concentration, particularly of K-salts, do depolarize the membrane potential at some point.

In pea epicotyl tissue and roots of *Avena*, the potential is buffered against increasing K-concentration (see chapter 7, page 188). This buffering breaks down eventually and the potential is reduced. Figure 10.12 shows some data from Pitman *et al.* (1971) working with excised barley roots of low salt status. The potential difference between the vacuole and the external medium remained constant over the 'system I' range of KCl concentrations but began to depolarize between 0·25 and 1·0 mM KCl. Pitman *et al.* found that high concentrations of NaCl and $CaCl_2$ had only minor effects on the membrane potential of barley roots, whereas in the experiments by Weigl (1970a) these salts were as effective as KCl in stimulating Cl^- efflux in maize roots. Although it does not explain the discrepancy, it should be pointed out that the ionic relations of maize roots differ from those of other cereal roots in several respects, e.g., in having an apparent 'active' calcium uptake (Maas, 1969). In barley, then, we must assume that the depolarization is due to the increased diffusion of potassium into the cell.

The depolarization of the membrane potential, whatever its cause, will have two consequences. Firstly, with cations, it will offset the increases in the inwardly directed electrochemical potential gradient caused by high external concentrations. Secondly, it will reinforce the effects of increasing concentration in reducing the electrochemical potential gradient up which anions move into the cell. In these circumstances, passive anion influx can assume significant proportions whereas at lower concentrations and more negative electropotentials it must be insignificant.

I have contrived some data in Table 10.4 which illustrates these effects. The assumed concentrations for K^+ and $H_2PO_4^-$ are not unreasonable and the membrane potentials are from Pitman *et al.* (1971). The columns marked 'equilibrium concentration' are calculated from the Nernst equation on the assumption that the observed potential is the Nernst potential. Departures from the 'assumed' values give a measure of how far the system is from equilibrium. In case I where the KH_2PO_4 is at the limit of the system I range it is evident that both ions, and in particular $H_2PO_4^-$, enter the cell against gradients of electrochemical potential. When the KH_2PO_4 solution is increased to 25 mM in case II there is a striking change in the steepness of the electrochemical potential gradient for $H_2PO_4^-$; about one-third of the assumed internal concentration could be accounted for by passive diffusion. This proportion will rise steeply as depolarization and concentration increase.

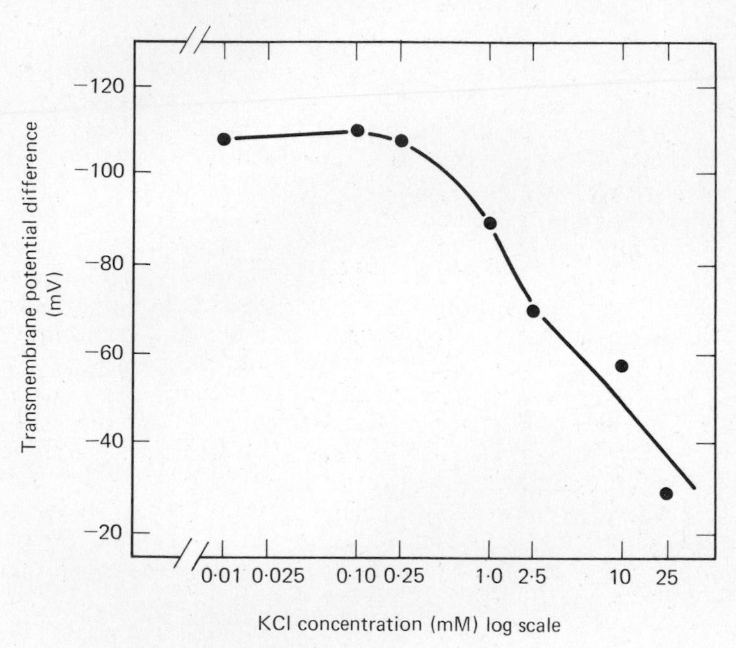

FIG. 10.12 Effect of KCl-concentration in the external medium on the electrical potential difference between the vacuole and the outside of the cells in excised barley roots. The roots in this treatment had been raised in dilute calcium sulphate solution and were therefore of low salt status. The potential difference seemed to be buffered against increasing KCl-concentration in the system I range, but at higher concentrations the cells depolarized. (Taken from Pitman* et al., *1971.)

In the absence of the fall in the membrane potential from −84 to −38 mV, the equilibrium concentration of K^+ in the cell interior would be 749 mM; the depolarization of 46 mV reduced this considerably, to 112 mM. The electrochemical potential gradient is, however, still directed inwards across the plasmalemma. Further depolarization will diminish this driving force and may be responsible for the gradual approach towards 'saturation' in system II.

***Table 10.4** Effect of hypothetical changes in concentration and membrane potential on the internal and equilibrium concentration of K^+ and $H_2PO_4^-$ in a plant cell*

Ion	Case I: KH_2PO_4 1·0 mM		Case II: KH_2PO_4 25·0 mM	
	Assumed concentration (mM)	Equilibrium concentration (mM)	Assumed concentration (mM)	Equilibrium concentration (mM)
K^+	66	27	99	112
$H_2PO_4^-$	10	0·04	15	5·6
Membrane potential	$E_{vo} = -84$ mV		$E_{vo} = -38$ mV	

The location of system I and system II

There seems to be little doubt that the carriers and ion pumps of system I reside at the plasmalemma of the cell. Such a location is consistent with all the physical evidence that, at system I concentrations, this membrane represents the major permeability barrier of the cell.

If, for the moment, we ignore the uncertainty about the nature of system II absorption, we should be aware that there are two opinions about its location in the cell. The first, and most straightforward, is that of Epstein who, in a number of papers, states that, without doubt, both system I and system II reside in the plasmalemma. This view is summarized with clarity in a short paper by Welch and Epstein (1969). The second interpretation is offered by Laties, who, with equal confidence, states that system II resides in the tonoplast. This view is also put forward with clarity in a review by Laties (1969). For convenience in the discussion I shall refer to the former scheme as the 'parallel' arrangement and the latter as the 'series' arrangement. Before I embark on the delicate matter of criticising these schemes, we should appreciate the profound differences that exist between them.

At high salt concentrations, in the 'parallel' arrangement the two systems operate together, each contributing ions to the cytoplasm across a membrane which 'does not become permeable to monovalent cations' (*sic*) Welch and Epstein (1969). In the 'series' arrangement the key assumption is that the plasmalemma becomes *highly permeable* to ions, more so than the tonoplast which then provides the rate limiting step in ion movement to the vacuole.

Evidence for the parallel arrangement. Laties (1969) points out that, if system I and system II are operating in parallel at the plasmalemma, the observed velocity of absorption will always be the sum of two processes. At concentrations above 1 mM, system I will be going at maximum velocity so that, as the concentration increases, any increase in rate will be due to system II. With one process at saturation and another one increasing it is quite obvious if we make a double reciprocal plot of I/V against I/S we would not expect to get a straight line. If the contribution (V_{max}) made by system I is deducted and the reciprocal of the remaining velocity plotted against I/S, a straight line should be obtained.

This is essentially what Epstein, Rains and Elzam (1963) found in their experiments and accordingly they proposed that at high salt concentrations both systems I and II contribute ions to the cytoplasm.

Welch and Epstein (1968, 1969) took this point a little further and specifically stated that the systems were operating in parallel at the plasmalemma in excised barley roots of low salt status. The evidence for this assertion came from two types of experiment.

In experiment I, Welch and Epstein (1969) subjected excised roots to one of two treatments; group 1 were treated continuously for 10 minutes with either 0·5 mM or 5·0 mM KCl labelled with ^{86}Rb; group 2 were treated for five separate periods of one minute each with the labelled solution, each period being separated by a one minute treatment with unlabelled solution of the same strength. Thus, the total exposure of group 2 to labelled potassium was only half that of group 1. Table 10.5 shows that roots in group 2 contained exactly half of the ^{86}Rb found in the continuous treatment. Similar results were obtained for the uptake of labelled chloride from 0·5 and 5·0 mM solutions of sodium chloride.

Table 10.5 ***Absorption of labelled potassium and chloride by excised low-salt barley roots in a continuous 10 min period and in five 1 min pulses separated by 1 min interruptions***

External solutions mM		Labelled ion absorbed in 10 min (μmoles g^{-1} tissue)		Labelled ion absorbed in five 1 min pulses (μmoles g^{-1} tissue)	Ratio
K (^{86}Rb) Cl	0·1	(K^+)	0·96	0·47	2:1
K (^{86}Rb) Cl	5·0	(K^+)	1·26	0·63	2:1
$Na^{36}Cl$	0·1	(Cl^-)	0·18	0·092	2:1
$Na^{36}Cl$	5·0	(Cl^-)	1·49	0·79	1·9:1

Data from Tables I and III, Welch and Epstein (1969).

This experiment showed that, at both concentrations, the absorption of tracer started and stopped as soon as the isotope was introduced to or withdrawn from the root. The label accumulated in each one minute pulse was additive and even at the higher concentration there was no appreciable efflux of ^{86}Rb from the tissue during the intervening period in unlabelled solution. Had the plasmalemma become highly permeable to potassium one might have expected this, unless all of the ^{86}Rb was actually in the vacuole, which seems unlikely. Efflux through the plasmalemma would have reduced the total labelled potassium accumulated in the five one minute pulses to less than half of that in the continuous treatment. They concluded, therefore, that at 5·0 mM KCl, when system II is contributing about 20 per cent of the total uptake, the passive permeability of the plasmalemma is not significant.

In the second experiment Welch and Epstein (1969) found that at higher concentrations (25 mM) of KCl both system I and system II start operating with no discernible lag when excised barley roots were placed in experimental solutions. Table 10.6 makes it clear that the amount of K^+ absorbed both at low (system I) and high (system II) concentrations after one minute was 1/60 of the amount absorbed after one hour. Thus, absorption by system I and system II at the higher concentration remained constant in the period from one minute to 60 minutes. The authors stated that, if equilibration of the cytoplasm with the external medium was necessary before system II absorption was maximal, the half-time for cytoplasmic exchange would have to be 10 seconds or less. This is improbably rapid when compared with measured values for the cytoplasm exchange in barley roots which is in the order of 5–6 minutes.

Table 10.6 ***Absorption of potassium (^{86}Rb-labelled) by low salt excised barley roots in 1 minute and 1 hour***

External solution KCl (mM)	Potassium absorbed (μmoles g^{-1} tissue) 1 min	60 min	Ratio 1 min:60 min
0·5	0·10	6·1	1:61
25·0	0·24	14·6	1:61

Data from Welch and Epstein (1969).

Welch and Epstein (1969) conclude that their evidence is consistent with the location of both systems at the plasmalemma and incompatible with a model (Laties, 1969) in which system II absorbs ions only after they have diffused through the plasmalemma and cytoplasm to reach the tonoplast.

Evidence for the 'series' arrangement. The experiments from which Laties (1969) deduces that system II is at the tonoplast are similar in many respects to those of Epstein. In both, low salt tissues have been used, but Laties' work differs because the element of time, and changes of absorption rates with time, have been studied in relation to the concentration of ions in the external solution.

The most significant observation in Laties (1969) is that, if excised roots or disks of beetroot tissue are treated with labelled solutions of KCl greater than 10 mM, there are important changes in the uptake kinetics over a period of 3–4 hours which cast very grave doubt on the 'parallel' arrangement. Two things happen:

(a) The uptake velocity declines to a slower rate but then remains constant.

(b) A simple plot of I/V against I/S yields a straight line without correction for system I.

The latter finding suggests that, somehow, system I has been lost and that the overall influx is controlled by system II alone. How can we explain this loss?

One explanation employs reasoning essentially the reverse of that used to interpret the results of efflux experiments in chapter 7. It depends rather heavily on the 'low salt' condition of the tissue at the point when it is first exposed to the labelled ions, and also on the marked change in the physical properties of the plasmalemma at high electrolyte concentrations.

Suppose we have a cell in which the cytoplasm contains little K^+ (e.g., $CaSO_4$-grown barley root), and this cell is suddenly placed in a concentrated solution of KCl (>10 mM). The permeability of the plasmalemma increases (see Fig. 10.11) and K^+ diffuses into the cytoplasm. The specific activity of labelled-K in the cytoplasm will quickly rise until it reaches equilibrium with the external solution. At this point the influx across the plasmalemma will be equal to the sum of the efflux back into the medium and the influx across the tonoplast. In these circumstances, the tonoplast influx will limit the overall rate of K^+ accumulation in the cell. A graphical plot of such a sequence of events might resemble the hypothetical example in Fig. 10.13.

We might ask how system I fits into this scheme and why it should disappear at later stages. In the 'cytoplasm filling' stage there is a net influx of labelled ions into the cytoplasm both *via* system I and passive diffusion—a dual mechanism. When the 'cytoplasm full' stage is reached there is no net influx of K^+ and system I at the plasmalemma can no longer be distinguished; it may be operating but in practical terms it has 'disappeared'. Laties (1969) claims that the ability of Epstein and his many co-workers to distinguish system I activity at high salt concentrations is a consequence of their experimental design, and suggests that had they examined the rate of uptake after 3–4 hours only system II would have been found.

If Laties (1969) is correct in his interpretation of the facts it is clear that Welch and Epstein (1969) were really observing the 'cytoplasm filling' stage and not system II at all. On this basis, however, it is difficult to explain away the results in Table 10.6 where the ^{86}Rb influx measured over one minute was the same as that measured over one hour. If the plasmalemma were as freely permeable to K^+ as Laties suggests, one might have expected an appreciable efflux of tracer from the cytoplasm during the first hour and a declining rate of apparent influx. One can hardly imagine 'cytoplasm filling' being linearly related to time in the presence of a leaky plasmalemma.

Perhaps the most powerful evidence against the 'parallel' arrange-

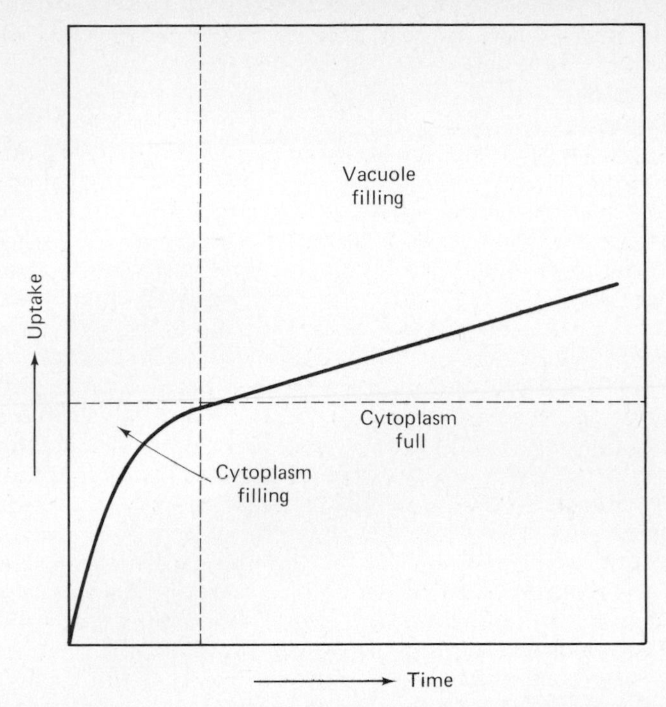

***FIG. 10.13** Hypothetical sequence of events when low salt roots are presented with a high concentration of a salt in the system II range. In the 'cytoplasm-filling' stage, net fluxes due to both system I and diffusion occur across the plasmalemma. When 'cytoplasm-full' stage is reached net uptake is governed by the flux across the tonoplast.*

ment comes from experiments on beet disks by Osmond and Laties (1968). Disks were pretreated, for a period of time equivalent to 'cytoplasm filling', with unlabelled solutions of KCl of 10 mM or greater and then transferred to a solution containing ^{36}Cl. A steady rate of Cl-influx was then measured for many hours and the rate was characteristic of the 'cytoplasm full' stage. In 'prefilled' plants there was no evidence for the operation of system I, the influx being controlled by the influx into the vacuole mediated by system II.

Although the 'series' interpretation is satisfying in many ways there are certain difficulties in its general application. One of these concerns the absorption of ions such as sodium, calcium and magnesium, high concentrations of which have not been shown to greatly alter the cation permeability of the plasmalemma or depolarize the membrane potential (Pitman *et al.*, 1970), and yet in each case there is good evidence for a dual pattern of absorption and the operation of system II.

Uptake by 'non-vacuolated' cells. In earlier papers, Torii and Laties (1966) examined the uptake of Cl^- and Rb^+ in cells of the root tip of maize and found evidence only for simple ion diffusion across the plasmalemma and not for system II absorption. This, they stated, was probably related to the 'non-vacuolated' condition of the unexpanded cells. This seems a curious statement since even the most casual inspection of cells in this region shows them to contain abundant small vacuoles. When these fuse, a central vacuole is formed just prior to cell expansion. The surface area of tonoplast surrounding them must be much greater in one of these 'non-vacuolated' cells than in the 'vacuolated' mature cells. Torii and Laties (1966) seem to assume that the properties of this tonoplast with respect to system II change as the cell matures; it seems no less reasonable to propose a similar change in the properties of the plasmalemma. In the unicellular alga, *Chlorella pyrenoidosa*, which also lacks a central vacuole, there is evidence that both system I and system II can operate at high salt concentrations (Kannan, 1971). Atkinson *et al.* (1972) have pointed out, however, that this and other species of *Chlorella* do have several small vacuoles in the cytoplasm the surface area of which may be as much as 35 per cent of that of the plasmalemma.

In concluding this discussion I can only observe that neither the 'series' arrangement nor the 'parallel' arrangement, in the rather restricted sense envisaged by Epstein (1972), is capable of explaining all of the experimental data on the dual pattern of ion transport. The attractive features of the 'series' arrangement which have been worked out for K^+ and Cl^- do not seem likely to explain the system II uptake of ions which do not render the plasmalemma 'leaky'.

It is, perhaps, worth commenting that 'dual' absorption mechanisms are exclusive neither to plant cells nor to electrolyte fluxes; Fishbarg *et al.* (1947) showed that chloride absorption, at increasing chloride concentrations, by frog skin resembled the situation we have described above, having two kinetically distinct components. Similar 'dual' mechanisms may be seen in Fig. 4 of Taylor (1960) illustrating glucose uptake by cells of *Scenedesmus quadricauda* (a green alga).

Relationships between the fluxes of ions and water in roots

In chapter 8 we learnt that excised roots actively pump ions into the xylem exudate, thus setting up a gradient of water potential down which water flows into the xylem. In this system it is evidently the ion pumping which causes the water flux and not the other way round. The situation is rather different in the intact plant where, *in certain circumstances*, the rate at which ions are absorbed may depend largely on the size of the water flux into the root xylem promoted by transpiration.

In plant roots there seem to be two pathways for ion movement into the stele, an active transport pathway which we will call the symplast component, and a second which depends directly on the movement of water across the root. This latter pathway is sometimes said to be the cell wall pathway but, as we shall learn, this may not be strictly accurate and for the moment we will call this the 'water linked component'. The proportion of the total delivered by either component depends very strongly on the physiological state of the plants being studied, the rate of transpiration and the ionic concentration of the external medium.

Low versus high salt status plants

The classic work of Broyer and Hoagland (1943) showed that plants grown in a dilute nutrient solution absorbed potassium and bromide at rates independent of the rate of water uptake, whereas in plants raised in relatively concentrated nutrient solution, the rate of ion absorption was strongly influenced by the water flux into the root (Fig. 10.14). In addition to having low salt status, the plants had a higher sugar content and a larger amount of root per unit shoot than the high salt status plants. When we talk of salt status it is as well to remember that there are more variables than simply the ionic condition of the plant cells. Broyer and Hoagland (1943) pointed out that the more rapid rate of ion uptake by the low salt status plants was correlated with their high sugar content and respiratory activity. This essentially metabolic interpretation of the data is consistent

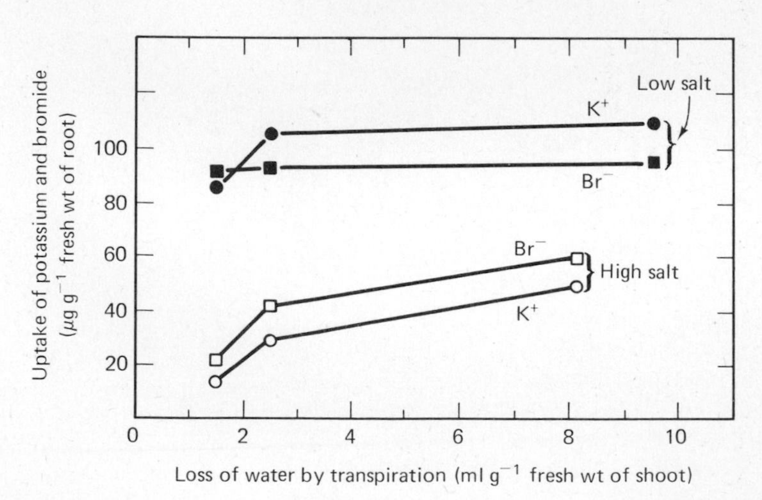

FIG. 10.14 The effect of transpiration on the uptake of potassium and bromide by low and high salt status roots. (Drawn from data in Broyer and Hoagland, 1943.)

with much that has been learnt subsequently about the effects of salt concentration on the properties of the plasmalemma.

System I and system II and ion movements into the xylem

We have seen that when plant roots are treated with high concentrations of certain ions the permeability of the plasmalemma increases very markedly and its electrolyte barrier properties are much reduced. In the opinion of Laties (1969) and many others, ions can diffuse more freely across the plasmalemma in the range of concentrations at which system II appears to operate. At lower ionic concentrations system I controls the ion fluxes into the cell and the plasmalemma represents a major permeability barrier.

Before we proceed further we should remember that free diffusion and mass flow into the stele are limited by the Casparian band of the endodermis and the firm attachment between the band and the plasmalemma of the endodermal cell. This combination of circumstances obliges water and ions travelling radially in the free space to cross a membrane barrier; their subsequent movement in the stele itself may be either in the symplast or in the cell walls. If we reduce the resistance of the endodermal plasmalemma to ions very considerably, the barrier set up by the Casparian band is largely eliminated.

Let us consider the possible sequence of events which might occur at the endodermis (and elsewhere in the root) when the solution entering the free space is changed from low to high concentration. In the initial low salt situation (<1 mM):

(a) System I is rate limiting.
(b) Plasmalemma flux determined by ion pumps or carriers which are dependent on metabolism in some way.
(c) Passive diffusion, in response to electrochemical potential gradients, negligible.
(d) Dilution of cytoplasmic phase by water flux has *little effect* on ion transport across the plasmalemma.

We now change the solution to one where the ions under consideration are more highly concentrated (>10 mM):

(a) System I no longer limits flux across the plasmalemma.
(b) Plasmalemma flux has large diffusive component not markedly dependent on metabolism.
(c) Passive flux responsive to electrochemical potential gradients.
(d) Dilution of cytoplasmic phase by water flux *will affect markedly* ion transport across the plasmalemma.

It is important to realize what this hypothesis does not say. It does not say that ions and water are carried together across the membrane in the same pathway or that they travel together towards the xylem exclusively in the cell walls (the properties of the Casparian band itself are likely to be unchanged by salt concentration). The ions do not, therefore, move by mass flow in the strict sense. The hypothesis says that the flux of water effectively sweeps ions out of the cytoplasm towards the xylem and, in the presence of a 'leaky' plasmalemma, this process will increase the rate of ionic diffusion across the plasmalemma by diluting the cytoplasm.

Evidence for water linked ion fluxes into the xylem

In many of the experiments where a relationship has been demonstrated between the flux of ions into the xylem and the rate of water movement across the root, the plants have been pretreated and examined in solutions in which ions are relatively concentrated. For instance, Lopushinsky (1964) pretreated tomato plants with Hoagland solution (Ca^{2+}, 5 mM; K^+, 6 mM; Mg^{2+}, 4 mM; NO_3^-, 15 mM; $H_2PO_4^-$, 3 mM) and conducted his experiments in a similar concentrated medium. Lopushinsky found that the application of hydrostatic pressure to the solution surrounding the excised root system

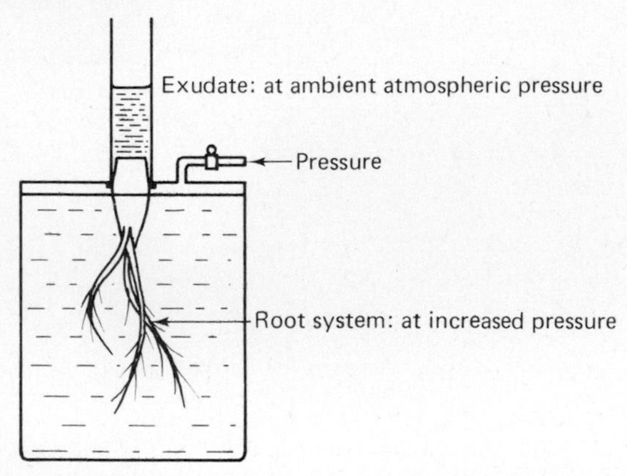

FIG. 10.15 Schematic view of an apparatus used to increase the water flux through an excised root system. A similar arrangement was used for studies on tomato roots by Lopushinsky (1964.)

of a tomato plant increased the water flux into the xylem and consequently the volume of exudate collected, at ambient pressure, from the stump (Fig. 10.15). An increase in pressure from ambient to two bar increased the volume of exudate 15-fold and the amount of labelled phosphate from the external solution by a factor of two (Fig. 10.16). In the absence of water flow induced by pressure, it was

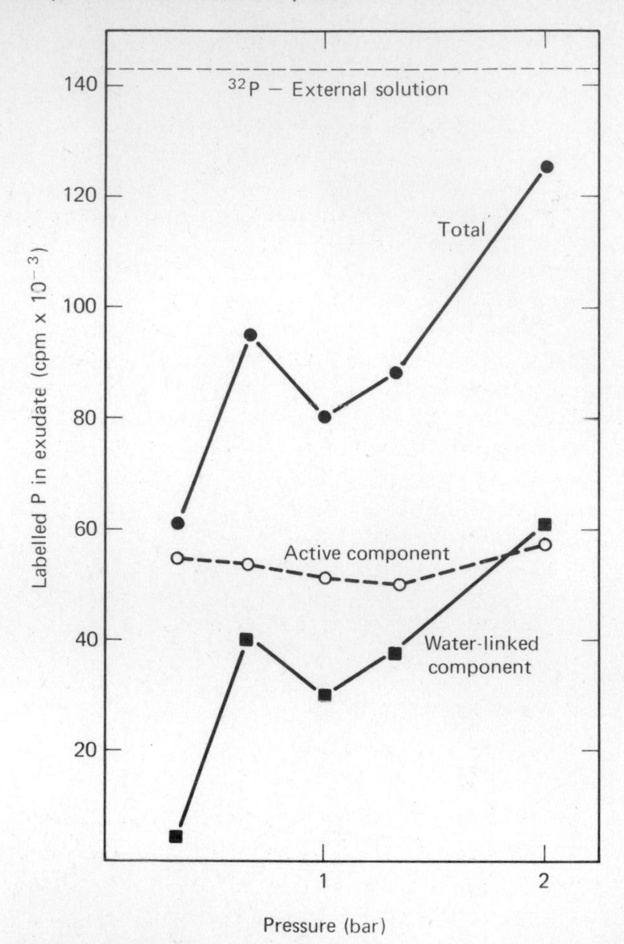

***FIG. 10.16** Effect of hydrostatic pressure on the amount of labelled phosphorus in the exudate from detopped tomato root systems. The active component was determined by measuring the ^{32}P-content of root exudates at ambient pressure; the same root was then pressurized and the value for 'Total' was obtained. The water-linked component was obtained by subtraction of the active component from the total. (From Lopushinsky, 1964.)*

assumed that the ^{32}P arriving in the xylem was due to active transport processes. The additional phosphate in the exudate from pressurized roots was linked to the movement of water if it is assumed that neither the pressure nor the increased water flux affected active phosphorus transport. Such evidence as there is on this point comes from experiments with cation transport (Jackson and Weatherley, 1962; Bowling and Weatherley, 1965) and may not be applicable in

Lopushinsky's work, but it is clear that for cations there can be interactions between pressure and transport processes even in the absence of extra water flow.

It is important to note that the fluxes of phosphate and water into the exudate did not increase by the same factor. There was a restriction on the entry of labelled phosphate relative to water even at high water fluxes and the ^{32}P concentration of the exudate was always lower than that of the external solution.

Some of the best known work showing a relationship between water flux and ion uptake by intact plants was reported by Hylmö (1953, 1955). At great length this author showed that, in pea seedlings, salt uptake was precisely related to water uptake but unfortunately he did not define the nutrient status of his plants. If, as seems likely (Russell and Barber, 1960), the pea plants were of high salt status there is really no conflict between his data and that of earlier workers who also showed that such a relationship exists (see Fig. 10.14). Hylmö (1955) attempted to elaborate his findings into a more general hypothesis in which he claimed that '. . . The linear relationship between water and ion transport means that the true concentration in the xylem in the transpiration stream is the same for all rates of water transport.' This manifestly untrue statement (see Fig. 10.14) gave rise to a heated and largely unnecessary controversy in the literature; this controversy was dealt with, and resolved to the satisfaction of most, in an excellent review by Russell and Barber (1960).

Water fluxes in the symplast and ion movement to the xylem

There is general agreement that some of the water reaching the xylem is carried in the symplast, even though there is some controversy about the proportion so carried (see Woolley, 1965; Tanton and Crowdie, 1972). As the water potential of the xylem decreases, flow *via* this pathway might increase and thus speed up the movement of ions through the symplast. We can think of this process as intercellular mass flow, the plasmodesmata providing the essential links between the cells and the xylem being the compartment into which the contents of the symplast are emptied.

Greenway and Klepper (1968) have shown convincingly that even in low salt status roots treated with dilute solutions of phosphate (*ca.* 0·1 mM) there is a marked effect of water flux, not on the rate of phosphate absorption, but on the rate at which previously absorbed phosphate is transferred to the xylem. In their experiments, intact tomato plants were pulse labelled for five minutes with ^{32}P and then returned to either dilute culture solution or to a culture solution containing mannitol with a water potential of −5·0 bar. The increased solute concentration of this solution reduced the osmotic potential difference between the xylem and the external solution and hence the water flux across the root was reduced by a factor of three; phosphate movement to the shoot was decreased by a factor of two. Since the ^{32}P had been absorbed before the water potential in the bathing medium was altered, the differences in performance are related not to differences in uptake but to differences in the pattern of long distance transport.

While this process is unlikely to affect transport the rate of phosphate transport across the plasmalemma in the short term, it may well influence the rate of movement into the symplast as opposed to the rate of movement into the vacuole, especially if these two processes can be thought of as competing for phosphate ions in some way. The rapid dilution of the symplast by water moving through it might increase the diffusion of phosphate into the symplast at the expense of phosphate moving into the vacuole.

Ion uptake in relation to plant 'demand'

Much of the information on this topic comes from the experimental work and theory developed by soil scientists and crop physiologists who, in their normal line of business, tend to take a broader view of mineral nutrition than the physiologist working on ion transport in the laboratory.

In relatively short term experiments, however, we often find that the nutritional pretreatment received by the experimental plants strongly influences the rate at which ions are absorbed from a solution of a given ionic strength (e.g., Fig. 10.14). It is generally true to say that a plant which has not received an adequate supply of a given ion will absorb that ion rapidly when it is provided during an experiment, especially if the growth of the plant has been limited by a deficiency of that ion. The reverse situation is also found. Rates of *net uptake* by salt saturated plants are low and determined almost entirely by the rate of growth of the plants (e.g., Pitman, 1965), thus although tracer experiments on such plants often indicate a fairly brisk influx we see that most of this must be matched by an efflux of unlabelled ions. We made this point earlier (page 192), but the significance of this fairly commonplace observation does not seem to have been grasped by those who insist that ion movement across the membranes of roots is unidirectional and that there is no efflux (see page 297). In some circumstances, we find that there is a *net efflux* from the plant; the data of Woodford and McCalla (1934) and others show that in the later stages of growth the K^+ content of cereal plants actually began to fall.

Evidence for absorption rate dependent on 'demand'

The picture which emerges from work on the mineral nutrition of whole plants is that the net ion absorption by roots is closely regulated by the growth of the plant in circumstances where the supply of ions is not limiting. Thus, the various parameters used in growth analysis have their counterparts in ion absorption. This idea has been developed by Nye and Tinker (1969) who derive a *root demand coefficient* which can vary depending on the ionic concentration at the root surface and the rate at which ions are needed by the plant to keep its expanding volume 'topped-up' with salt. The latter component is determined by the growth rate. Notice that this concept of root 'demand' makes no assumptions about the nature of the absorbing process itself. The idea that the activity of system I carriers described earlier can be directly controlled by growth is not one which emerges from the laboratory studies of Epstein and others but the results of quite simple experiments show that this is, in fact, the case.

I shall use some results in a paper by Drew and Nye (1969) to illustrate the effect of demand on the rate of ion uptake. Each root from seedlings of *Lolium multiflorum* (Italian ryegrass) was grown through a narrow column divided into three horizontal compartments (see inset to Fig. 10.17). The middle compartment, 1 cm in depth, was filled with soil which, in case 2, was labelled with ^{86}Rb as a tracer for potassium, while the compartments on either side were filled with either nutrient free sand or soil. When the roots had grown through the middle compartments the initial potassium content of some of the replicates was measured. The plants were then harvested in groups on each of the three following days and the weights and potassium content recorded. In case 1 (soil/sand) all of the potassium taken by the plant must have come from the 1 cm length of root in the middle compartments since the sand on either side contained negligible amounts of potassium. In the second case (soil/soil) the whole root system took up potassium but that coming from the middle compartment could be distinguished by the ^{86}Rb label.

Figure 10.17 shows that, when the 1 cm treatment zone was the only source of potassium for the plant, the rate of absorption by the root in the middle compartment was more than eight times greater than when all of the root system was supplied with potassium.

The initial concentration of K^+ in the soil solution was the same in both cases (approx. 1·0 mM) but may have been appreciably less than this at the surface of the more rapidly absorbing root in case 1. The potassium concentration was, therefore, at a level where we would normally find that system I carriers were operating at maximum velocity and system II was negligible by comparison. Applying a conventional interpretation to the data we see that the activity of the

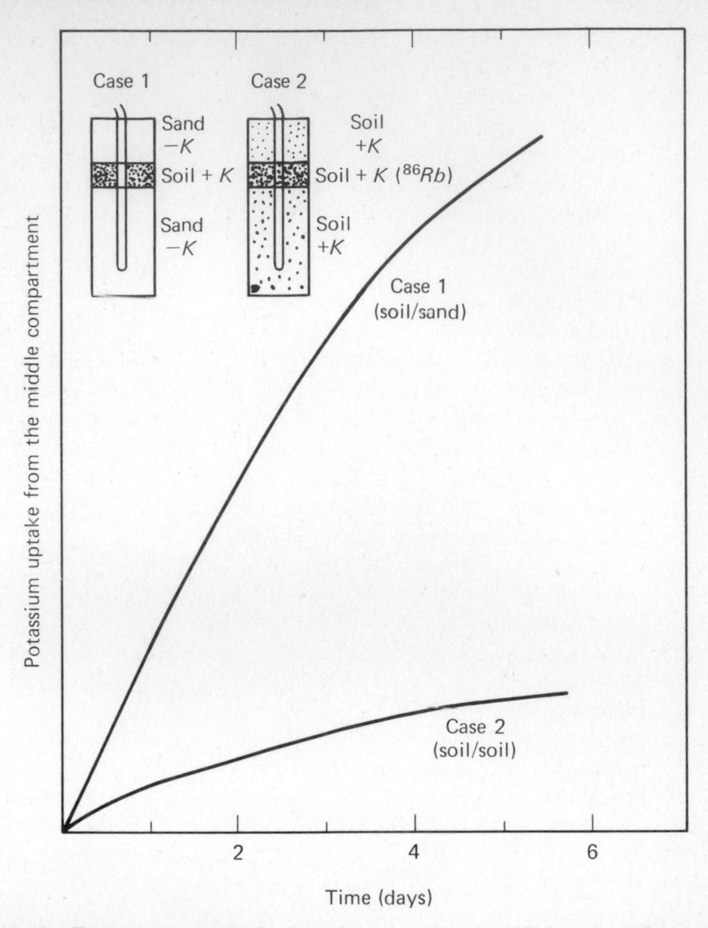

FIG. 10.17 Potassium uptake by 1 cm lengths of roots of* Lolium multiflorium *where the remaining part of the root system was treated either with nutrient-free sand, or with soil. Inset to the figure shows the experimental arrangement in the two cases examined. (Adapted from Drew and Nye, 1969, with the authors' kind permission.)

K^+-transport system can be adjusted over quite a wide range at a given potassium concentration. It would be interesting to know, in circumstances like these, what happens to the apparent K_m for K-transport. It need not alter, but if it does not, some factor is causing considerable shifts in the V_{max} of the system; possibly the number of carrier sites is greater in case 1.

Conclusions which are broadly similar to those above were drawn from an experiment in which most of the adventitious roots developing on the tillers of *Lolium perenne* were continually excised (Brouwer and Kleinendorst, 1965). Although this reduced the size of the tillers, the plants appeared to be healthy and grew steadily in spite of the very large shoot: root ratio. Mineral absorption per gram of the roots remaining on the plant integrated over the course of the experiment was nearly five times that of the control plants and the concentration of ions in tillers of root pruned and control plants was similar. Here again we see a response by a limited amount of root to meet a relatively constant 'demand' in the developing shoot.

Nutrient uptake in relation to supply

The ability of plants to take what they need from a solution is quite remarkable. In a mammoth experiment, Williams (1961) grew barley plants for 11 weeks in very large containers and attempted to keep the potassium concentration in a given treatment relatively constant. The nominal concentrations covered an enormous range of concentration from 0·25 to 1260 μM potassium and were adjusted twice daily to counter depletion of the solution during absorption. Table 10.7 shows that the final dry weight of the plants and their shoot potassium concentration were increased only by factors of 1·7 and 1·5 respectively by this 5000-fold range of K^+-concentration.

Table 10.7 ***Shoot dry weight and potassium concentration in 11-week-old barley plants grown in a wide range of potassium concentrations***

Potassium concentration (μM)	Shoot dry weight (g)	Shoot potassium concentration mmoles g^{-1} (dry weight)
0·25	46·1	1·03
2·5	61·8	1·18
25	89·1	1·26
1260	79·2	1·53

(Data recalculated from Williams, 1961.)

Provided with a constant low supply of phosphate, a wide range of species were found by Asher and Loneragan (1967) to be able to grow and maintain an adequate internal phosphorus concentration from very dilute solutions. In order to control depletion of the culture solution they used a system in which 2800 litres (!) of solution were continuously circulated through the pots in which the plants were grown. The phosphate concentration in the solution was checked and adjusted daily. Eventually in the most dilute solutions it was necessary to provide a continuous drip-feed of phosphate to the solution. The nominal concentrations were thus approximately maintained.

At the lowest concentration (0·04 μM) all species grew slowly (Table 10.8), showed signs of severe phosphorus deficiency and had very low phosphorus contents. Increasing the phosphate concentration to 0·2 μM caused two- to four-fold increases in the shoot phosphorus concentration of all species and eliminated the visible symptoms of phosphorus deficiency. Above this concentration there were increases in the growth rate of some species but the changes in the shoot phosphorus concentration of the shoot tissue were small or absent.

Table 10.8 Shoot phosphorus concentration and relative growth rate of species grown in culture solutions with low concentrations of phosphate

Species		Nominal phosphate concentration in uptake solution (μM)				
		0·04	0·20	1	5	25
Hypochoeris glabra	$[P]_{shoot}$[1]	0·09	0·20	0·59	0·53	0·57
(Flatweed)	Growth rate[2]	70	115	113	156	164
Bromus rigidus	$[P]_{shoot}$	0·14	0·25	0·55	0·70	0·64
(Brome grass)	Growth rate	66	98	113	127	128
Trifolium subterraneum	$[P]_{shoot}$	0·14	0·36	0·89	1·33	1·43
(Clover)	Growth rate	68	95	109	115	109
Cryptostemma calendula	$[P]_{shoot}$	0·08	0·19	0·41	0·35	0·34
(Cape weed)	Growth rate	85	137	147	179	183

[1] The units of shoot phosphorus concentration are mg P g^{-1} fresh weight.
[2] Growth rate is the relative growth rate (RGR) which has the units mg g^{-1} day. The RGR is calculated from the following formula:

$$RGR = \frac{\log_e W_2 - \log_e W_1}{t_2 - t_1}$$

where W_1 and W_2 are the initial and final weights of the plant and $t_2 - t_1$ is the time interval over which growth was measured.

(These data were abstracted from Tables 4 and 6 in Asher and Loneragan, 1967.)

The results show that plants can sustain their growth from very dilute phosphate solutions if care is taken to ensure a relatively constant composition and that, in the range 1 to 25 μM, the phosphate absorption per unit root changed little, if at all. Quite a different picture might have emerged from uptake kinetic studies from the excised roots of low salt plants of the same species. In the lower concentrations it seems that the root can be made to work harder to get the same amount of phosphate to the shoot. I introduce the word *work* to emphasize that it requires a greater expenditure of free energy to transport the same amount of phosphate into the plant from a dilute solution than from a more concentrated one.

Nutrient content of 'deficient' plants

Even when the growth of the plant is being restricted by the supply of a nutrient the concentration in the tissues of the plant may remain constant. Where this is the case growth is adjusting itself to the availability of some essential component of the cell. Such results are often found in experiments on nitrate, which exerts a very strong influence on growth, particularly of cereal plants and grasses. In Table 10.9 I have assembled some data from my own unpublished work. The growth of barley was clearly reduced when the nitrate fell below a nominal concentration of 1·0 mM, but the gross nitrogen concentration in the shoots was relatively constant. It was only in severely deficient levels that the nitrogen concentration of the shoot began to fall. Thus, even when growth begins to be limited by nitrate supply the amount of nitrogen per unit shoot remains constant. There comes a point, however, when the internal concentration and the yield fall abruptly. At this point, recognizable deficiency symptoms develop.

Table 10.9 ***Effect of nitrate supply on the growth and nitrogen concentration of the shoot of barley seedlings, 14 days old***

Nitrate concentration (mM)	Shoot fresh weight (g)	Shoot N Conc. mg N g^{-1} (fresh wt)
10	0·56	4·7
5	0·61	4·4
2·5	0·57	5·0
1	0·53	4·5
0·5	0·41	5·7
0·25	0·42	4·8
0·10	0·35	5·1

(Clarkson, unpublished data.)

Concluding remarks

Although the data on ion absorption by whole plants have been presented at such a late stage in this book, I hope that the impact of the results has been properly felt and that the student appreciates that the inherent properties of the transport system are not the only factors controlling the rate at which roots absorb ions. It is evident that these properties can be greatly modified and to some extent suppressed by the growth processes in the plant.

We should see the valuable work on ion transport mechanisms in its proper context. Quite obviously, we must know how ions negotiate membrane barriers if we hope to have a proper understanding of the ionic relations of plants but I hope that, by now, we are convinced that this is not the end of the matter. A great deal remains to be learnt about the overall control of ion uptake by the growing, intact plant.

Bibliography

a Further reading

Neame, K. D. and Richards, T. G. (1972) *Elementary Kinetics of Membrane Carrier Transport*. Blackwell (Oxford).

This excellent book is written and illustrated with great clarity. It assumes that the student has very little prior acquaintance with the subject; thoroughly recommended as background reading in relation to the 'dual' carrier systems proposed by Epstein and others.

b References in the text

Asher, C. J. and Loneragan, J. F. (1967) Response of plants to phosphate concentration in solution culture: I. growth and phosphorus content. *Soil Sci.* **103**, 225–233.

Atkinson, A. W., Jr, Gunning, B. E. S., John, P. C. L. and McCullough, W. (1972) Dual mechanisms of ion absorption. *Science* **176**, 694–695.

Barber, D. A. (1968) Micro-organisms and the inorganic nutrition of higher plants. *Ann. Rev. Pl. Physiol.* **19**, 71–88.

Barber, D. A. (1972) 'Dual isotherms' for the absorption of ions by plant tissues. *New Phytol.* **71**, 255–262.

Barber, D. A. and Frankenburg, U. C. (1971) The contribution of micro-organisms to the apparent absorption of ions by roots grown under non-sterile conditions. *New Phytol.* **70**, 1027–1034.

Barber, D. A., Sanderson, J. and Russell, R. S. (1968) Influence of micro-organisms on the distribution in roots of phosphate labelled with Phosphorus-32. *Nature*, **217**, 644.

Bowen, J. E. (1969) Absorption of copper, zinc and manganese by sugar-cane leaf tissue. *Plant Physiol.* **44**, 255–261.

Bowling, D. J. F. and Weatherley, P. E. (1965) The relationship between transpiration and potassium uptake in *Ricinus communis*. *J. Exp. Bot.* **16**, 732–741.

Brouwer, R. and Kleinendorst, A. (1965) Effect of root excision on growth phenomena in perennial ryegrass. *Jaarb. I.B.S.* 1965, 11–20.

Broyer, T. C. and Hoagland, D. R. (1943) Metabolic activities of roots and their bearing on the relation of upward movement of salts and water in plants. *Am. J. Bot.* **30**, 261–283.

Carter, O. G. and Lathwell, D. J. (1967) Effects of temperature on orthophosphate absorption by excised corn roots. *Plant Physiol.* **42**, 1407–1412.

Drew, M. C. and Nye, P. H. (1969) The supply of nutrient ions by diffusion to plant roots in soil. II. The effect of root hairs on the uptake of potassium by roots of Rye Grass (*Lolium multiforum*). *Pl. Soil* **31**, 407–424.

Elzam, O. E. and Epstein, E. (1969) Salt relations of two grass species differing in salt tolerance. II. Kinetics of absorption of K, Na and Cl by their excised roots. *Agrochimica* **13**, 196–206.

Epstein, E. (1972) *Mineral Nutrition of Plants: Principles and Perspectives.* Wiley (N.Y.).

Epstein, E. and Hagen, C. E. (1952) A kinetic study of the absorption of alkali cations by barley roots. *Plant Physiol.* **27**, 457–474.

Epstein, E., Rains, D. W. and Elzam, O. E. (1963) Resolution of dual mechanisms of potassium absorption by barley roots. *Proc. Nat. Acad. Sci.* (US) **49**, 684–692.

Ferrari, G. and Renosto, F. (1972) Comparative studies of the active transport by excised roots of inbred and hybrid maize. *J. Agric. Sci.* **79**, 105–108.

Fischbarg, J., Zadunaisky, J. A. and De Fisch, F. W. (1967) Dependence of sodium and chloride transports on chloride concentration in isolated frog skin. *Am. J. Physiol.* **213**, 963–969.

Greenway, H. and Klepper, B. (1968) Phosphorus transport to the xylem and its regulation by water flow. *Planta* (Berlin) **83**, 119–136.

Hylmö, B. (1953) Transpiration and ion absorption. *Physiol. Plant.* **6**, 333–405.

Hylmö, B. (1955) Passive components in the ion absorption of the plant. I. The zonal ion and water absorption in Brouwer's experiments. *Physiol. Plant.* **8**, 433–449.

Jackman, R. H. (1965) The uptake of rubidium by roots of some graminaceous and leguminous plants. *New Zealand J. Agric. Res.* **8**, 763–777.

Jackson, J. E. and Weatherley, P. E. (1962) The effect of hydrostatic pressure gradients on the movement of sodium and calcium across the root cortex. *J. Exp. Bot.* **13**, 404–413.

Jacobson, L., Cooper, B. R. and Volz, M. G. (1971) The interaction of pH and aeration of Cl uptake by barley roots. *Physiol. Plant.* **25**, 432–435.

Jeschke, W. D. (1970) Der Influx von Kaliumionen bei blättern von *Elodea densa*, Abhängigkeit vom Licht, von kalium-konzentration und von der Temperatur. *Planta* (Berlin) **91**, 111–128.

Jeschke, W. D. and Simonis, W. (1965) Über die Aufnahme von Phosphat- und Sulfationen durch Blätter von *Elodea densa* und ihre Beeinflussung durch Licht, Temperatur und Aussenkonzentration. *Planta* (Berlin) **67**, 6–32.

Kannan, S. (1971) Plasmalemma: The seat of dual mechanisms of ion absorption in *Chlorella pyrenoidosa*. *Science* **173**, 927–929.

Laties, G. G. (1969) Dual mechanisms of salt uptake in relation to compartmentation and long distance transport. *Ann. Rev. Pl. Physiol.* **20**, 89–116.

Lopushinsky, W. (1964) Effect of water movement on ion movement into the xylem of tomato roots. *Plant Physiol.* **39**, 494–501.

Lüttge, U. and Laties, G. G. (1966) Dual mechanisms of ion absorption in relation to long distance transport in leaves. *Plant Physiol.* **41**, 1531–1539.

Maas, E. V. (1969) Calcium uptake by excised maize roots and interactions with alkali cations. *Plant Physiol.* **44**, 985–999.

Nissen, P. (1971) Uptake of sulfate by roots and leaf slices of barley: Mediated by single, multiphasic mechanisms. *Physiol. Plant.* **24**, 315–324.

Nye, P. H. and Tinker, P. B. (1969) The concept of a root demand coefficient. *J. Appl. Ecol.* **6**, 293–300.

Osmond, C. B. and Laties, G. G. (1968) Interpretation of the dual isotherm for ion absorption in beet tissue. *Plant Physiol.* **43**, 747–755.

Pitman, M. G. (1965) Sodium and potassium uptake by seedlings of *Hordeum vulgare*. *Aust. J. Biol. Sci.* **18**, 10–24.

Pitman, M. G., Mertz, S. M., Graves, J. S., Pierce, W. S. and Higinbotham, N. (1971) Electrical potential differences in cells of barley roots and their relation to ion uptake. *Plant Physiol.* **47**, 76–80.

Polle, E. O. and Jenny, H. (1971) Boundary layer effects in ion absorption by roots and storage organs of plants. *Physiol. Plant.* **25**, 219–224.

Robinson, J. B. and Smith, F. A. (1970) Chloride influx into citrus leaf slices. *Aust. J. Biol. Sci.* **23**, 953–960.

Russell, R. S. and Barber, D. A. (1960) The relationship between salt uptake and the absorption of water by intact plants. *Ann. Rev. Pl. Physiol.* **11**, 127–140.

Smith, R. C. and Epstein, E. (1964) Ion absorption by shoot tissue: kinetics of potassium and rubidium absorption by corn leaf tissue. *Plant Physiol.* **39**, 992–996.

Smith, F. A. and Robinson, J. B. (1971) Sodium and potassium influx into citrus leaf slices. *Aust. J. Biol. Sci.* **24**, 861–871.

Spanswick, R. M. (1970) Electrophysiological techniques and magnitude of membrane potentials and resistances in *Nitella translucens*. *J. Exp. Bot.* **21**, 617–627.

Tanton, T. W. and Crowdy, S. H. (1972) Water pathways in higher plants. II. Water pathways in roots. *J. Exp. Bot.* **23**, 600–618.

Taylor, F. J. (1960) The absorption of glucose by *Scenedesmus quadricauda*. I. Some kinetic aspects. *Proc. Roy. Soc.* **B151**, 400–418.

Torii, K. and Laties, G. G. (1966) Dual mechanisms of ion uptake in relation to vacuolation in corn roots. *Plant Physiol.* **41**, 863–870.

Van der Honert, T. H. and Hooymans, J. J. M. (1955) On the absorption of nitrate by maize in water culture. *Acta. Bot. Neerl.* **4**, 376–384.

Weigl, J. (1970a) Die Wirkung hoher Salzkonzentration auf die Permeabilität der Pflanzlichen Zellmembran. *Z. Naturfors. B.* **25**, 96–100.

Weigl, J. (1970b) Mechanismus der *cis*- und *trans*-stimulierung der Anionen Fluxe durch die Plasmamembran der Pflanzenzelle. *Z. Naturfors. B.* **25**, 631–636.

Welch, R. M. and Epstein, E. (1968) The dual mechanism of alkali cation absorption by plant cells: their parallel operation across the plasmalemma. *Proc. Nat. Acad. Sci.* (USA) **61**, 447–453.

Welch, R. M. and Epstein, E. (1969) The plasmalemma: seat of the type 2 mechanisms of ion absorption. *Plant Physiol.* **44**, 301–314.

Williams, D. E. (1961) The absorption of potassium as influenced by its concentration in the nutrient medium. *Pl. Soil.* **15**, 387–399.

Woodford, E. K. and McCalla, A. G. (1936) The absorption of nutrients by two varieties of wheat grown on black and gray soils of Alberta. *Can. J. Res.* **C14**, 245–266.

Woolley, J. T. (1965) Radial exchange of labelled water in intact maize roots. *Plant Physiol.* **40**, 711–717.

eleven

Interactions between the soil, micro-organisms and ion uptake by roots : an introduction to the real world

Several interactions of major interest in the ionic relations of plants growing in soil were not considered in chapter 10. The soil is, of course, an extremely variable environment and it is difficult to avoid enumerating large numbers of specific cases when we consider how its properties can affect the absorption of ions. While this is properly a subject for a separate book, by way of an epilogue we will consider two factors of major importance.

It will be shown that, in some circumstances, the flow of nutrients to the root surface can be the limiting step in ion uptake by the plant, irrespective of its metabolic activity. This situation arises because the soil solution, unlike laboratory nutrient cultures, is an unstirred medium in which gradients of ionic concentration develop frequently.

The second factor concerns the microflora associated with the root surface (the *rhizosphere*). Every root in natural conditions, and even in non-sterile conditions in the laboratory, has, in its rhizosphere, an attendant population of bacteria, actinomycetes and fungi which derive some benefit from the plant, and which may, in return, confer benefit on the host plant in several ways. In many instances, particularly in perennial species, there are closer structural links between micro-organisms and the host as in the various types of mycorrhiza and in bacteria-containing root nodules. These associations, as well as the less formal ones in the rhizosphere, can become the dominating factors in the mineral nutrition and growth of the plant.

It should become apparent that the innate capabilities of the plant may not necessarily control all aspects of its mineral nutrition, and that it is unwise to interpret the ionic relations of the plant with its environment solely in terms of uptake systems, ionic concentrations, transpiration rates, fine structure, etc. This does not mean that I have been wasting the student's time in discussing these subjects at length; the soil/micro-organism/plant system is a complex one which cannot be understood while we are ignorant of the properties of its component parts. At some time in the past a similar kind of statement may have seemed necessary to justify the *in vitro* study of enzymes or of nucleic acids.

Nutrient flows in the soil around plant roots

The rate of ion absorption by a length of root growing in the soil is determined by a combination of soil and plant-dependent factors. The latter have been our concern in much of this book but we did see how, in chapter 10, page 287, the unstirred layer of solution around the roots suspended in a static nutrient solution can limit the rate of uptake of ions. We can think of the soil as a completely unstirred medium in which roots may be regarded as nutrient sinks. Ions will move towards these sinks by two processes, by diffusion and by mass flow in the water extracted from the soil by the root.

Diffusion. As ions are removed by absorption from the root/soil interface, their concentration in the solution close to the root will be lowered setting up a gradient of potential down which ions, from the bulk solution, will diffuse. If the rate at which ions are absorbed greatly exceeds the rate at which the source of ions can be replenished from the bulk soil, a shell of soil will develop around the root where the ion is depleted; this shell will gradually increase its diameter. In the case of an ion which is not adsorbed to any great extent on soil particles, e.g., NO_3^-, the radial spread of the diffusion shell is equal to $\sqrt{(Dt)}$ where D is the diffusion coefficient in the soil in question and t

is the elapsed time (see Fig. 11.1). The value of D for a given ion can be expected to be influenced strongly by the nature of the soil (particularly its ion exchange properties, water content and particle size); for NO_3^-, values around 1×10^{-6} cm^2 s^{-1} are usually reported (Nye, 1969) which is an order of magnitude slower than values for self-diffusion in a simple salt solution. With ions such as $H_2PO_4^-$ and K^+, where there are invariably complex interactions with ion exchange materials, the values for D are much lower; for $H_2PO_4^-$ the values range from 1×10^{-8} to 1×10^{-7} cm^2 s^{-1}, and for K^+ $0{\cdot}5 \times 10^{-7}$ cm^2 s^{-1}.

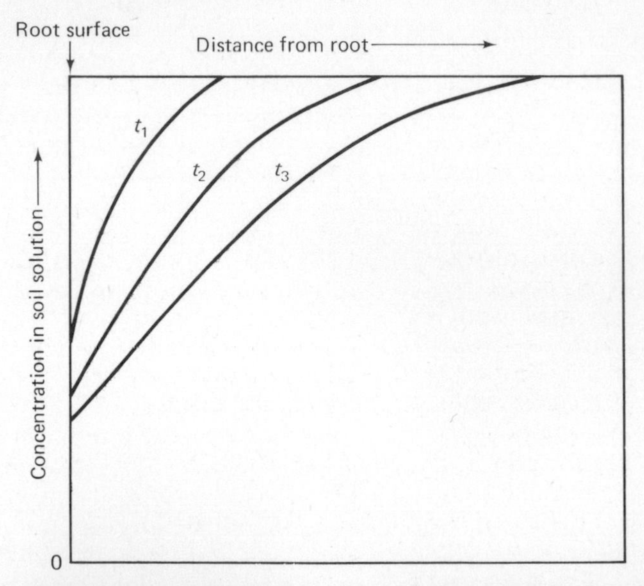

FIG. 11.1 Development of the depletion zone of nitrate ions at the root surface. The distance from the root surface is calculated from $\sqrt{Dt}$ (see text). The shape of the depletion zone alters with time (t_1–t_3); changes in the soil solution concentration at the root/soil boundary gradually approach a steady state while the width of the depletion zone increases. (Redrawn from Nye, 1969.)

When an ion in the bulk soil solution is dilute and its diffusion is slow, its concentration at the root surface can be reduced very quickly (in the space of a few hours) to almost zero if plant demand is high. This is particularly true for phosphate but can apply to K^+ and NO_3^-, especially in non-agricultural soils. Situations like this can be visualized by means of autoradiography. Walker and Barber (1962) grew maize (*Zea*) plants in soil which was initially labelled uniformly with ^{86}Rb as a tracer for soil potassium. The soil was contained in a box with an inclined front covered in thin polyethylene film. Roots growing down this inclined front could be observed for several days and at daily intervals an X-ray film was exposed for a few hours to the labelled soil and roots. Their plates showed that the roots were strongly radioactive but were surrounded by a very weakly radioactive zone extending radially for as much as 0·5 to 1 cm from the root surface; the bulk soil itself remained strongly radioactive. By making a series of densitometer tracings from the autoradiographs of a given location on the root, Walker and Barber (1962) were able to observe the development of the diffusion shell and we are able to calculate a practical value of 5×10^{-8} cm^2 s^{-1} for the diffusion coefficient for potassium in this particular soil.

Mass flow. The absorption of water by plant roots creates the opportunity for ions to move to the root surface by mass flow. While this movement must occur with all ions, the nutritional significance of the process will be heavily dependent on the rate of delivery relative to the rate of absorption at the root surface. Since the rate of delivery will itself depend on soil solution concentration we might safely assume that mass flow will be of greatest significance with ions such as calcium, magnesium and nitrate (in agricultural soils) although there is evidence that the potassium supply to plants in well fertilized soil may also be influenced significantly by mass flow (Barber, Walker and Vasey, 1963; Brewster and Tinker, 1970).

The significance of the mass flow component can be roughly assessed by measuring three things: (a) the mean concentration of an ion in the soil solution, (b) the total volume of water taken up (transpiration plus growth) in a given time interval, (c) the amount of the ion entering the plant during that given time interval. Brewster and Tinker (1970) showed that the amount of calcium taken up by the roots of *Allium porrum* (leek) was much less than would have been delivered to the root surface by the measured volume of soil water absorbed. In this situation we might anticipate that the concentration of calcium at the root surface was actually in excess of that in the bulk soil and that calcium ions would tend to diffuse away from the root. Working with several other species Barber and Ozanne (1970) produced autoradiographs from labelled soil showing that this situation does, in fact, occur. Brewster and Tinker (1970) also calculated that the concentration of Mg^{2+} and Na^{+} would have been slightly greater in the vicinity of the root since the delivery of these slowly absorbed ions by mass flow was greater than their flow into the plant.

In the case of phosphate the soil solution is usually dilute so that the quantity delivered by mass flow will be small. Even when plants are transpiring rapidly, it has been estimated that only five to 10 per cent of the measured phosphate uptake from the soil could be supplied by mass flow (Olsen, Kemper and Jackson, 1962). Lewis and Quirk (1965), using measured values of *D* for phosphate, calculated that diffusion to the root could supply all the phosphate required by growing wheat plants.

It is worth remembering that experimental assessments of the importance of mass flow are usually made in well fertilized agricultural soils where major nutrients are relatively concentrated (see Table 11.1). In many soils supporting natural vegetation the situation may be quite different (Table 11.1). For instance, in acidic, podsolized soils, which cover a large proportion of Northern Europe and Scandinavia, the concentration of calcium in the soil solution may be low and the concentration of aluminium and hydrogen ions quite high. It is possible that diffusion may become the factor controlling the overall uptake of calcium by the plant. On the other hand, mass flow may deliver to the root injurious amounts of aluminium and hydrogen ions, a fact of life which must be countered by physiological adaptation.

Table 11.1 ***Comparison of the ionic composition of the soil solution in a well fertilized agricultural soil and an acidic wet heathland soil***

Soil type	Ionic concentration in soil solution (mM)							
	K	Na	Mg	Ca	NO_3	Cl	P	SO_4
Silty-loam well fertilized (Brewster, 1971)	0·43	1·65	0·62	15·1	20·6	9·1	0·021	2·5
Soil water from wet heath (Clymo, 1962)	0·04	0·48	0·19	0·23	0·03	0·84	0·008	0·50

Absorption zones in soil grown roots. From what has been said, it must be apparent that portions of a given root member which have been in the soil for some time may be bathed in very dilute solutions of ions such as phosphate or potassium and, perhaps, nitrate. Irrespective of the innate capabilities of this length of root as far as ion absorption is concerned, diffusion through the soil will determine its rate of uptake. If we now consider the growing apical region of the root it is clear that the situation is different. By cell division and extension, the apical region is being pushed continually into unexploited parts of the soil volume where nutrient ions have not been depleted. In the short period of time (a few hours) before the depletion shell develops the properties of the vigorous absorbing system of the young root tissues may determine a rate of uptake considerably greater than in the 'mature' zone whose ion supply is limited. Thus, having decided in chapter 8 (page 229) that there seemed no sound physiological or anatomical basis for 'absorbing zones' for P and K in barley and marrow roots, I must now concede that in soil such zones may be a reality because of the time dependent development of depletion shells.

Except for the few millimeters at the root apex the limitation placed on ion uptake by diffusion may contribute to a relatively constant rate of absorption of P and K per unit length. This has often been assumed by those who produce mathematical models of the uptake of ions by roots in soil; the physiological evidence cited in chapter 8 supports this view.

The role of root hairs. We must also reassess the role of root hairs in the soil. In water culture conditions their role is negligible, but in the soil, theory predicts and experiment has confirmed that they may be of importance.

When the uptake of P and K is determined by the diffusion of ions through the soil towards a cylindrical sink, it is obvious that an expansion of the effective root radius will increase the surface area for delivery and absorption. In practical terms, root hairs are a very economical method of achieving this objective. They can effectively double or treble the root radius in return for a relatively minor expenditure of metabolic capital in cell growth; the dry weight of root hairs is small in relation to the root which bears them. Let us envisage a root surrounded by hairs as two concentric cylinders; the core is the central root and the outer layer is the root hair cylinder. Ions can be absorbed by root hairs both within the cylinder and at its surface. Each hair will, of course, become surrounded by its own depletion shell within the root hair cylinder and, in time, shells between adjacent hairs may come to overlap. If the soil solution concentration and the exchangeable fraction of an ion are known, it is possible to compare observed uptake by the root with the quantity available for absorption in the root hair cylinder. When this was done by Drew and Nye (1969), they found that only 0·8 to 6·3 per cent of the K absorbed by the roots of *Lolium multiflorum* (rye grass) could have come from within the root hair cylinder. Nevertheless the presence of root hairs was calculated to enhance K uptake by 77 per cent due to the effective increase in root diameter indicating that diffusion of K^+ from the soil beyond the root hair cylinder supplied most of the potassium absorbed. In circumstances where root demand is low, i.e., plants growing slowly in a soil heavily fertilized with K, diffusion will cease to be a limiting factor in absorption and the presence or absence of root hairs will be of little consequence.

One evident advantage in possessing root hairs is that ion movement towards the root core is speeded up. Once across the plasmalemma of the root hair, ions can enter the symplast, where, as we have seen in chapter 8, page 213, rates of transport are rapid in comparison with diffusion in simple salt solutions ($D \simeq 1 \times 10^{-5}$ cm^2 s^{-1}) and are very rapid indeed when compared with diffusion through soil ($D \sim 1 \times 10^{-8}$ to 1×10^{-6} cm^2 s^{-1}). If we consider a root hair, 0·1 cm in length and calculate the average time for a potassium ion to travel this distance, in the symplast, towards the root core we find that only a few seconds are necessary to complete the journey whereas in the soil an equivalent movement might take one day (taking D as 1×10^{-7} cm^2 s^{-1}).

Influence of the rhizosphere in ion uptake

Soil-grown roots are surrounded by large numbers of micro-organisms which tend to concentrate in their immediate vicinity. The species composition of the microflora is dependent on the host plant and probably reflects the range substrates for bacterial and fungal growth which are present in root exudates (Rovira, 1965). Although we should not be too impressed simply by numbers (bacteria are very small), it is worth noting that Berezova (1965) estimated there to be nearly 10^8 bacteria per gram of soil in the immediate vicinity of the roots of potato while in the soil 1 cm from the roots the bacterial numbers fell to 3×10^6 per gram of soil (Table 11.2).

While, in many cases, it is not exactly clear how these micro-organisms interact with the mineral nutrition of the host plant there is no doubt that in certain conditions their influence is very considerable.

Mineralization and nutrient mobilization. Bacteria play a vital role in the cycling of nitrogen in the biosphere. In almost all natural vegetation and in many agricultural situations, the nitrate and ammonium concentration in the soil solution is chiefly determined by the combined activity of bacteria which fix atmospheric nitrogen and those which oxidize the resultant NH_4^+ to NO_3^-. These activities influence the nitrogen supply but do not interact directly with the plant root except in special circumstances; the exceptions are roots of legumes and certain other plants (e.g., *Alnus*, the alder) which enter into symbiotic associations with N-fixing micro-organisms. The micro-organisms involved in the familiar root nodules of legumes are of the genus *Rhizobium* and have been the subject of a great deal of research, some of which is reviewed in a book by Stewart (1966).

Micro-organisms may also benefit the plant by transforming mineral and organic constituents of the soil into forms from which roots can

Table 11.2 Bacterial numbers from the rhizosphere of three species and the surrounding soil

Plant	Zone	Bacterial numbers in millions g^{-1} soil				
		Bacteria solubilizing $Ca_3(PO_4)_2$	Bacteria mineralizing organic P	Clostridium (N-fixer)	Cellulose decomposers	Nitrifiers
Rye	Rhizo.	22	27	0·7	0·04	0·08
(*Secale*)	Soil	3	4	0·1	0·001	0·008
Potato	Rhizo.	62	20	2	0·09	0·07
(*Solanum*)	Soil	2	1	0·1	0·004	0·04
Clover	Rhizo.	8	26	trace	0·02	0·005
(*Trifolium*)	Soil	0·08	0·09	trace	0·001	trace

Data taken from Berezova (1965).

obtain nutrients. This is frequently of great significance in phosphate nutrition because much of the soil phosphorus is tied up in unavailable forms. Of the bacteria in the rhizosphere of potatoes, Berezova (1965) found that 60 million per gram of soil could release $H_2PO_4^-$ from calcium triphosphate (virtually insoluble) and that 20 million per gram of soil could break down a range of organic phosphorus compounds. The net effect of these organisms on the phosphate supply to the root surface must have been considerable. Table 11.3 shows some data of Gerretsen (1948) who added soil micro-organisms to previously sterile cultures containing insoluble phosphorus compounds. In the inoculated cultures plants grew more strongly and absorbed more phosphorus than plants which grew throughout the experiment in sterile cultures.

Table 11.3 ***The effect of rhizosphere bacteria on the growth and phosphorus uptake of* Sinapis alba *(mustard) supplied as $Ca_3(PO_4)_2$ at pH 6·8***

Treatment	Total dry weight g ± S.E.	P absorbed mg ± S.E.
Sterile	11·3 ± 1·4	37·8 ± 5·8
Inoculated	18·6 ± 1·8	58·0 ± 5·2
Increase (%)	+64%	+56%

(Data from Gerretsen, 1948.)

The activity of soil micro-organisms is not, however, always beneficial. For instance, a number of common soil bacteria, e.g., *Pseudomonas*, *Aerobacter*, can oxidize soil manganese from its divalent form to the trivalent form which is less soluble and less readily absorbed by plants. Gerretsen (1937) showed this type of microbial activity was responsible for the development of manganese deficiency in oats (*Avena*); if the microbes were killed by sterilization, the same soil supported manganese sufficient plants.

Competition for nutrients. It seems possible that, when the supply of an ion to the root is severely limited, organisms in the rhizosphere may actively compete with the root for the incoming ions and, being at the root surface, they seem well placed to do so. Much of the laboratory work of Barber (1968) and Barber and Frankenburg (1971) suggests that in non-sterile nutrient solutions where ionic concentrations are less than 10 μM, bacteria absorb appreciable amounts of phosphate and rubidium, reducing the amount translocated to the shoot. The micro-organisms present in these experiments were general laboratory contaminants of unknown species, certainly different from the microflora which might be found surrounding soil grown plants. An attempt to get nearer the soil situation can be seen in the work of Bowen and Rovira (1966) where a mixed suspension of soil micro-organisms was added to previously sterile water cultures of tomato and clover plants. Plants in the inoculated cultures both absorbed more phosphate than those in sterile ones and, in the case of tomato, there was a highly significant increase in the amount of phosphorus reaching the shoots (Table 11.4).

Table 11.4 The influence of soil micro-organisms on phosphate uptake by plants of tomato and subterranean clover

Species	Condition of plants	Phosphate uptake	
		Roots	Shoots
		(pmoles $H_2PO_4^-$)	
Tomato	Sterile	78	4
	Non-sterile	125*	20*
Subterranean clover	Sterile	97	12
	Non-sterile	172*	13

** Significantly different from corresponding sterile plant value. (Data from Bowen and Rovira, 1966.)*

The direct relation of these results to the soil situation is made difficult by the fact that experimental observations cover only short periods of time (20 minutes).

Barber (1969) has pointed out that the results of all experiments designed to test interactions between micro-organisms have one common feature: whether the effects are stimulatory or inhibitory they are most prominent when the ion under consideration is in short supply.

Microbial secretions and ion uptake. We have learnt (chapter 4) that soil micro-organisms, e.g., *Fusarium orthoceras*, can release cyclic polypeptides which act as ion carriers across membranes (e.g., Valinomycin, Enniatin, Gramicidin, etc.). These substances may increase either the influx or the efflux of the ions they carry, depending on the direction of the diffusion gradient. Many fungi, known to be pathogenic, produce 'toxin' molecules which greatly increase the general permeability of cell membranes (Page, 1972); their effect is, of course, generally deleterious and causes a loss of both electrolytes and organic metabolites from the cells of the host. In some cases, however, the effect of the toxin appears to be quite specific. In the roots of susceptible varieties of *Zea mays*, dilute solutions of the 'toxin' from *Helminthosporium carbonum* appear to selectively increase the absorption of nitrate ions (Yoder and Scheffer, 1971).

Examples such as these have come to light because of the intense interest in antibiotics and plant pathology, but there are reasons for believing that they may represent extremes of more general relationships between micro-organisms and plant roots.

Barber and Lee (1972) have found that the absorption and translocation of manganese by barley plants is markedly promoted by some factor diffusing out of the micro-organisms which contaminates laboratory cultures. If the solution from a non-sterile culture of barley roots was passed through a sterilizing filter and the filtrate added to a culture of sterile plants, manganese uptake was markedly stimulated, although the performance fell short of that of plants in non-sterile cultures (Table 11.5).

Table 11.5 Effect of microbial products on the uptake of manganese during 24 hours by barley plants

Origin of solution	Condition of plants	Uptake of manganese (μg Mn g^{-1} dry weight)	
		Roots	Shoots
Sterile plants	Sterile	241 ± 25	9·4 ± 0·5
Non-sterile plants	Sterile	804 ± 52	13·4 ± 0·7
Non-sterile plants	Non-sterile	1695 ± 106	18·6 ± 1·3

(Data from Barber and Lee, 1972.)

It has also been found that the presence of micro-organisms increases the amount of iron transported to the shoot of barley seedlings (Clarkson and Sanderson, in press). In neither case is it clear how the microbial product or the micro-organisms interact with the host tissue, although the results for manganese suggest that a chelate may be formed which is absorbed more readily than manganese ions from the culture solution.

The rate at which the root system explores the unexploited parts of the soil volume is clearly an important factor in determining the rate at which certain ions enter the plant (page 322). It is most interesting to learn, therefore, that rhizosphere micro-organisms in culture can release considerable amounts of substances such as auxin (Libbert *et al.*, 1966) and gibberellins (Domsch, 1963; Jackson *et al.*, 1964) which are known to influence root growth. These observations are clearly in accord with the claim, frequently made in publications from the Soviet Union (e.g., Samtsevich, 1962), that improvements in the growth of crop plants in inoculated cultures are due to the secretion of growth promoting substances by the micro-organisms. There is, however, some experimental evidence that root growth can be less vigorous in the presence of soil micro-organisms than in their absence. Bowen and Rovira (1961) observed that, when sterile sand/agar cultures supporting subterranean clover, tomato, phalaris and radiate pine seedlings were inoculated with soil micro-organisms, the rate of root extension and the development of root hairs was reduced in comparison with sterile controls.

Mycorrhizal associations

Mycorrhizas are formed by the association of fungi and plant roots. They are of very widespread occurrence in both trees and herbaceous species, and are of two kinds. Those which form a complete sheath of tightly organized mycelium over the root surface are known as *Ectotrophic* and are characteristic of forest trees, e.g., *Fagus sylvatica* (Beech), *Betula* sp. (Birch), *Pinus* sp. (Pine). A second type embraces a much wider range of associations in which the hyphae of the fungus penetrates the cells of the host root extensively, unlike the ectotrophic mycorrhizas where the penetration is less evident; they are collectively referred to as *endotrophic* mycorrhiza. It is quite impossible to review the very large published literature on this subject and the few examples below are given simply to draw attention to their significant role in the ionic relations between roots and the soil.

Ectotrophic mycorrhizas. In ectotrophic mycorrhizas, there is what may be regarded as an extra cortex formed by the fungal sheath (see Fig. 11.2). This fungal layer has a greater capability for accumulating some nutrients, particularly phosphate, than the cortical cells of the host beneath it. Harley (1969) regards it as an 'extra accumulating zone of high efficiency'. If mycorrhiza from beech are labelled for two hours with a dilute solution of $KH_2{}^{32}PO_4$ (10 μM) most of the ^{32}P is present in the fungal sheath. This phosphate becomes mobilized and passed on to the plant, by a mechanism which is dependent on metabolism, in conditions where the external supply of P becomes deficient (Harley and Brierley, 1954). Where the P supply remains abundant there seems to be little interchange between the fungal sheath and the host. The chief benefit from this association, as far as the plant is concerned, seems to be the provision of a reservoir of phosphorus, a kind of annuity which can be drawn upon when times are lean. This has an immediate relevance if we realize that in mature forest vegetation there may be considerable fluctuations in the available nutrient supply at different seasons of the year. In the spring

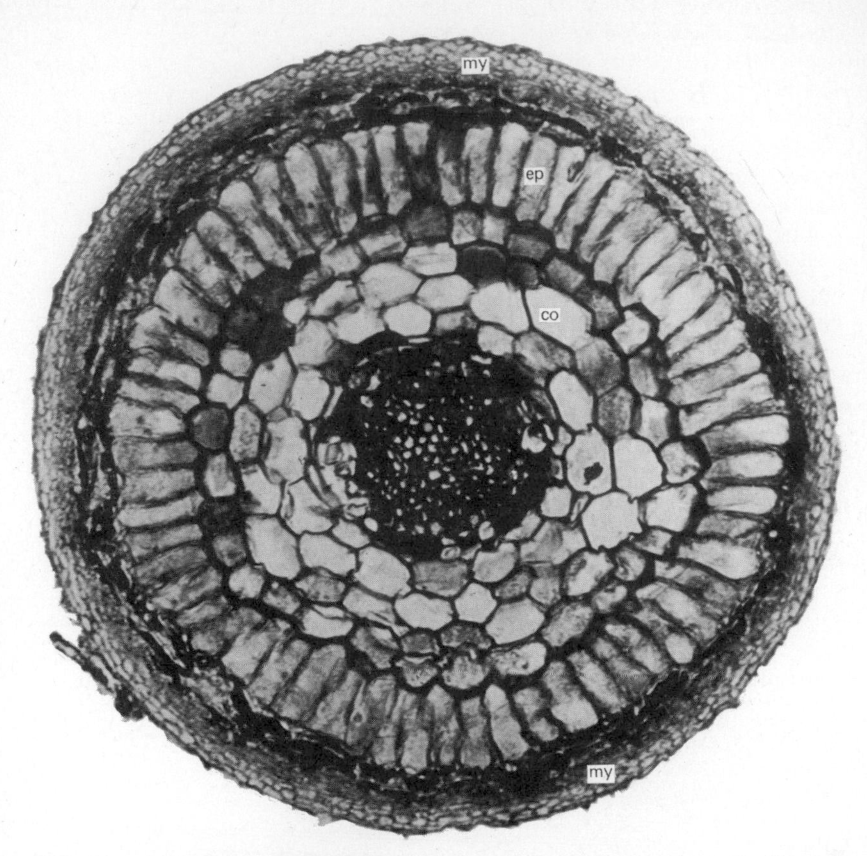

FIG. 11.2 A transverse section from a mycorrhiza of Fagus *(Beech). Note the sheath of fungal mycelium, my, and the radially elongated epidermal cells, ep, of the beech root. Hyphae from the fungus ramify between the epidermal cells and through the free space of the outer layers of the cortex, co, occasionally penetrating the cortical cells. Hyphae are never found internal to the endodermis in healthy roots. (Plate by courtesy of Professor J. L. Harley.)*

there is a large return of phosphorus to the forest floor by leaching of inorganic phosphate from expanding leaves. The presence of the efficient mycorrhizal collection system, which is present in the upper layers of the soil profile, ensures that much of this suddenly released material is reabsorbed and eventually recycled when the soil becomes depleted of P later in the season.

Endotrophic mycorrhizas. Of the diverse associations in this group, those caused by fungi with aseptate hyphae (Phycomycetes) have been receiving considerable attention of late. They are referred to as *vesicular-arbuscular mycorrhizas* and have been reported for nearly every family of flowering plant. The important features of such fungi are: (a) that their infection of the host is usually benign, causing little disturbance to the growth of the root, (b) unlike the ectotrophic mycorrhizas, there is little specificity between the fungus and the host, so that a given species of *Endogone* may infect plants as diverse as *Zea*, *Fragaria* (strawberry) and *Trifolium* (clover), (c) the hyphae enter the protoplasts of cells in the root, (d) the hyphae extend considerable distances into the soil beyond the root surface. In some respects, they may be thought of as 'super' root hairs exploiting the soil solution well beyond the normal depletion shells and the rhizosphere and their effects on mineral nutrition are striking.

Gerdemann (1964) showed that the growth and the uptake of phosphate by maize plants was far superior when infected with *Endogone fasciculata* than when they were uninfected. The influence of the fungus was particularly pronounced in conditions of phosphorus deficiency. A similar result has been shown unequivocally by Daft and Nicolson (1966) working with tomato and tobacco in addition to maize (Table 11.6).

Table 11.6 *The influence of infection by* Endogone *on the dry weight and phosphorus content of tomato plants in sand culture containing bone meal as the sole source of phosphorus*

Relative P level	Condition of plants	Mean dry weight (g)	Total P content µmoles/100 mg dried plant material	
			Roots	Shoots
0·25	Mycorrhizal	0·393	6·2	4·7
	Control	0·088	3·3	3·2
1·0	Mycorrhizal	0·334	9·3	5·4
	Control	0·115	3·6	3·3
4·0	Mycorrhizal	0·730	6·1	3·3
	Control	0·707	1·8	3·0

(Abridged from Table 4 in Daft and Nicolson, 1966.)

More recently, Sanders and Tinker (1971) showed that the increased uptake of phosphate from ^{32}P-labelled soil by onion roots infected with *Endogone mosseae* was due to increased absorption per unit root length; the increase was substantial, being on average five times greater. Both infected and uninfected plants drew on the same source of P in the soil since the specific activity of the phosphate absorbed was the same. This observation largely precludes the possibility that the fungal hyphae were mobilizing some insoluble form of P unavailable to the control plants. Thus, the enhanced uptake by mycorrhizal roots was explained by the more efficient exploration of the soil volume by the hyphae of *Endogone*.

In the discussion of the relationships between the rhizosphere organisms and the nutrition of roots I am afraid that apparently contradictory statements are often found. This reflects both the small amount of good work published and the fact that the system has so many variables. Since so much needs to be learnt, this subject offers the research worker many opportunities to advance our knowledge of the mineral nutrition of whole plants in their natural environment.

Bibliography

a Further reading

Barber, S. A. (1962) A diffusion and mass flow concept of soil nutrient availability. *Soil Sci.* **93**, 39–49.

Brewster, J. L. and Tinker, P. B. H. (1972) Nutrient flow rates into roots. *Soils Fertilizers* **35**, 355–359.

The two short papers provide, in turn, a valuable introduction to the ideas of nutrient flow in soil and an up-to-date review of published work.

Soil Biology: Reviews of Research (Multi-author) Unesco Publications 1969.

This well produced paperback volume contains two reviews of particular relevance to the matters discussed in the present chapter.

125–161 *Ecological Associations among Soil Organisms* by F. E. Clark.

163–208 *Biology and Soil Fertility* by E. G. Mulder, T. A. Lie and J. W. Woldendorp.

Bowen, G. D. and Rovira, A. D. (1969) The influence of micro-organisms on growth and metabolism of plant roots. In *Root Growth* (Ed. J. W. Whittington). Butterworth, London.

Harley, J. L. (1959) *The Biology of Mycorrhiza*. Leonard Hill, London.

This remains the standard work in English on the subject and deals with all aspects from ecology and taxonomy to biochemistry in a concise and readable manner.

b References in the text

Barber, D. A. (1968) Micro-organisms and the inorganic nutrition of higher plants. *Ann. Rev. Pl. Physiol.* **19**, 71–88.

Barber, D. A. (1969) The influence of the microflora on the accumulation of ions by plants. In: *Ecological Aspects of the Mineral Nutrition of Plants* (Ed. I. B. Rorison), 191–200. Blackwell, London.

Barber, D. A. and Frankenburg, U. C. (1971) The contribution of micro-organisms to the apparent absorption of ions by roots grown under non-sterile conditions. *New Phytol.* **70**, 1027–1034.

Barber, D. A. and Lee, R. B. (1972) The effect of micro-organisms on the uptake of manganese by barley plants. ARC Letcombe Laboratory Report 1971, 9–10.

Barber, S. A., Walker, J. M. and Vasey, E. H. (1963) Mechanisms for the movement of plant nutrients from the soil and fertilizer to the plant root. *J. Agr. Food. Chem.* **11**, 204–207.

Barber, S. A. and Ozanne, P. G. (1970) Autoradiographic evidence for the differential effect of four plant species in altering the calcium content of the rhizosphere soil. *Soil Sci. Soc. Am. Proc.* **34**, 635–637.

Berezova, F. F. (1965) Significance of root system micro-organisms in plant life. In: *Plant Microbe Relations* (Ed. J. Macura and V. Vančura), 171–177. Czech. Acad. Sci., Prague.

Bowen, G. D. and Rovira, A. D. (1961) The effects of micro-organisms on plant growth. I. Development of roots and root hairs in sand and agar. *Pl. Soil* **15**, 166–188.

Bowen, G. D. and Rovira, A. D. (1966) Microbial factor in short-term phosphate uptake studies with plant roots. *Nature* **211**, 665–666.

Brewster, J. L. (1971) Some factors affecting the uptake of plant nutrients from the soil. D. Phil. Thesis, Oxford.

Brewster, J. L. and Tinker, P. B. (1970) Nutrient cation flows in soil around plant roots. *Soil Sci. Soc. Amer. Proc.* **34**, 421–426.

Clarkson, D. T. and Sanderson, J. (In press) The response of barley plants to iron-deficiency. I. Characterization of the enhanced capability for iron translocation.

Clymo, R. S. (1962) An experimental approach to part of the calciole problem. *J. Ecol.* **50**, 707–731.

Daft, M. J. and Nicolson, T. H. (1966) Effect of *Endogone* mycorrhiza on plant growth. *New Phytol.* **65**, 343–350.

Domsch, K. H. (1963) The effect of saprophytic fungi in soil on the early development of higher plants. *Z. Pflkrankh.* **70**, 470–475.

Drew, M. C. and Nye, P. H. (1969) The supply of nutrient ions by diffusion to plant roots in soil. II. The effect of root hairs on the uptake of potassium by roots of Rye Grass (*Lolium multiflorum*). *Pl. Soil* **31**, 407–424.

Gerdemann, J. W. (1964) The effect of mycorrhiza on the growth of maize. *Mycologia* **56**, 342–349.

Gerretsen, F. C. (1937) Manganese deficiency of oats and its relation to soil bacteria. *Ann. Bot.* **1**, 208–230.

Gerretsen, F. C. (1948) The influence of micro-organisms on phosphate intake by the plant. *Pl. Soil* **1**, 51–81.

Harley, J. L. (1969) A physiologist's viewpoint. In: *Ecological Aspects of the Mineral Nutrition of Plants* (Ed. I. H. Rorison), 437–447. Blackwell, London.

Harley, J. L. and Brierley, J. K. (1954) Uptake of phosphate by excised mycorrhiza of beech. VI. *New Phytol.* **53**, 240–252.

Jackson, R. M., Brown, M. E. and Burlingham, S. K. (1964) Similar effects on tomato plants of *Azotobacter* inoculation and application of gibberellins. *Nature* **203**, 851–852.

Lewis, D. G. and Quirk, J. P. (1965) Diffusion of phosphate to plant roots. *Nature* **205**, 765–766.

Libbert, E., Wichner, S., Schiewer, U., Risch, H. and Kaiser, W. (1966) The influence of epiphytic bacteria on auxin metabolism. *Planta* (Berlin) **68**, 327–334.

Nye, P. H. (1969) The soil model and its application to plant nutrition. In: *Ecological Aspects of the Mineral Nutrition of Plants* (Ed. I. H. Rorison), 105–114. Blackwell, London.

Olsen, S. R., Kemper, W. D. and Jackson, R. D. (1962) Phosphate diffusion to plant roots. *Soil Sci. Soc. Am. Proc.* **26**, 222–227.

Page, O. T. (1972) Effect of phytotoxins on the permeability of cell membranes. In: *Phytotoxins in Plant Diseases* (Ed. R. K. S. Wood, A. Ballio and A. Graniti), 211–247. Academic Press, London.

Rovira, A. D. (1965) Plant root exudates and their influence on soil microorganisms. In: *Ecology of Soil-borne Pathogens* (Ed. K. F. Baher and W. C. Snyder), 170–186. University of California Press.

Samtsevich, S. A. (1962) Preparation, use and effectiveness of bacterial fertilizers in the Ukrainian SSR. *Mikrobiologiya* **31**, 923–933.

Sanders, F. E. T. and Tinker, P. B. (1971) Mechanism of absorption of phosphate from soil by *Endogone* mycorrhizas. *Nature* **223**, 278–279.

Stewart, W. D. P. (1966) *Nitrogen Fixation in Plants*, Athlone Press, London.

Walker, J. M. and Barber, S. A. (1962) Absorption of potassium and rubidium from the soil by corn roots. *Pl. Soil* **17**, 242–259.

Yoder, O. C. and Scheffer, R. P. (1971) Selective stimulation of ion uptake by *Helminthosporium carbonum* toxin. *Phytopathology* **61**, 918.

Author index

(*Page numbers in italic refer to the bibliography.*)

Subject index

MADE AND PRINTED IN GREAT BRITAIN BY WILLIAM CLOWES & SONS, LIMITED, LONDON, BECCLES AND COLCHESTER